Map Projection Transformation

Map Projection Transformation

Principles and Applications

QIHE H. YANG

JOHN P. SNYDER

WALDO R. TOBLER

TAYLOR & FRANCIS

Founded 1798

First published 2000 by Taylor & Francis
11 New Fetter Lane, London EC4P 4EE

Simultaneously published in the USA and Canada
by Routledge 29 West 35th Street, New York, NY 10001

Taylor & Francis is an imprint of the Taylor & Francis Group

© Taylor & Francis 2000

Typeset in Times 10/12 pt by Graphicraft Limited, Hong Kong

Every effort has been made to ensure that the advice and information in this book is true and accurate at the time of going to press. However, neither the publisher nor the authors can accept any legal responsibility or liability for any errors or omissions that may be made. In the case of drug administration, any medical procedure or the use of technical equipment mentioned within this book, you are strongly advised to consult the manufacturer's guidelines.

British Library Cataloguing in Publication Data
A catalogue record for this book is available from the British Library

Library of Congress Cataloging in Publication Data
A catalogue record for this book has been requested
ISBN 0–7484–0667–0 (cased)
ISBN 0–7484–0668–9 (limp)

Contents

Symbols

These symbols are used frequently enough to merit general listing. Occasionally, some are used for other purposes. Symbols used temporarily are not listed here.

a as a linear value: (1) equatorial radius or semimajor axis of the ellipsoid of revolution; (2) maximum scale factor at point on a map; or (3) element of coefficients in series. As an angle: azimuth in spherical (or spheroidal) polar coordinate system.

b (1) for an ellipsoid of revolution, the polar radius or semiminor axis; (2) minimum scale factor at point on map; or (3) element of coefficients in series. For triaxial ellipsoid, the equatorial radius at a right angle to the semimajor axis a.

B north geodetic or geographic latitude on the ellipsoid (if latitude is south, apply a minus sign).

B_0 central or initial geodetic or geographic latitude on the ellipsoid at origin of y-coordinates along central meridian (if latitude is south, apply a minus sign).

C projection constant.

c (1) for triaxial ellipsoid, the polar radius or semiaxis; or (2) element of coefficients in series.

e (1) for ellipsoid of revolution, the (first) eccentricity of the ellipse defined by dimensions a and b, where $e = [(a^2 - b^2)/a^2]^{1/2}$; for triaxial ellipsoid, the eccentricity of the ellipse forming the equator; or (2) a Gaussian coefficient.

e' for an ellipsoid of revolution, the second eccentricity, where $e' = e/(1 - e^2)^{1/2}$.

e_1 base of natural logarithm $e_1 = 2.718281828 \ldots$

F (1) area of a portion of the surface of an ellipsoid or sphere; or (2) equivalent latitude function.

f (1) function of the following parenthetical argument such as φ; or (2) a Gaussian coefficient.

h (1) elevation of a given point above the surface of the reference ellipsoid; or (2) a Gaussian coefficient.

L longitude east of Greenwich on a map of the ellipsoid (for longitude west of Greenwich, use a minus sign), or east of the central meridian in many formulae.

L_0 inital longitude east of Greenwich on a map of the ellipsoid (for longitude west of Greenwich, use a minus sign), or east of the central meridian in many formulae.

ln   natural logarithm, or logarithm to base e (e_1 here) where $e_1 = 2.718281828\ldots$

M radius of curvature of meridian at a given point on the ellipsoid.

m linear scale factor along a meridian.

N radius of curvature of ellipsoid surface in a plane orthogonal to the meridian.

n linear scale factor along parallel of latitude.

n' an auxiliary function of the ellipsoid semiaxes a and b, namely $n' = (a - b)/(a + b)$.

p area scale factor.

q (1) isometric latitude; or (2) conformal latitude function.

R radius of sphere.

r radius of parallel of surface of ellipsoid or sphere, for ellipsoid $r = N \cos B$; for sphere $r = R \cos \varphi$.

S (1) arc length of meridian; or (2) equidistant latitude function.

s linear distance along surface of sphere or ellipsoid.

u, v coordinate system for intermediate use.

u reduced latitude on the surface of ellipsoid.

v distortion variable.

X,Y,Z three-dimensional axes with center at a given point on the surface of the sphere or ellipsoid, the Y-axis in the meridian plane pointing to the North Pole, the Z-axis coinciding with the normal to the ellipsoid surface, and the X-axis positive east from this point.

X_g, Y_g, Z_g three-dimensional axes for the sphere or ellipsoid, with the center at the center of the ellipsoid, and the X_g-, Y_g-, and Z_g-axes increasing in the direction of the Greenwich meridian in the equatorial plane, the meridian 90°E in the equatorial plane, and the North Pole, respectively.

x rectangular coordinate: distance to the vertical line (y-axis) passing through the origin or center of a projection (if negative, it is distance to the left).

x_φ partial derivative $\partial x / \partial \varphi$; similarly for y_λ, etc.

y rectangular coordinate: distance above the horizontal line (x-axis) passing through the origin or center of a projection (if negative, it is distance below).

z (1) spherical angle from selected point on the surface of a sphere or ellipsoid to some other point on the surface, as viewed from center; or (2) rectangular coordinate in direction of z-axis.

α cone constant for conic projection.

γ convergence of meridians (deviation from direction of y-axis on a map).

δ azimuth in plane polar coordinates.

ε deviation of graticule intersection from a right angle on map.

λ longitude east of Greenwich on a map of the sphere (for longitude west of Greenwich, use a minus sign), or east of the central meridian in many formulae.

λ_0 longitude east of Greenwich of the central meridian of a map of the sphere, or of the origin of rectangular coordinates (for west longitude, use a minus sign), often assumed to be zero.

μ linear scale factor (not restricted by direction if not subscripted).

μ_1 linear scale factor along a vertical (or great circle passing through the pole of a projection).

μ_2 linear scale factor along a small circle centered on the pole of a projection.

ρ	radius in plane polar coordinates.
ϕ	geocentric latitude on the surface of ellipsoid.
φ	(1) north geodetic or geographic latitude on the surface of sphere (if latitude is south, apply a minus sign); or (2) conformal latitude.
φ_0	central latitude or origin of y-coordinates at central meridian.
ω	maximum angular deformation.
ψ	equidistant latitude.
τ	equivalent latitude.

Preface

'Map Projections' is also called 'Mathematical Cartography'. It is one of the fundamental theories of cartography – a comprehensive science.

With the advance of science and technology, there have been breakthroughs in the field of classical research and methods of map projection. Among these, computer science and space science have had the greatest influence upon the field in research and in the formation of the working methods of map projection, developing these in breadth and depth. The most apparent characteristics of modern mathematical cartography are: research on the theory and methods of map projection is being deepened and combined with many special fields; new areas of research and application are being developed; and new concepts are constantly appearing. This book is an attempt to reflect several aspects of the development of modern map projections, especially the theory and methods of map projection transformation.

Map projection transformation during the past 20 years is a newly developed area of research in map projections. It is widely used in surveying and computer-assisted cartography, in data processing for information systems, and in the transformation of data from space, remote sensing, and other space sciences. The development of map projection transformation not only expands new areas of research on map projections, but also further develops the applied area with the creation and application of map projection transformation software and mapping mathematics based on the use of computers.

Research into the theory and applications of map projection transformation began in the 1970s in China. Now it has attained a high level both in theory and practice. A series of papers and works have been published in recent years. It has become an important postgraduate course in cartography. This book systematically sums up the results of research on the theory and application of map projection transformation in recent years, most of them being the fruits of the author's research. It is available for use and reference by scientists and engineers who engage in computer-assisted cartography and in spatial data processing for information systems such as Geographic Information Systems (GIS), Land Information Systems (LIS), and the like. It can also be used as a textbook or reference book for senior students and postgraduates specializing in geography, remote sensing, surveying, and computer applications.

The Western edition of this book was largely rewritten based on the book *Ditu touying bianhuan yuanli yu fangfa* (*Map Projection Transformation – Principles and Methods*), by Qihe Yang[1], published by the Chinese People's Liberation Army Press in 1990, with editorial and technical revisions by John P. Snyder[2] and Waldo R. Tobler[3]. The Western edition is divided into 13 chapters, with six appendices. Chapter 1 outlines 50 years' of Chinese advances in the subject of map projections; Chapter 2 introduces many kinds of commonly used equations for map projections and some projections updated in China; Chapters 3–8 systematically and thoroughly discuss principles and methods of map projection transformation; Chapters 9–13 discuss applications of map projection transformation and some theories and methods that mark advances in modern map projections. Throughout, the positions of the *x*- and *y*-axes have been reversed from the Chinese to the American orientation.

The first author gives a great deal of attention in the content of the book to explaining the principles of creating mathematical models for map projection transformation with examples of calculated results. Some application examples are given for typical transformation methods, such as several types of numeric transformation, and so on. In this book, many chapters are devoted to discussing the theory and methods of conformal projection transformation. This is because they are methods used widely and studied thoroughly in China. The most apparent characteristic is that it adopts formulae for, or tables of, constant coefficients that are suitable for computer calculation instead of the classical method of variable coefficients in conformal projection transformation.

There are a series of innovative theoretical results in this book. The models of analytical transformation and numerical transformation have been systematically constructed. Several thematic subjects reflecting the advances in map projections have been explored. Widely used calculating examples are given. These make map projection transformation more theoretical, and systematized for practice. This is an important subdiscipline in the subject of map projections. Map projection transformation has become a basis for teaching postgraduates in the first author's institute since the 1980s. Based on it, this English version of the book, entitled *Map Projection Transformation – Principles and Applications*, has been compiled by collecting the results of the first author's research for the past 15 years and using the results of his fellow researchers for reference in this area.

The book, based on the original, was revised and rewritten by the first author, who also managed to prepare the Chinese material – translating, typing, plotting figures and editing. Gratefully acknowledged are Dr Yang Xiaomei, Engineer Lan Rongqin, Dr Ren Liucheng, and others, who provided valuable help in the success of this effort. The second author John Snyder, a notable map projection scientist in the USA, initiated the publication plan for the book, made technical corrections and edited Chapters 1, 3–9, 12, Appendix 1, and the first five sections of Chapter 2; he also wrote Section 11 in Chapter 2 and sorted out Appendix 6. He died before he had finished compiling the full book. We cherish the memory of him. Professor Waldo Tobler, a famous cartographer in the USA, continued John's work and made revisions and corrections to the remaining chapters and sections, arranged an index, and assembled the whole book. The publication of this book is due to the fruitful cooperative effort between Chinese and American scholars in map projection science.

Q.H. YANG

J.P. SNYDER

W.R. TOBLER

NOTES

1 Professor, Department of Cartography, Zhengzhou Institute of Surveying and Mapping, Zhengzhou, Henan Province 450052, People's Republic of China.
2 Formerly the US Geological Survey, Reston, Virginia.
3 Professor Emeritus, Geography Department, University of California, Santa Barbara, CA 93106-4060 USA.

Introduction

RESEARCH OBJECTS AND TASKS OF MAP PROJECTION SCIENCE

'Map projections' is also called 'mathematical cartography.' It principally studies the theories and methods of building the basic mathematics for maps. This research object is the result of the characteristic contradiction between the non-developable earth's surface and the map plane. As we know, the earth has a natural surface that is complicated. For the purpose of measurement and calculation on the earth's surface, the earth should be replaced by an ellipsoid or sphere that approximates the earth's natural body in overall shape. This is because an ellipsoid and a sphere can be expressed by a simple mathematical equation. In geometry such an ellipsoid or sphere is also known to be a non-developable surface. But it is necessary to expand part of or the entire terrestrial ellipsoidal (or spherical) surface onto a plane surface according to some required specification for map use. To expand the non-developable earth's surface into a plane forms the characteristic contradiction of map science.

This contradiction can be solved by means of a map projection. The so-called map projection problem is, firstly, to project the graticule of meridians and parallels of an ellipsoid (or sphere) onto a plane using a specified mathematical method according to certain projection conditions, and then to use the graticule as a control to transfer images of all of the elements from the earth's surface onto this plane. For this reason the theory and method of projecting the graticule of meridians and parallels of the earth ellipsoid (or sphere) onto a plane constitutes one of the major research objects of map projection science.

In addition, it delves into general, special and celestial map projections. The application of map projections to navigation has a long history. Especially with the development of radio navigation and the technology of radio positioning, a series of new research topics is presented to the subject of map projections. Map projections, as a means of establishing the relation between space and a plane, have been widely used to solve some geometric problems of spherical geometry, astronomy, crystallography, and geology in graphical form. The application of Landsat to mapping introduced completely new concepts for map projections. Time has now become a parameter in mapping. This is quite different from conventional static mapping in which the relations among earth shape, perspective center and projection plane are fixed. It is a dynamic one. An entirely new research topic of studying a projection fit for satellite mapping confronts map projection science. In recent years, the electronic computer, especially the personal computer, has been widely applied to all aspects of map projections and has thoroughly changed the look of map projection science. Examples are the applications of computers to the calculation of coordinates; to the automatic creation of the mathematical foundation of maps; and to the automatic plotting of thematic mathematical elements on a map. Computer-aided map projection transformation is even more of a leap for cartography. To meet the need of computer-aided cartography, it is a pressing task to study the theory and methods of map projection transformation, to study topographic data processing, spatial information positioning, and transformation in information systems.

As described above, the major tasks before map projection science can be summed up as follows:

- To explore the theory and method of creating geographic meshes for different purpose maps, their calculating and plotting, as well as to solve some problems about map disposition and sheet separation.

- To study the projection of selected curves from an ellipsoid (or sphere) to a plane and their expression thereon, as well as problems involving the measurement of elements on a map.

- To delve into the theory and method of map projection transformation meeting the requirements of automatic cartography and the transformation between different projections.

- To develop a projection method fit for solving the problems of planar and linear geometry in order to meet the needs of other sciences such as astronomy, crystallography, and geology.

- To research celestial space projections and the Landsat projection to meet the need of developing of space technology.

- To probe digital data processing in map databases, spatial information positioning systems, and map projection transformation systems to meet the requirement of building different specific information systems and developing spatial information science.

SEVERAL CHARACTERISTICS OF DEVELOPMENT OF MODERN MAP PROJECTION SCIENCE

With the advance of science and technology, there have been breakthroughs in the field of classical research objects and methods of map projection. We now present several characteristics of this development.

- The first characteristic is that research on the theory and methods of map projection is being deepened and combined with many special fields.

In the following we give some references to documents published after the 1980s as examples to illustrate these characteristics.

For example, in a paper 'On the classification of map projections' (Li 1979) the author gave his new views on the classification of map projections. He summed up projections as belonging to three types – elliptic projections, parabolic projections and hyperbolic projections – according to whether or not the equilength direction of arbitrary points in the projection area is the sole one. As well developed as the azimuthal projection is, the study of its theory and method has been deepened in recent years. As an example, the paper 'The perspective azimuthal projection under variable view points' (Yang 1987c) combined perspective azimuthal projections with non-perspective azimuthal projections and unified the treatment of perspective azimuthal projection systems using variable view points.

Research into the theory and method of conformal projections has also been further developed. As an example, the monograph *Map Projections: a Reference Manual* (Bugayevskiy and Snyder 1995) summed up and systematized the conditions of optimum and ideal conformal projection exploration methods and developed the method of the Chebyshev projection. The monograph *The Theory of a Single Space Photograph* (Bugayevskiy and Portnov 1984) is an example of the integration of two research fields: cartography and space photography. With the advance of modern science and technology, it is increasingly important that scientists in different scientific and research fields be united to overcome this difficulty and develop new technology situated between scientific frontiers.

- The second characteristic is that new research and application areas are continuously being developed.

The study of map projection transformation not only opens up new research areas for map projections but also, with the computer-aided creation of the map's mathematical foundation and the construction of map projection transformation software systems, develops new application areas for map projections.

The traditional map's mathematical basis is the geographic mesh but this incompletely meets the needs of newer practical uses. This is because the advance of modern science and technology needs increased spatial positioning data, including such information as the position of points, or distance, azimuth, and track data. Adding these thematic mathematical items to an existing map will be commensurate with a set of newly edited thematic maps. So it can be said that the development of thematic mathematical elements for a map enlarges the application area of map projection science.

- The third characteristic is that the continuous presentation of new concepts promotes the advance of map projection science.

To meet the needs of developing the technology of remote sensing from space, a proposal was put forward in the 1970s by the US Geological Survey, suggesting the study of a space projection (Colvocoresses 1974). The mathematical model for a space oblique Mercator projection was derived by John P. Snyder in the late 1970s and was used for satellite series mapping. The theory and method for a space conformal projection was presented in the beginning of 1990s by a Chinese cartographer (Cheng 1992). With the emergence of space projections, there have been breakthroughs in the field of classical research on map projections. The functional relation between a spatial point (in three dimensions) and a plane point has been extended to a functional relation between a spatial point (in four dimensions) with the parameters of a spatial point's position in time and on a plane.

Variable scale projection is a newly developed and widely used map projection in recent years. There is a classical objective for a general map, namely to choose a projection that has limited deformation. But with a variable scale projection, according to the map's uses and requirements we can choose a projection that retains only the map graphic's topological relations. Chinese cartographers have been engaged in research into variable scale projection and have achieved many results. For example, *Methods for*

Variable Scale Map Projections (Wang *et al.* 1996) introduced many methods for variable scale projections. In 'Composite projections' (Section 11.2 of this book), the principles and methods of obtaining a new kind of variable scale map projection by composite projection obtained from conventional map projections are studied and presented.

- The fourth characteristic is that map projection and its transformation, as a spatial information positioning model, has become an important component of space information science.

No activities of humanity can take place without geographical space. The spatial information rich digital map and remote sensing image as geographical foundations has been combined with positioning models (map projections and their transformation) to form a spatial information positioning system (SIPS) (Yang and Yang 1997). This is the carrier of all spatial information and is the foundation of positioning. The result is spatial information with a united geographical and plane coordinate system. Also it renders a point's position measurable. Today, spatial information graphical positioning systems based on digital maps have permeated all aspects of GIS and have become an important component thereof. Space information science is a comprehensive discipline while the spatial information graphic positioning system is one of its important theoretical bases and an important component.

THE RESEARCH CONTENTS OF THE BOOK

This book is divided into 13 chapters and six appendices. Chapter 1 outlines 50 years' of advances in map projections in China; Chapter 2 introduces several commonly used equations of map projections and some projections updated in China. They are the basis of map projection transformation. Chapter 3 considers the general theory of map projection transformation, and systematically and thoroughly discusses some common theoretical problems and methods of map projection transformation. Chapters 4 and 5 systematically discuss several kinds of analytical transformation models, as well as analytical models for conformal projections. Chapters 6 and 7 systematically discuss numerical methods for general polynomial approximation and for conformal polynomial approximation. Chapter 8 discusses in detail the principle, method, and application of the third type of coordinate transformation of map projections. As can be seen, the foregoing chapters organize the principles of map projection transformation. They include the correct theory, plentiful detail, and comprehensive methods. Later chapters specialize in the application of map projection transformation and in the development of modern map projections. Chapter 9 presents the principal, methods, and applications of zone transformation for the Gauss–Krüger projection; Chapter 10 discusses the problem of developing new map projections with coordinate transformations. Chapter 11 covers variable-scale map projections. Chapter 12 describes position lines. These last three chapters separately discuss some theories and methods that mark the advance of modern map projections. Chapter 13, on spatial information positioning systems, discusses the relationship between a map projection (and its transformation) and spatial information science and the accompanying research contents.

    The appendices of this book introduce several applied and basic principles of map projection transformation, and include mathematical methods, constant coefficient tables, typical examples, the mathematical element transformation of topographic maps and an introduction to a software system. Appendix 6 consists of a map projection bibliography of Chinese literature on map projections, and additional references.

CHAPTER ONE

Fifty Years' Advancement of Map Projection Study in China

Before 1949 China was lacking in activity in the discipline of map projections. It had limited researchers and few applications. Since the founding of the People's Republic of China, just like in the other sciences and technology, great advances have been achieved in this field.

Beginning in the late 1950s, a learned journal, *Acta Geodaetica et Cartographica Sinica*, published by the Chinese Society of Surveying and Mapping, started to carry articles on the subject of map projections. The other journals, such as the *Bulletin of Surveying and Mapping* and *Translations of Surveying and Mapping*, often carried articles about map projections. They gave Chinese cartographers an opportunity for further learning, discussion and exchange.

To meet the requirements of teaching and production, many educational materials, such as *Mathematical Cartography* (Wu 1961), *Map Projections* (Li *et al.* 1993), *Mathematical Basis for Charts* (Hua 1985), and *Principles of Mathematical Cartography* (Wu 1989), were edited and published by Chinese colleges and universities of surveying and mapping, and by secondary technical schools of surveying and mapping. These monographs systematically expounded the general theory of map projection and gave a representation of new research results on map projection at home and abroad. They included the practical Chinese experience in the designing of the mathematical basis of maps. They continue to play a positive role in improving the teaching quality of the course of map projection in colleges and universities.

From the late 1950s to the late 1990s, Chinese cartographers working on map projections have made great achievements in both the study of the theory of map projection and the exploring of new types of projections, as well as their applications and the development of new directions in map projection. Many theses and monographs have been published. As a result, the theory of this discipline has advanced further both in depth and in breadth in China. The results can be summed up in the following six sections.

1.1 BASIC MATHEMATICS FOR TOPOGRAPHIC MAPS AND FOR ATLASES

The topographic map is the basic national scale map. The fundamental mathematics for the topographic map are important for the scientific character and uses of the map. In the early post-Liberation period China began to adopt the commonly used Gauss–Krüger projection as the mathematical basis for surveying and mapping. Many calculation tables fit for our country were translated and compiled. A new zone transformation formula for the Gauss–Krüger projection was derived by a Chinese engineer Jin Yangshan. As a chief editor, Mr Jin also compiled a Gauss–Krüger zone coordinate transformation table between the 6° zones and the 3° zones. It is one of the simplest and the most convenient tables for use at home and abroad.

The distortion of the Gauss–Krüger projection is very large at some latitudes. For this reason, Chinese cartographers developed the 'Cylindrical transverse conformal projection with two standard meridians' (Li 1981) and conducted 'Research on the Gauss–Krüger projection family' (Yang 1983c) to raise the precision of topographic maps. The former improved the distribution of distortion of the projection and also reduced the distortion of the Gauss–Krüger projection. The latter made a thorough study of the Gauss–Krüger projection and derived generalized calculating formulae for all conformal projections that are zone-divided along meridians, and discussed some included conformal projection systems.

In order to improve the precision of coordinates of points in the grid of adjacent zones, to raise work efficiency, and to change the current manual operational method, Chinese cartographers developed and compiled the *Tables of Coordinate Transformation for Gridlines of Adjacent Zones of the Gauss Projection* (Yang 1980a).

Atlases are an important centralized product that reflect the developmental level of the science and technology of cartography. Since the Liberation, China has achieved remarkable successes in compiling a whole China atlas and also provincial (or district) atlases. In order to solve the problem of the basic mathematics for the atlas, Chinese cartographers put forward many theses and monographs, for example, 'Study on the projections for the Provinces' Atlas of China' (Meng 1959), 'Several problems on the basic mathematics for the common atlas' (Yang 1962), etc. These theses and articles laid a good foundation for the design of map projections for compiling a large atlas and raised the scientific value of this kind of map product.

1.2   GENERAL THEORY OF MAP PROJECTION

In the last 50 years Chinese cartographers have made many investigations and studies to complete and broaden map projection theory and have obtained many valuable results.

Firstly, it must be pointed out that the article 'Problems with the discovery of new map projections with the distortion ellipse on hand' (Zhou 1957) issued in *Acta Geodaetica et Cartographica Sinica* presented a general condition $a = \varphi(b)$ as a projection function and gave examples to illustrate how to explore new types of azimuthal projection, cylindrical projection, and conical projection according to the projection conditions $a = b^k$, $a = Kb$. This method expands the conventional projection condition to a functional condition. For this reason it is an original theory.

Secondly, in the 1960s articles, such as 'Comments about functions for cartographic projections' (Dang 1960), 'On the law of conversions of distortions in cylindrical, azimuthal and conic projections' (Liu 1965), and 'The use of a numerical methods for establishing optional conical projections' (Yang 1965), were published in rapid succession. The first item substituted the relation of the length ratio into the condition equation of a projection $k_1 = q(\varphi)k_2^n + p(\varphi)k_2$ and obtained a differential equation of the Bernoulli type, and then achieved some basic formulae that generalize the three commonly used projection types. The last two papers, respectively, delved into the change rule of distortion of the three commonly used projection types, and the nature of conical projections. They not only contributed to the deepening of the knowledge of these projections but also provided a theoretical basis for drawing up the plans for projections of arbitrary character. A significant paper, 'On the classification of map projections' (Li 1979) published in the late 1970s analyzed the inherent law of distortion of projections and gave the author's views on the classification of map projection based on the functional expressions $a = b^k$, $a = Kb$.

Besides the exploration into basic theory, another important driving force to develop the subject of map projection is the research into the changing trends in the field of map projection and the dialectics of nature. In this respect there are several published papers, such as 'The condition and tendency of development of mathematical cartography' (Wu 1963), 'Elementary analysis of conflicting movements in map projection' (Wu 1965), and 'Contents and tasks of mathematical cartography' (Hu 1980). In recent years there have also been some notable theses in research on map projection theory, such as 'On probing into the graph topology in map projection' (Hu 1990), and 'The topological model and classification framework of map projections' (Hu 1996).

1.3 EXPLORATION OF NEW TYPES OF MAP PROJECTION

Within the last 50 years, in view of the need for teaching and production, and to raise the level of the discipline, Chinese cartographers have made encouraging progress in exploring new types of map projections, and have put forward a series of creative results in a wide range of subjects. Hence, they have obtained many interesting products.

For example, 'On graphic interpretation methods in seeking azimuthal projection' (Hu 1962), 'Graphic interpretation methods in seeking cylindrical map projections with pre-defined distortion' (Gong 1964), and 'The use of numerical methods for establishing optional conical projections' (Yang 1965), are among the better application examples of inverse solutions for map projections developed in modern times. In the first paper an expression for finding the solution of the radius of an azimuthal projection by the desired distribution of distortion was presented. As a result, the regular form of this kind of projection radius was changed, and the azimuthal projection has achieved a more flexible form for applications. Basing itself on a pre-defined distribution of distortion, the latter paper delved into theory and method of creating arbitrary conical projections, using a numerical method and an analytic method, thereby leading to the solution of the problem of creating arbitrary conical projections for the surface of ellipsoid. A new method of determining the constant of the cone for a conical projection was given in 'Conformal or equal-spaced conical projection with total-equal area' (Hu and Zhou 1958) and the essence of this unusual method was further researched in the paper on 'Pseudoazimuthal projections and their application for a whole map of China' (Liu and Li 1963). The inner structure was revealed and developed into a composite method. These enabled distortion isograms to be near, as far as possible, to the outline of the mapping region. The papers 'Cylindrical transverse conformal projection with two standard meridians' (Li 1981) and 'A modified transverse cylindrical conformal projection for 1:1 000 000-scale topographic maps' (Li 1983b) discussed two conformal projection schemes for zone division along meridians. The former reduces and improves the distortion distribution of the Gauss–Krüger projection. The latter presents a new projection scheme for separate sheets on the millionth scale, in which a novel method was given to make the length of the border meridian within 0.1 (mm).

A highlight in exploring new types of projection is the emergence of many summary formulae. For example, 'Double azimuthal projection' (Li 1963) comprehensively summarized almost all common azimuthal projections, systematized this type of projection and brought to light the inner relationship among them. In addition, 'Research on azimuthal projections' (Yang 1983b) summarized some well-known azimuthal projections and revealed certain new characteristics of the gnomonic projection for navigation. In two papers, 'Research on the Gauss–Krüger projection family' (Yang 1983c) and 'Research on the polyconic projection family' (Yang 1983e), the author delved into the combination of conformal projections with zone division along meridians and the combination of approximately conformal polyconic projections. They respectively revealed the inner relation and laws among conformal projections with zone division along meridians and the geometry characteristic of some conformal projection schemes. The paper 'A discussion of three modified projections – modified equidistant azimuthal, conic and cylindrical projections' (Yang 1983d) presented three modified equidistant projections whose distortion distribution differs from three regular equidistant projections. Among them the modified equidistant azimuthal projection is quite applicable to a map of all of China. In 'Mathematical model of direct and inverse solutions to coordinate transformation for the modified polyconic projection' (Yang 1989a) and 'Analytical computation method for two-point azimuthal projection and two-point equidistant projection' (Yang 1990c), some suggestions were made to replace a traditional graphical computing method. The two papers 'The perspective azimuthal projection under variable view points' (Yang 1987c) and 'On $m = n^k$ orthogonal projections' (Li 1987) enable us to explore arbitrary azimuthal projections, cylindrical projections and conic projections more flexibly and actively. The development of new map projections with coordinate transformation (see Chapter 10) is a new method studied and presented recently. There have also been other theses, including 'A family of new pseudocylindrical equal-area projections' (Hu 1994).

Exploration of the projection scheme suitable for a world map has also been fruitful. 'The design and analytical calculating method of polyconic projection in unequal graticules' (Zhong et al. 1965), 'A polyconic projection with unequally spaced parallel lines and its applications' (Fang 1983), and 'A polyconic projection with two spherical arcs as border meridians' (Zhong 1983a) were put forward in succession and they redesigned and improved the polyconic projection as applied to a world map. Some of the projection schemes are widely used in the world maps published in China and formed a design system for a polyconic projection.

1.4   SELECTION OF MAP PROJECTION AND ITS APPLICATION

In addition to the monographs and theses concerning the selection of map projections mentioned in Section 1.1, Chinese cartographers also edited and published *Map and Projections for Small-Scale Maps* (Hu and Gong 1974) and *Tables of Regional Map Projections* (Yang 1979b) to meet the requirement of compiling all types of internal or external regional maps. These two collections of tables for projections not only summed up Chinese cartographers' experience in using projections for mapping but also possessed their own characteristic. The former provided an all China map and provincial maps with a projection scheme and coordinate results, as well as coordinate results for a conic conformal projection of a separate map on the millionth scale. The latter provided a projection scheme and coordinate results not only for an entire map of China, provincial maps, and an ocean map, but also for a political map of world, continental maps and some special maps. It demonstrates how one can overcome region limitations and provide coordinate results of projections for arbitrary mapping regions stretching along a parallel or meridian. *Tables for Calculating Map Projections* (Fang 1979), made for calculation by the Chinese was published; it has a more detailed and complete content than that of the earlier translated *Table for Cartography*.

With regard to the scientific application of maps, from qualitative analysis to quantitative analysis, a series of specialized articles on map cartometry, description of position lines, and the usage of special grids on map projections were presented. To solve the geometry problem of a spherical surface on a global scale, 'A thematic world map projection and its measurement and annotation methods' (Hu and Wu 1983) was presented, thus breaking through the useful limit of the projection mesh of the transverse stereographic projection, by extension to the entire world, and directly relating to geographical position. To correct data measurement error caused by projection distortion, and improve its precision, a nomograph was developed and applied. *Map Cartometry* (Gong 1989) is a monograph on map applications, and systematically discusses the theory and method of map cartometry; it enabled Chinese map applications to achieve a new standard.

1.5 THEORY AND APPLICATION OF MAP PROJECTION TRANSFORMATION

Map projection transformation is a research field of map projection, exploring the theory and method of transforming one kind of map projection point's coordinates to another map projection point's coordinates. It is an important problem to be solved in the course of developing automatic cartography. If it is not necessary to delve into the problem in traditional map compilation work, it will be an important theoretical and practical problem for automated cartography. In the 1980s, Chinese cartographers made a thorough study of the theory and method of map projection transformation and have achieved significant results. A range of theses have been presented. These include 'How to transform coordinates of points from one kind of map projection to another' (Wu 1979), 'A research on the transformation of map projections in computer-aided cartography' (Wu and Yang 1981), 'On the numerical method for transforming the zones of Gauss projection' (Yang 1982a), 'The semi-numerical method for the map projection transformation' (Hu 1982), 'A research on numerical transformation between conformal projections' (Yang 1982b), 'A numerical method for the transformation of map projections by densifying with spline function' (Yuan 1983), 'A research on transformation of finite element between conformal projections' (Li 1985), 'The discussion and application of the constant coefficient and the variable coefficient formulae of direct and inverse solution of Gauss–Krüger projection' (Yang 1986b), 'BASIC programs for 1:10 000, 1:5 000, and larger-scale topographic sheet elements' (Yang 1986d), and 'A study of the software for automatically setting up map mathematical foundation' (Yang 1986c). They probed the theory and method of map projection transformation and illustrated commonly used projection transformations with examples. Two monographs, *The Principles and BASIC Program for Conformal Projection Transformation* (Yang 1987a) and *The Principles and Methods of Map Projection Transformation* (Yang 1990a), were published. They were characterized by merging transformation theory, transformation models, algorithms, computer programs, and examples of transformation into an integral whole. Thus map projection transformation became even more theoretically developed, systematized, and practical. Chinese map projection science was raised to a new high level.

1.6   CONTINUOUS DEVELOPMENT OF NEW RESEARCH AND APPLICATIONS TO THE FIELD OF MAP PROJECTION

With the advance of science and technology, there have been breakthroughs in the field of classical research and methods of map projection science. Among these, computer science and space science have had an especially profound influence upon the field of research and the formation of a working body of map projection science, developing them both in depth and breadth.

Chinese cartographers paid a good deal of attention to these new growth areas in map projection science and have made fruitful efforts, such as the research into map projection transformation introduced in Section 1.5. In addition, from 1980s they began to study the principle and method of four-dimension space projection with spatial position and time as parameters to meet the requirement of the positioning of remote sensing images. In the 1990s they promoted research to create the theory and method of the Conformal Space Projection. They are also currently carrying out research for the Creation of the Mathematical Foundation for Satellite Images and its Application (research topic funded by National Natural Science Fund of China), which integrates pattern recognition with space projection to build the mathematical foundation for satellite images.

After the emergence of the theory and method of variable-scale projection such as the polyfocus projection in the 1970s, from the 1980s Chinese cartographers have been engaged in research into the method of variable-scale map projection and have attained a high level. They have published 'Map projections for variable scale city maps' (Huang 1985), 'Cartographic projection system for variable scale maps' (Hu 1987), 'A kind of adjustable map projection with "magnifying glass" effect' (Wang and Hu 1993a), and *Methods for Variable Scale Map Projections* (Wang et al. 1996). Chapter 11 of this book discusses research in composite projection.

More spatial positioning information on a map, for example the position, distance, azimuth, and locus of points, is required with the advance of science and technology. For this purpose, Chinese cartographers, beginning in the 1980s, undertook studies of the theory of position lines and its application and have achieved a great many results in both theory and application, for example, 'A study of the software for automatically setting up map mathematical foundation' (Yang 1986c), 'The transforming and measuring of position information of map' (Yang 1987d), 'Computation method for positioning by long-range hyperbolic navigation system' (Hua 1990), and 'The theory of position line and positioning navigation software system' (Yang and Lu 1995).

By the 1990s, Geographic Information Systems (GIS) have been used widely in various national departments and every profession. Map projection is however the mathematical foundation for the spatial information positioning system in a GIS. Thus Chinese cartographers attached great importance to the application of map projections in this new field and attended to the study of spatial information and graph positioning systems and their application. New concepts of spatial information and graph positioning systems based on a digital map or based on a remote sensing image were presented (see Chapter 13 of this book). 'Space information positioning system and its application' (Yang et al. 1997) is one item that was put forward.

In China, *Map Projection Papers* (Wu and Hu 1983a), *Mathematical Basis for Charts* (Hua 1985), *Principles of Mathematical Cartography* (Wu and Yang 1989) and *Map Cartometry* (Gong 1989) were published in the 1980s. *The Principles and Methods of Map Projection Transformation* (Yang 1990a) and two publications, both entitled *Map Projections* (Hu and Jianwen 1992a; Li et al. 1993), were published in the 1990s. These monographs in combination give a representation of the Chinese achievements and development level of map projection science in the last 50 years. This English edition of *Map Projection Transformation – Principles and Applications*, published by the Taylor and Francis publishing house, is a representative work that opens up the Chinese achievements and development level on map projection science to the eyes of the readers of the world.

Chinese achievements in map projection science in the last 50 years are due to numerous industrious Chinese cartographers (see Appendix 6), of whom the following scholars made the greatest contribution: Prof. Fang Jun, Prof. Wu Zhongxing, Prof. Hu Yuju, Prof. Yang Qihe, Prof. Li Guozhao, Prof. Gong Jianwen and senior engineer Hua Tang.

NOTE

This chapter has been updated and completed, based on the Chinese essay 'Thirty years' development of map projection study in China' (Wu and Hu 1983b).

CHAPTER TWO

Map Projection Equations

2.1 EARTH ELLIPSOID AND SPHERE

The earth has a natural surface that is complicated. In surveying and cartography, it is replaced by an earth ellipsoid or sphere that approximates the natural surface of the earth in overall shape, especially the general curvature. In the computation of map projections, the earth ellipsoid or sphere is called the earth datum; it is also called the datum used for the related map projections. Now we will introduce some of the basic elements and fundamental formulae.

2.1.1 Earth ellipsoid

The earth ellipsoid is obtained as a curved surface in which a plane ellipse is rotated about its minor axis. It has two semimajor axes a and one semiminor axis b. The equation is expressed as

$$\frac{x^2 + y^2}{a^2} + \frac{z^2}{b^2} = 1 \tag{2.1.1}$$

The following five basic elements are often used to represent the shape and size of the earth ellipsoid:

semimajor axis $\quad a$
semiminor axis $\quad b$
flattening $\quad \alpha = (a - b)/a$
first eccentricity $\quad e^2 = (a^2 - b^2)/a^2$
second eccentricity $\quad e'^2 = (a^2 - b^2)/b^2$

Of the above basic elements, (a, b) are in units of length, and α, e, e', are all calculated from a and b, and represent the flattening of the earth ellipsoid. That is, in order to determine the shape and size of the earth ellipsoid, it is sufficient to have only two basic elements, but one of them must be a length element, a or b. Table 2.1 lists the earth ellipsoid parameters used in various countries.

Before 1952, China had adopted the Hayford (or International) ellipsoid. After 1953, it used the Krasovskiy ellipsoid, but in 1978 it decided to adopt the IUGG/IGA 1975

Table 2.1　Principal ellipsoids used throughout the world

Ellipsoid	Year	Semimajor axis a (m)	Flattening α	Use
Delambre	1800	6 375 653	1:334.0	Belgium
Airy[1]	1830	6 377 491	1:299.32	Great Britain
Everest[1]	1830	6 377 276.3	1:300.801	India; Pakistan; Burma
Bessel	1841	6 377 397.2	1:299.152	Mid Europe; Indonesia
Clarke	1866	6 378 206.4	1:294.978	North America
Clarke[1]	1880	6 378 249	1:293.459	France; most of Africa
Helmert	1907	6 378 200	1:298.30	Egypt
Hayford[2]	1909	6 378 388	1:297.00	Much of world
Krasovskiy	1940	6 378 245	1:298.30	Soviet Union; China
IUGG	1975	6 378 140	1:298.257	China
GRS 80	1980	6 378 137	1:298.257	Newly adopted

[1] Also used in some regions with other constants.
[2] Also called the International ellipsoid.

ellipsoid. Accordingly, China is building a new and independent geodetic coordinate system.

There are conversion relationships among the basic elements as follows:

$$\left.\begin{array}{l} e^2 = e'^2/(1 + e'^2), \quad e'^2 = e^2/(1 - e^2) \\ e^2 = 2\alpha - \alpha^2 \approx 2\alpha, \quad b^2 = a^2(1 - e^2) \end{array}\right\} \tag{2.1.2}$$

Basic formulae in common use for the earth ellipsoid are expressed as follows:

- radius of curvature along a meridian:

$$M = a(1 - e^2)/(1 - e^2 \sin^2 B)^{3/2} \tag{2.1.3}$$

- radius of curvature of the prime vertical:

$$N = a/(1 - e^2 \sin^2 B)^{1/2} \tag{2.1.4}$$

- mean radius of curvature:

$$R = \sqrt{MN} \tag{2.1.5}$$

- radius of a parallel circle:

$$r = N \cos B \tag{2.1.6}$$

- meridian arc length from equator to lat. B:

$$S_m = \int_0^B M dB \tag{2.1.7}$$

For the calculation formula for a meridian arc length see (3.5.20); for the Krasovskiy ellipsoid computing formula see (3.8.14).

Parallel arc length:

$$S_p = rl = lN \cos B \tag{2.1.8}$$

where the longitude difference l is in units of radians.

The area of a trapezoidal area of the ellipsoid:

$$F = \int_0^B M r dB = a^2(1 - e^2)\left[\frac{\sin B}{2(1 - e^2 \sin^2 B)} + \frac{1}{4e}\ln\frac{1 + e \sin B}{1 - e \sin B}\right] \tag{2.1.9}$$

This is the trapezoidal area formula on the ellipsoid for a longitude difference of one radian, with latitude extending from 0 to B.

Isometric latitude on an ellipsoid converts the geodetic latitude B to a non-dimensional number proportional to the distance of B from the equator on the Mercator projection. It has numerous other uses in conformal mapping:

$$q = \int \frac{M}{r} dB = \ln \left[\tan \left(\frac{\pi}{4} + \frac{B}{2} \right) \left(\frac{1 - e \sin B}{1 + e \sin B} \right)^{e/2} \right] = \ln U \qquad (2.1.10)$$

2.1.2 The earth as a sphere

In cartographic practice, for small-scale maps, the ellipsoidal flattening of the earth is often neglected, and the earth is regarded as a sphere with a particular radius R. Now we introduce some different earth radii in common use:

■ mean radius of sphere:

$$R_e = \frac{a + b + a}{3} \qquad (2.1.11)$$

■ equivalent radius of sphere:

$$R_f = \sqrt{\frac{a^2}{2} + \frac{b^2}{4e} \ln \frac{1 + e}{1 - e}} \qquad (2.1.12)$$

■ equidistant radius of sphere along meridian:

$$R_S = a \left(1 - \frac{1}{4} e^2 - \frac{3}{64} e^4 - \frac{5}{256} e^6 - \frac{175}{16\,384} e^8 \right) \qquad (2.1.13)$$

■ equivolume radius of sphere:

$$R_v = \sqrt[3]{a^2 b} \qquad (2.1.14)$$

■ mean radius of curvature:

$$R = \sqrt{M_0 N_0} \qquad (2.1.15)$$

For the Krasovskiy ellipsoid, we have:

$$R_e = 6\,371\,118 \text{ m}; \quad R_f = 6\,371\,116 \text{ m}; \quad R_s = 6\,367\,558 \text{ m}; \quad R_v = 6\,371\,110 \text{ m}.$$

From the basic formulae (2.1.3)–(2.1.10) of the ellipsoid, when $e = 0$, the elemental formulae of the sphere are obtained as follows (although some calculations using the above equations are indeterminate unless the equations are revised):

$$M = N = R = a \qquad (2.1.16)$$

Radius R instead of semimajor axis a and (φ, λ) instead of (B, l) are often used for a sphere. Thus we have:

$$\left. \begin{array}{l} r = R \cos\varphi, \quad S_m = R\varphi, \quad S_p = \Delta\lambda R \cos\varphi \\[2mm] F = R^2 \sin\varphi, \quad q = \ln \tan \left(\frac{\pi}{4} + \frac{\varphi}{2} \right) \end{array} \right\} \qquad (2.1.17)$$

2.2 AZIMUTHAL PROJECTIONS

2.2.1 Azimuthal projections in common use

For azimuthal projections, from a spherical coordinate system, almucantars (circles of equal distance) can be projected onto circles concentric about the center, and verticals can be projected onto straight lines passing through the center of these circles. The angle separating any two verticals is equal to the corresponding separation angle on the sphere. Formulae for rectangular coordinates for azimuthal projections have the form:

$$x = \rho \sin \delta, \quad y = \rho \cos \delta \tag{2.2.1}$$

where $\rho = f(Z)$, $\delta = a$, (Z, a) are spherical polar coordinates.

Expressions for conversions between spherical polar coordinates (Z, a) and geographic coordinates (φ, λ) can be written in the form.

$$\left. \begin{aligned} \cos Z &= \sin\varphi \sin\varphi_0 + \cos\varphi \cos\varphi_0 \cos(\lambda - \lambda_0) \\ \sin Z \cos a &= \sin\varphi \cos\varphi_0 - \cos\varphi \sin\varphi_0 \cos(\lambda - \lambda_0) \\ \sin Z \sin a &= \cos\varphi \sin(\lambda - \lambda_0) \end{aligned} \right\} \tag{2.2.2}$$

and:

$$\cot a = \tan\varphi \cos\varphi_0 \csc(\lambda - \lambda_0) - \sin\varphi_0 \cot(\lambda - \lambda_0) \tag{2.2.3}$$

Distortion formulae for azimuthal projections take the form:

$$\left. \begin{aligned} &\text{linear scale factor along a radial } \mu_1 = \frac{d\rho}{R\,dZ} \\[2mm] &\text{linear scale factor along an almucantar } \mu_2 = \frac{\rho}{R \sin Z} \\[2mm] &\text{area scale factor } P = \mu_1 \mu_2 \\[2mm] &\text{maximum angular deformation } \sin\frac{\omega}{2} = \left| \frac{\mu_2 - \mu_1}{\mu_2 + \mu_1} \right| \end{aligned} \right\} \tag{2.2.4}$$

It becomes clear that the difference between different azimuthal projections is the different form taken by the $\rho = f(Z)$ function.

For the gnomonic projection, we have:

$$\rho = R \tan Z \tag{2.2.5}$$

For the stereographic projection, we have:

$$\rho = 2R \tan\frac{Z}{2} \tag{2.2.6}$$

For the azimuthal equivalent projection, we have:

$$\rho = 2R \sin\frac{Z}{2} \tag{2.2.7}$$

For the azimuthal equidistant projection, we have:

$$\rho = RZ \tag{2.2.8}$$

For the orthographic projection, we have:

$$\rho = R \sin Z \qquad\qquad\qquad\qquad\qquad\qquad\qquad\qquad (2.2.9)$$

If the datum surface is an ellipsoid, a normal conformal azimuthal projection has the form:

$$\rho = \frac{C}{U}, \quad \delta = l \qquad\qquad\qquad\qquad\qquad\qquad\qquad (2.2.10)$$

where:

$$C = n_0 \cdot 2 N_{90^\circ} \left(\frac{1-e}{1+e}\right)^{e/2}, \quad U = \tan\left(\frac{\pi}{4} + \frac{B}{2}\right)\left(\frac{1 - e\sin B}{1 + e\sin B}\right)^{e/2}$$

2.2.2 Perspective azimuthal projection of the ellipsoid from an exterior point

Projection from an exterior point of the ellipsoid onto a perpendicular pictorial plane

From Bugayevskiy and Snyder (1995), this projection has the formulae:

$$X = \frac{H N_0' \sin Z \sin a}{D - N_0' \cos Z}, \quad Y = \frac{H N_0' \sin Z \cos a}{D - N_0' \cos Z} \qquad\qquad (2.2.11)$$

where:

$$N_0' \sin Z \cos a = N[\sin B \sin B_0 - \cos B \cos B_0 \cos(L - L_0)] + e^2(N_0 \sin B_0 - N \sin B)\cos B_0$$
$$N_0' \sin Z \sin a = N \cos B \sin(L - L_0)$$
$$N_0' \cos Z = N[\sin B \sin B_0 + \cos B \cos B_0 \cos(L - L_0)] + e^2(N_0 \sin B_0 - N \sin B)\sin B_0$$
$$D = N_0 + H$$

and H is the distance of the camera above the ellipsoid, in the same units as N.

Perspective azimuthal projection from an exterior viewpoint of the ellipsoid onto an inclined pictorial plane

For space photography, if the elements of inner and outer orientation are known, when determining coordinates (X, Y) on a horizontal pictorial plane, the values $a' = a - t$ are used instead of spherical values a (t is the direction of 'the plane of the base vertical' or the plane of photography). Then we calculate the abscissas and ordinates (X, Y) rotated by angle t in azimuth, and then the coordinate transformation matrix; using the other angles (ε_0, χ) will take the form:

$$A = \begin{pmatrix} \cos\varepsilon_0 \cos\chi & -\cos\varepsilon_0 \sin\chi & -\sin\varepsilon_0 \\ \sin\chi & \cos\chi & 0 \\ \sin\varepsilon_0 \cos\chi & -\sin\varepsilon_0 \sin\chi & \cos\varepsilon_0 \end{pmatrix} = \begin{pmatrix} a_1 & a_2 & a_3 \\ b_1 & b_2 & b_3 \\ c_1 & c_2 & c_3 \end{pmatrix} \qquad (2.2.12)$$

The formulae relating coordinates (x, y) of points on the inclined pictorial plane, i.e. on a photograph (with the origin of the coordinates at the optical center), and the coordinates (X, Y) of points on a horizontal pictorial plane (with the origin of the coordinates at the pole Q_0, i.e., the photographic nadir) take the form:

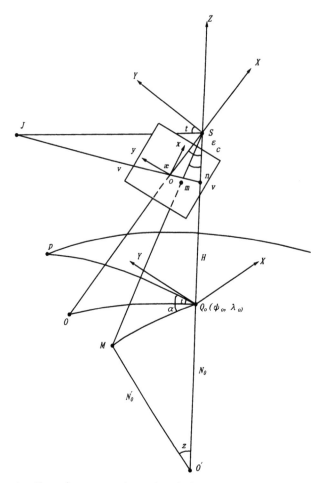

Figure 2.1 Construction of a perspective azimuthal projection with positive transformation of an included pictorial plane

$$y = f\frac{-(X\cos\varepsilon_0 - H\sin\varepsilon_0)\sin\chi + Y\cos\chi}{X\sin\varepsilon_0 + H\cos\varepsilon_0}$$

$$x = f\frac{(X\cos\varepsilon_0 - H\sin\varepsilon_0)\cos\chi + Y\sin\chi}{X\sin\varepsilon_0 + H\cos\varepsilon_0}$$

(2.2.13)

$$Y = H\frac{y\sin\chi + x\cos\chi}{-(\sin\varepsilon_0\cos\chi)y + (\sin\varepsilon_0\sin\chi)x + f\cos\varepsilon_0}$$

$$X = H\frac{(\cos\varepsilon_0\cos\chi)y - (\cos\varepsilon_0\sin\chi)x + f\sin\varepsilon_0}{-(\sin\varepsilon_0\cos\chi)y + (\sin\varepsilon_0\sin\chi)x + f\cos\varepsilon_0}$$

(2.2.14)

where H is the height above photographic nadir Q_0, f is the focal length of the camera lens, ε_0 is the angle between the principal optical beam and the normal to the ellipsoid from the point of view of S in the principal vertical plane, and χ is the angle between the X-axis of the photograph and the principal vertical line, i.e., between the plane of intersection of the principal vertical and that of the photograph.

2.2.3 Roussilhe stereographic projection

This projection has the following properties:

1. The projection is conformal.
2. The projection is symmetrical about the central meridian.
3. The projection is a modification of the oblique stereographic projection adapted to the ellipsoid.
4. The relation of points with coordinate y_0 along the central meridian has the form:

$$y_0 = 2R_0 \tan \frac{\Delta S}{2R_0} \tag{2.2.15}$$

The above may be expanded to a series as follows:

$$y_0 = \Delta S + \frac{\Delta S^3}{12R_0^2} + \frac{\Delta S^5}{120R_0^4} + \ldots \tag{2.2.16}$$

where $R_0 = \sqrt{M_0 N_0}$.

From the theory of map projections, equations for conformal projections take the basic form:

$$y + ix = f(q + il) \tag{2.2.17}$$

and the inverse takes the form:

$$q + il = F(y + ix) \tag{2.2.18}$$

where (q, l) are isometric coordinates.

In the Roussilhe stereographic projection, according to condition 3, when $l = x = 0$, then we have:

$$y_0 = f(q) \tag{2.2.19}$$

and:

$$q = F(y_0) \tag{2.2.20}$$

In order to obtain constant coefficients for a formula directly or inversely solving for the Roussilhe stereographic projection, Equations (2.2.19) and (2.2.18) need to be expanded in a Taylor series from an initial point along the central meridian.

$$\Delta B = B - B_0, \quad \Delta y = y - y_0, \quad \Delta q = q - q_0, \quad l = L - L_0$$
$$\Delta y + ix = a_1(\Delta q + il) + a_2(\Delta q + il)^2 + a_3(\Delta q + il)^3 + a_4(\Delta q + il)^4 + \ldots \tag{2.2.21}$$

and:

$$\Delta q + il = b_1(\Delta y + ix) + b_2(\Delta y + ix)^2 + b_3(\Delta y + ix)^3 + b_4(\Delta y + ix)^4 + \ldots \tag{2.2.22}$$

where:

$$a_n = \frac{1}{n!} \left(\frac{d^n(y + ix)}{d(q + il)^n} \right)_0, \quad b_n = \frac{1}{n!} \left(\frac{d^n(q + il)}{d(y + ix)^n} \right)_0$$

Noting that $l = x = 0$, we have:

$$a_n = \frac{1}{n!}\left(\frac{d^n y_0}{dq^n}\right)_0 \tag{2.2.23}$$

and:

$$b_n = \frac{1}{n!}\left(\frac{d^n q}{dy_0^n}\right)_0 \tag{2.2.24}$$

After separating real and imaginary parts of (2.2.21) and (2.2.22), constant coefficient formulae for direct and inverse solution of conformal projections are obtained:

$$\left.\begin{aligned}
\Delta y &= a_1\Delta q + a_2(\Delta q^2 - l^2) + a_3(\Delta q^3 - 3\Delta q l^2) \\
&\quad + a_4(\Delta q^4 - 6\Delta q^2 l^2 + l^4) + a_5(\Delta q^5 - 10\Delta q^3 l^2 + 5\Delta q l^4) \\
x &= a_1 l + a_2 \cdot 2\Delta q l + a_3(3\Delta q^2 l - l^3) + a_4(4\Delta q^3 l - 4\Delta q l^3) \\
&\quad + a_5(5\Delta q^4 l - 10\Delta q^2 l^3 + l^5)
\end{aligned}\right\} \tag{2.2.25}$$

and:

$$\left.\begin{aligned}
\Delta q &= b_1\Delta y + b_2(\Delta y^2 - x^2) + b_3(\Delta y^3 - 3\Delta y x^2) \\
&\quad + b_4(\Delta y^4 - 6\Delta y^2 x^2 + x^4) + b_5(\Delta y^5 - 10\Delta y^3 x^2 + 5\Delta y x^4) \\
l &= b_1 x + b_2 \cdot 2\Delta y x + b_3(3\Delta y^2 x - x^3) + b_4(4\Delta y^3 x - 4\Delta y x^3) \\
&\quad + b_5(5\Delta y^4 x - 10\Delta y^2 x^3 + x^5)
\end{aligned}\right\} \tag{2.2.26}$$

By the derivation, the constant coefficients for the Roussilhe projection are obtained:

$$\left.\begin{aligned}
a_1 &= N_0\cos B_0, \quad a_2 = \frac{1}{2}N_0\cos^2 B_0 t_0(-1) \\
a_3 &= \frac{1}{6}N_0\cos^3 B_0\left(-\frac{1}{2} + t_0^2 - \frac{1}{2}\eta_0^2\right) \\
a_4 &= \frac{1}{24}N_0\cos^4 B_0 t_0(2 - t_0^2 + 6\eta_0^2) \\
a_5 &= \frac{1}{120}N_0\cos^5 B_0\left(1 - \frac{1}{2}t_0^2 + t_0^4\right)
\end{aligned}\right\} \tag{2.2.27}$$

and:

$$\left.\begin{aligned}
b_1 &= \frac{1}{N_0\cos B_0}, \quad b_2 = \frac{1}{2N_0^2\cos B_0}t_0 \\
b_3 &= \frac{1}{6N_0^3\cos B_0}\left(\frac{1}{2} - 12t_0^2 + \frac{1}{2}\eta_0^2\right) \\
b_4 &= \frac{1}{24N_0^4\cos B_0}(3 + 6t_0^2 - \eta_0^2) \\
b_5 &= \frac{1}{120N_0^5\cos B_0}\left(\frac{3}{2} + 18t_0^2 + 24t_0^2\right)
\end{aligned}\right\} \tag{2.2.28}$$

where:

$$t_0 = \tan B_0, \quad \eta_0^2 = e'^2 \cos^2 B_0.$$

2.3 CYLINDRICAL PROJECTIONS

2.3.1 Cylindrical projections in common use

By definition, in developing a cylindrical projection from a geographic coordinate system, parallels are projected into horizontal parallel lines and meridians are projected into vertical parallel lines, the interval between two meridians being in direct ratio to the corresponding difference in longitude. The formulae for rectangular coordinates of cylindrical projections have the form:

$$x = Cl, \quad y = f(B) \tag{2.3.1}$$

The distortion formulae for cylindrical projections take the form:

$$m = \frac{dy}{MdB}, \quad n = \frac{C}{r}, \quad p = mn, \quad \sin\frac{\omega}{2} = \left|\frac{m-n}{m+n}\right| \tag{2.3.2}$$

where m is the linear scale factor along a meridian, n is the linear scale factor along a parallel, p is the area scale factor, and ω is the maximum angular distortion.

The direct conformal cylindrical projection (Mercator projection) has the form:

$$x = r_0 l, \quad y = r_0 q, \quad \mu = \frac{r_0}{r} \tag{2.3.3}$$

where:

$$q = \int_0^B \frac{M}{r} dB$$

The direct equivalent cylindrical (or cylindrical equal-area) projection has the form:

$$x = r_0 l, \quad y = \frac{1}{r_0} F \tag{2.3.4}$$

where:

$$F = \int_0^B MrdB$$

The direct equidistant cylindrical projection has the form:

$$x = r_0 l, \quad y = S \tag{2.3.5}$$

where:

$$S = \int_0^B MdB$$

In order to obtain oblique and transverse cylindrical projections, we initially consider the earth as a sphere to build the spherical coordinate system. In the spherical coordinate system, the general formulae for cylindrical projections have the form:

$$x = Ca, \quad y = f(Z)$$
(2.3.6)

The coordinate formulae for transverse tangent conformal cylindrical projections take the form:

$$x = R \ln \cot \frac{Z}{2}, \quad y = R\left(\frac{\pi}{2} - a\right)$$
(2.3.7)

The above takes the tangent central meridian as the y-axis, so the (x, y) coordinates of formula (2.3.3) are respectively interchanged. Coordinate formulae for an oblique conformal cylindrical projection have the form:

$$x = C(\pi - a), \quad y = R \ln \cot \frac{Z}{2}$$
(2.3.8)

The coordinate formulae for a transverse equivalent cylindrical projection have the form:

$$x = R \cos Z, \quad y = R\left(\frac{\pi}{2} - a\right)$$
(2.3.9)

The coordinate formulae for the transverse equidistant cylindrical (Cassini) projection have the form:

$$x = R\left(\frac{\pi}{2} - Z\right), \quad y = R\left(\frac{\pi}{2} - a\right)$$
(2.3.10)

2.3.2 Gauss–Krüger (ellipsoidal transverse Mercator) projection

The Gauss–Krüger projection is geometrically a transverse tangent ellipsoidal cylindrical conformal projection. It is also called a transverse Mercator projection, but its datum is an ellipsoid.

The Gauss–Krüger projection has the following requirements and characteristics:

1. The projections of meridians and parallels are curves symmetrical about the central meridian.

2. No angular distortion is introduced by the projection transformation.

3. No distortion of length occurs along the central meridian of the projection.

According to projection requirement 3, if $l = x = 0$, then $y_0 = S$. Thus the coefficients for the direct and inverse coordinate transformation formulae (2.2.25) and (2.2.26) for conformal projection are obtained:

$$a_n = \frac{1}{n!}\left(\frac{d^n S}{dq^n}\right)_0$$
(2.3.11)

and:

$$b_n = \frac{1}{n!}\left(\frac{d^n q}{dS^n}\right)_0$$
(2.3.12)

Through computation, constant coefficients for the formulae for direct and inverse coordinate transformation for the Gauss–Krüger projection are obtained respectively:

$$a_1 = N_0 \cos B_0, \quad a_2 = -\frac{1}{2}N_0 t_0(-1)\cos^2 B_0$$

$$a_3 = -\frac{1}{6}N_0 \cos^3 B_0(-1 + t_0^2 - \eta_0^2)$$

$$a_4 = \frac{1}{24}N_0 \cos^4 B_0 \cdot t_0(5 - t_0^2 + 9\eta_0^2 + 4\eta_0^4)$$

$$a_5 = \frac{1}{120}N_0 \cos^5 B_0(5 - 18t_0^2 + t_0^4 + 14\eta_0^2 - 58t_0^2\eta_0^2)$$

$$a_6 = -\frac{1}{720}N_0 \cos^6 B_0 \cdot t_0(-61 + 58t_0^2 - t_0^4 - 270\eta_0^2 + 330t_0^2\eta_0^2)$$

(2.3.13)

and:

$$b_1 = \frac{1}{N_0 \cos B_0}, \quad b_2 = \frac{1}{2N_0^2 \cos B_0}t_0$$

$$b_3 = \frac{1}{6N_0^3 \cos B_0}(1 + 2t_0^2 + \eta_0^2)$$

$$b_4 = \frac{1}{24N_0^4 \cos B_0}t_0(5 + 6t_0^2 + \eta_0^2 - 4\eta_0^4)$$

$$b_5 = \frac{1}{120N_0^5 \cos B_0}(5 + 28t_0^2 + 24t_0^4 + 6\eta_0^2 + 8t_0^2\eta_0^2)$$

$$b_6 = \frac{1}{720N_0^6 \cos B_0}t_0(61 + 180t_0^2 + 120t_0^4 + 46\eta_0^2 + 48t_0^2\eta_0^2)$$

(2.3.14)

where:

$$t_0 = \tan B_0, \quad \eta_0^2 = e'^2 \cos^2 B_0$$

Now the formula with variable coefficients of the Gauss–Krüger projection can be solved. For the direct solution formula for the Gauss–Krüger projection, suppose point 0 and point Q coincide, i.e.:

$$B_0 = B, \Delta B = 0;$$
$$y_0 = S, \Delta y = y - S;$$
$$q_0 = q, \Delta q = 0$$

(2.3.15)

Substituting the above into formula (2.2.25), we have:

$$\left.\begin{array}{l} y - S = -a_2 l^2 + a_4 l^4 - a_6 l^6 + \dots \\ x = a_1 l - a_3 l^3 + a_5 l^5 - \dots \end{array}\right\}$$

(2.3.16)

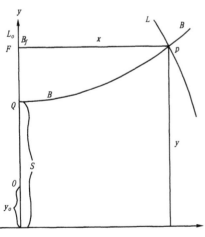

Figure 2.2 Coordinate system for Gauss–Krüger projection

Noting formula (2.3.13) and using B instead of B_0, then we have:

$$y = S + \frac{1}{2}Nt\cos^2 B \cdot l^2 + \frac{1}{24}Nt(5 - t^2 + 9\eta^2 + 4\eta^4)\cos^4 B \cdot l^4$$

$$+ \frac{1}{720}Nt(61 - 58t^2 + t^4 + 270\eta^2 - 330t^2\eta^2)\cos^6 B \cdot l^6$$

$$+ \frac{1}{40\,320}Nt(1\,385 - 3\,111t^2 + 543t^4 - t^6)\cos^6 Bl^8$$

$$x = N\cos B \cdot l + \frac{1}{6}N(1 - t^2 + \eta^2)\cos^3 B \cdot l^3$$

$$+ \frac{1}{120}N(5 - 18t^2 + t^4 + 14\eta^2 - 58t^2\eta^2)\cos^5 B \cdot l^5$$

$$+ \frac{1}{5\,040}N(61 - 479t^2 + 179t^4 - t^6)\cos^7 B \cdot l^7 + \ldots$$

(2.3.17)

where:

$$t = \tan B, \quad \eta^2 = e'^2\cos^2 B$$

and S is the meridian arc length from the equator to latitude B.

Inverse solution formula for the Gauss–Krüger projection:

Suppose point 0 and point F coincide, i.e.,

$$B_0 = B_f, \quad \Delta B = B - B_f; \quad y_0 = y, \quad \Delta y = 0; \quad q_0 = q_f, \quad \Delta q = q - q_f \tag{2.3.18}$$

Substituting the above into formula (2.2.26), we have:

$$q - q_f = -(b_2)_f x^2 + (b_4)_f x^4 - (b_6)_f x^6 + \ldots$$
$$l = (b_1)_f x - (b_3)_f x^3 + (b_5)_f x^5 - \ldots$$

(2.3.19)

Noting formula (2.3.14) and using B_f instead of B_0, after transformation, we have:

$$B = B_f + \frac{1}{2N_f^2}t_f(-1 - \eta_f^2)x^2$$

$$+ \frac{1}{24N_f^4}t_f(5 + 3t_f^2 + 6\eta_f^2 - 6t_f^2\eta_f^2 - 3\eta_f^4 - 9t_f^2\eta_f^4)x^4$$

$$+ \frac{1}{720N_f^6}t_f(-61 - 90t_f^2 - 45t_f^4 - 107\eta_f^2 + 162t_f^2\eta_f^2 + 45t_f^4\eta_f^2)x^6 + \ldots$$

$$l = \frac{1}{N_f\cos B_f}x + \frac{1}{6N_f^3\cos B_f}(-1 - 2t_f^2 - \eta_f^2)x^3$$

$$+ \frac{1}{120N_f^5\cos B_f}(5 + 28t_f^2 + 24t_f^4 + 6\eta_f^2 + 8t_f^2\eta_f^2)x^5 + \ldots$$

(2.3.20)

where B_f is the footpoint latitude, i.e., $y = S_f$.

The expressions of the above direct and inverse solution formulae use an expansion point along the central meridian, and belong to the category of variable coefficients, i.e. coefficients (a_n, b_n) change along latitude B (or y).

The Universal Transverse Mercator projection (UTM projection) is closely related to the Gauss–Krüger projection, where (x_u, y_u) are UTM coordinates and (x_G, y_G) are GK coordinates:

$$x_u = m_0 x_G, \quad y_u = m_0 y_G \tag{2.3.21}$$

and $m_0 = 0.9996$.

2.3.3 Transverse cylindrical conformal projection with two standard meridians (Li 1981)

A transverse cylindrical conformal projection with two standard meridians is also called a transverse Mercator projection with two standard meridians, or a secant transverse Mercator. The characteristics of this projection are as follows:

1. The projections of meridians and parallels are curves symmetrical about the central meridian.

2. No angular distortion is introduced by the projection transformation.

3. The proportion of length or scale factor along the central meridian is:

$$m_0 = 1 - \frac{l_1^2}{2} \cos^2 B (1 + \eta^2).$$

According to projection characteristic 3, we have:

$$y_0 = \int_0^B m_0 M dB = \int_0^B M \left[1 - \frac{l_1^2}{2} \cos^2 B (1 + \eta^2) \right] dB \tag{2.3.22}$$

The formulae for coordinates on the transverse cylindrical conformal projection with double standard meridians are obtained after derivation:

$$
\left.
\begin{aligned}
y &= y_0 + \frac{1}{2} N t \cos^2 B \left[1 - \frac{l_1^2}{2} \cos^2 B (3 + 7\eta^2) \right] l^2 \\
&\quad + \frac{1}{24} N t \cos^4 B \left[(5 - t^2 + 9\eta^2 + 4\eta^4) \right. \\
&\quad \left. - \frac{l_1^2}{2} \cos^2 B (33 - 27 t^2 + 182\eta^2) \right] l^4 + \frac{1}{720} N t \cos^6 B (61 - 58 t^2 + t^4) l^6 \\
x &= N \cos B \left[1 - \frac{l_1^2}{2} \cos^2 B (1 + \eta^2) \right] l \\
&\quad + \frac{1}{6} N \cos^3 B \left[(1 - t^2 + \eta^2) - \frac{l_1^2}{2} \cos^2 B (3 - 9 t^2 + 10\eta^2 - 41 t^2\eta^2) \right] l^3 \\
&\quad + \frac{1}{120} N \cos^5 B \left[(5 - 18 t^2 + t^4 + 14\eta^2 - 58 t^2\eta^2) \right. \\
&\quad \left. - \frac{l_1^2}{2} \cos^2 B (33 - 246 t^2) \right] l^5
\end{aligned}
\right\} \tag{2.3.23}
$$

$$y_0 = S - \frac{a l_1^2}{2} \left[A' \mathrm{arc} B + \frac{B'}{2} \sin 2B + \frac{C'}{4} \sin 4B + \frac{D'}{6} \sin 6B + \frac{E'}{8} \sin 8B \right] \tag{2.3.24}$$

where S is the meridian arc length and l_1 is the difference in longitude between the two standard meridians.

$$A' = \frac{1}{2} + \frac{1}{16}e^2 + \frac{3}{128}e^4 + \frac{25}{2\,048}e^6 + \frac{245}{32\,768}e^8$$

$$B' = \frac{1}{2} - \frac{3}{256}e^4 - \frac{5}{512}e^6 - \frac{345}{32\,768}e^8$$

$$C' = -\frac{1}{16}e^2 - \frac{3}{128}e^4 - \frac{5}{512}e^6 - \frac{35}{8\,192}e^8 \qquad (2.3.25)$$

$$D' = \frac{3}{256}e^4 + \frac{5}{512}e^6 + \frac{455}{65\,536}e^8$$

$$E' = -\frac{5}{2\,048}e^6 - \frac{105}{32\,768}e^8$$

2.3.4 Gauss–Krüger projection family (Yang 1983c)

The Gauss–Krüger projection family may also be called the transverse Mercator projection family. The projection characteristics of the Gauss–Krüger projection family are as follows:

1. The projection of meridians and parallels are curves symmetrical about the central meridian.

2. No angular distortion is introduced by the projection transformation.

3. The proportion of length along the central meridian is $m_0 = f(B)$.

The Gauss–Krüger projection family consists of sets of conformal projections within a zone about a chosen meridian, its general coordinate formulae having the form:

$$\left.\begin{array}{l} y = y_0 - a_2 l^2 + a_4 l^4 - a_6 l^6 \\ x = a_1 l - a_3 l^3 + a_5 l^5 \end{array}\right\} \qquad (2.3.26)$$

where $y_0 = \displaystyle\int_0^B m_0 M\, dB$, $m_0 = f(B)$, and M is the radius of curvature along a meridian at a given latitude.

$$M = \frac{a(1 - e^2)}{(1 - e^2 \sin^2 B)^{3/2}}, \qquad a_n = \frac{1}{n!}\left(\frac{d^n x_0}{dq^n}\right)$$

After derivation, coefficients a_n are obtained as follows:

$$a_1 = m_0 r, \quad a_2 = \frac{1}{2}F(m_0 r)', \quad a_3 = -\frac{1}{3}F' a_2 + \frac{1}{6}F^2(m_0 r)''$$

$$a_4 = -\frac{1}{4}F' a_3 + \frac{1}{24}F^2[F''(m_0 r)' + 2F'(m_0 r)'' + F(m_0 r)''']$$

$$\begin{aligned} a_5 &= \frac{1}{5}F' a_4 + \frac{1}{120}F^2[(3F'F'' + FF''')(m_0 r)' + (4F'^2 + 4FF'')(m_0 r)'' \\ &\quad + 5FF'(m_0 r)''' + F^2(m_0 r)^{IV}] \end{aligned}$$

$$\begin{aligned} a_6 &= -\frac{1}{6}F' a_5 - \frac{1}{720}F^2[(7F'^2 F'' + 4FF''^2 + 6FF'F''' + F^2 F^{IV})(m_0 r)' \\ &\quad + (8F'^3 + 26FF'F'' + 5F^2 F''')(m_0 r)'' + (19FF'^2 + 10F^2 F'')(m_0 r)''' \\ &\quad + 9F^2 F'(m_0 r)^{IV} + F^3(m_0 r)^V] \end{aligned} \qquad (2.3.27)$$

where:

$$F = \cos B(1 + \eta^2), \quad F' = -\sin B(1 + 3\eta^2), \quad F'' = -\cos B(1 - 6e'^2 + 9\eta^2),$$
$$F''' = \sin B(1 - 6e'^2 + 27\eta^2), \quad F^{IV} = \cos B(1 - 60e'^2 + 81\eta^2)$$
$$(m_0 r)' = m_0 r' + m_0' r, \quad (m_0 r)'' = m_0 r'' + 2m_0' r' + m_0'' r$$
$$(m_0 r)''' = m_0 r''' + 3m_0' r'' + 3m_0'' r' + m_0''' r$$
$$(m_0 r)^{IV} = m_0 r^{IV} + 4m_0' r''' + 6m_0'' r'' + 4m_0''' r' + m_0^{IV} r$$
$$(m_0 r)^{V} = m_0 r^{V} + 5m_0' r^{IV} + 10m_0'' r''' + 10m_0''' r'' + 5m_0^{IV} r' + m_0^{V} r$$
$$r = a \cos B/(1 - e^2 \sin^2 B)^{1/2}$$
$$r' = -M \sin B = -a(1 - e^2)\sin B \cdot G$$
$$r'' = -a(1 - e^2)(\cos B \cdot G + \sin B \cdot G')$$
$$r''' = -a(1 - e^2)(-\sin B \cdot G + 2 \cos B \cdot G' + \sin B \cdot G'')$$
$$r^{IV} = -a(1 - e^2)(-\cos B \cdot G - 3 \sin B \cdot G' + 3\cos B \cdot G'' + \sin B \cdot G''')$$
$$r^{V} = -a(1 - e^2)(\sin B \cdot G - 4 \cos B \cdot G' - 6\sin B \cdot G'' + 4\cos B \cdot G''' + \sin B \cdot G^{IV})$$
$$G = A' - B'\cos 2B + C'\cos 4B - D'\cos 6B + E'\cos 8B$$
$$G' = 2B'\sin 2B - 4C'\sin 4B + 6D'\sin \cos 6B - 8E'\sin 8B$$
$$G'' = 4B'\cos 2B - 16C'\cos 4B + 36D'\cos 6B - 64E'\cos 8B$$
$$G''' = -8B'\sin 2B + 64C'\sin 4B - 216D'\sin 6B + 512E'\sin 8B$$
$$G^{IV} = -16B'\cos 2B + 256C'\cos 4B - 1\,296D'\cos 6B + 4\,096E'\cos 8B$$

For A', B', C', D', E' see formula (3.5.20).

For the conformal projection system in which $m_0 = 1 - q \cos^2 KB$ we have:

$$\left. \begin{array}{l} m_0' = Kq \sin 2KB, \quad m_0'' = 2K^2 q \cos 2KB, \quad m_0''' = -4K^3 q \sin 2KB \\ m_0^{IV} = -8K^4 q \cos 2KB, \quad m_0^{V} = 16K^5 q \sin 2KB \end{array} \right\} \qquad (2.3.28)$$

$$y_0 = \int_0^B m_0 M dB = \int_0^B (1 - q\cos^2 KB) M dB = \int_0^B \left(1 - \frac{q}{2} - \frac{q}{2}\cos 2KB\right) M dB$$

$$= \left(1 - \frac{q}{2}\right) S - \frac{q}{8} a(1 - e^2)\left\{ A'\left[\frac{1}{K}\sin 2KB + \frac{1}{K}\sin 2KB\right] \right.$$

$$- B'\left[\frac{1}{K+1}\sin 2(1 + K)B + \frac{1}{1-K}\sin 2(1 - K)B\right]$$

$$+ C'\left[\frac{1}{2+K}\sin 2(2 + K)B + \frac{1}{2-K}\sin 2(2 - K)B\right]$$

$$- D'\left[\frac{1}{3+K}\sin 2(3 + K)B + \frac{1}{3-K}\sin 2(3 - K)B\right]$$

$$\left. + E'\left[\frac{1}{4+K}\sin 2(4 + K)B + \frac{1}{4-K}\sin 2(4 - K)B\right]\right\}$$

When $K = 0$,

$$\frac{1}{K}\sin 2KB = 2B;$$

and when $K = n$,

$$\frac{1}{n-K}\sin 2(n-K)B = 2B, \quad (n = 1, 2, 3, 4)$$

In a conformal projection system where $m_0 = 1 - q\cos^2 KB$, with properly selected parameters (q, k), several conformal projection arrangements can be obtained. For example, with $q = 0$, i.e. $m_0 = 1$, we have the Gauss–Krüger projection; with $q = 0.000609$ and $K = 1$, the transverse cylindrical conformal projection with double standard meridian is obtained; if $q = 0.0004$ and $K = 0$, the Universal Transverse Mercator projection is obtained with $m_0 = 0.9996$.

The value of $y_0 = \displaystyle\int_0^B m_0 M dB$ can be calculated by the method of numerical integration.

In addition, there is the conformal projection system:

$$m_0 = \sum_{i=0}^{n} a_{2i} B^{2i}$$

This is a conformal projection system with great flexibility. Parameter a_{2i} of this projection can be determined by applying a given distortion distribution along the central meridian. In addition, there is a conformal projection system in which $m_0 = 1 - q\cos^2 KB$.

This projection system is suitable for the projection plan for sheet maps. When parameters (q, k) are properly selected, the meridian length stretch at the margin of the sheet is less than 0.1 mm. For example, $m_0 = 0.99875 - 4.524 \times 10^{-5}\sin^2 45B$ is used for the 1:1 000 000-scale quadrangle or sheet projection plan. Then the maximum vector length of the outer parallel has a scale factor $h = 0.0998$ mm, and the relative distortion of length on the projection varies from -0.00125 to $+0.0012675$.

2.4   CONIC PROJECTIONS

2.4.1   Conic projections in common use

By definition, from a geographic coordinate system parallels can be projected as concentric circular arcs, and meridians can be projected as a set of equally spaced lines passing through the center of the circles. The angle of separation of two meridians is in direct ratio to the corresponding difference in longitude. This produces conic projections, for which the formulae for rectangular coordinates have the form:

$$x = \rho\sin\delta, \quad y = \rho_s - \rho\cos\delta \qquad (2.4.1)$$

where $\rho = f(B)$, $\delta = al$.

The distortion formulae for conic projections take the form:

$$m = -\frac{d\rho}{MdB}, \quad n = \frac{\alpha\rho}{r}, \quad P = mn, \quad \sin\frac{\omega}{2} = \left|\frac{m-n}{m+n}\right| \qquad (2.4.2)$$

The minimum scale factor along the central parallel on a conic projection has the form:

$$n_0 = \frac{m_0\alpha}{\sin B_0} \qquad (2.4.3)$$

and the radius of the projected parallel with the least scale exaggeration is along the parallel B_0:

$$\rho_0 = m_0 N_0 \cot B_0 \qquad (2.4.4)$$

On the conformal conic projection, developed by Lambert and often called the Lambert conformal conic projection, usually with two standard parallels, although one standard parallel may also be used,

$$\rho = \frac{C}{U^{\alpha}}, \quad \delta = al \tag{2.4.5}$$

where:

$$U = \tan\left(\frac{\pi}{4} + \frac{B}{2}\right)\left(\frac{1 - e\sin B}{1 + e\sin B}\right)^{e/2}, \quad \alpha = \sin B_0, \quad \rho_0 = n_0 N_0 \cot B_0, \quad C = \rho_0 U_0^{\alpha} \tag{2.4.6}$$

The following methods may be used to determine parameters (α, C).

1. From (B_0, n_0):

$$\alpha = \sin B_0, \quad C = n_0 N_0 \cot B_0 \cdot U_0^{\alpha} \tag{2.4.7}$$

2. From the two standard parallels (B_1, B_2):

$$\alpha = \frac{\ln r_1 - \ln r_2}{\ln U_2 - \ln U_1}, \quad C = \frac{r_1 U_1^{\alpha}}{\alpha} = \frac{r_2 U_2^{\alpha}}{\alpha} \tag{2.4.8}$$

3. Equalizing the absolute values of the distortion along the outer and central parallels:

$$\alpha = \frac{\ln r_S - \ln r_N}{\ln U_N - \ln U_S}, \quad C = \frac{2 r_m r_N U_m^{\alpha} U_N^{\alpha}}{\alpha(r_m U_m^{\alpha} + r_N U_N^{\alpha})} \tag{2.4.9}$$

If the latitude difference is not large in the mapped area, we have:

$$C = \frac{1}{\alpha}\sqrt{r_N r_m U_N^{\alpha} U_m^{\alpha}} \tag{2.4.10}$$

For the equivalent conic projection (often called the Albers equal-area (or equivalent) conic projection), usually with two standard parallels,

$$\rho^2 = \frac{2}{\alpha}(C - F), \quad \delta = al \tag{2.4.11}$$

where:

$$F = \int_0^B Mr\,dB, \quad \alpha = n_0^2 \sin B_0, \quad \rho_0 = m_0 N_0 \cot B_0, \quad C = F_0 + \frac{\alpha}{2}\rho_0^2 \tag{2.4.12}$$

Determining parameters (α, C):

1. From (B_0, n_0):

$$\alpha = \sin B_0, \quad C = n_0 N_0 \cot B_0 \cdot U_0^{\alpha} \tag{2.4.13}$$

2. From the two standard parallels (B_1, B_2):

$$\alpha = \frac{1}{2} \cdot \frac{r_1^2 - r_2^2}{F_2 - F_1}, \quad C = \frac{r_1^2 F_2 - r_2^2 F_1}{r_1^2 - r_2^2} \tag{2.4.14}$$

For the equidistant conic projection,

$$\rho = C - S, \quad \delta = al \tag{2.4.15}$$

where:

$$S = \int_0^B M\,dB, \quad \alpha = n_0 \sin B_0, \quad \rho_0 = N_0 \cot B_0, \quad C = \rho_0 + S_0 \tag{2.4.16}$$

Determining parameters (α, C):

1. From (B_0, n_0):

$$\alpha = n_0 \sin B_0, \quad C = N_0 \cot B_0 + S_0 \tag{2.4.17}$$

2. From the two standard parallels (B_1, B_2):

$$\alpha = \frac{r_1 - r_2}{S_2 - S_1}, \quad C = \frac{r_1 S_2 - r_2 S_1}{r_1 - r_2} \tag{2.4.18}$$

To obtain oblique conic projections, taking the earth as a sphere, the spherical coordinate system is utilized. In the spherical coordinate system, polar coordinate formulae for conic projections have the form:

$$\rho = f(Z), \quad \delta = \alpha(\pi - a) \tag{2.4.19}$$

For example, formulae for the oblique conformal conic projection take the form:

$$\rho = C \cdot \tan^\alpha \frac{Z}{2}, \quad \delta = \alpha(\pi - a) \tag{2.4.20}$$

where:

$$\alpha = \cos Z_0, \quad \rho_0 = n_0 R \tan Z_0, \quad C = \rho_0 \cot^\alpha \frac{Z_0}{2} \tag{2.4.21}$$

Oblique equivalent conic projections have the form:

$$\rho^2 = \frac{2}{\alpha}(C - R^2 \cos Z), \quad \delta = \alpha(\pi - a) \tag{2.4.22}$$

where:

$$\alpha = n_0^2 \cos Z_0, \quad \rho_0 = \frac{1}{n_0} R \tan Z_0, \quad C = R^2 \cos Z_0 \left(1 + \frac{1}{2}\tan^2 Z_0\right) \tag{2.4.23}$$

Oblique equidistant conic projections have the form:

$$\rho = C - R\left(\frac{\pi}{2} - Z\right), \quad \delta = \alpha(\pi - a) \tag{2.4.24}$$

where:

$$\alpha = n_0 \cos Z_0, \quad \rho_0 = R \tan Z_0, \quad C = R \tan Z_0 + R\left(\frac{\pi}{2} - Z_0\right) \tag{2.4.25}$$

2.4.2 Arbitrary conic projection (Yang 1965)

From (2.4.2), the integration formula for $r(\rho)$ for different characteristics in conic projections can be developed as follows. For the relation of the linear scale factor m and $r(\rho)$ along a meridian:

Given $d\rho = -mMdB$, we have:

$$\rho = C - \int_{B_s}^{B} mMdB \tag{2.4.26}$$

The relation of area scale factor P and $r(\rho)$:

Given $P = -\dfrac{d\rho}{MdB} \cdot \dfrac{d\rho}{r}$, we have:

$$\rho^2 = \frac{2}{\alpha}\left(C - \int_0^B PMrdB\right) \tag{2.4.27}$$

The relation of maximum angular distortion ω and $r(\rho)$:

Given $\tan^2\left(\dfrac{\pi}{4} \pm \dfrac{\omega}{4}\right) = \dfrac{m}{n}$; if $\left(\dfrac{m}{n} < 1 \text{ take the '}-\text{' sign}\right)$, i.e. $\tan^2\left(\dfrac{\pi}{4} \pm \dfrac{\omega}{4}\right) = -\dfrac{d\rho}{MdB} \cdot \dfrac{r}{\alpha\rho}$,

we have:

$$\ln\rho = \ln C - \alpha\int_{B_S}^B \tan^2\left(\frac{\pi}{4} \pm \frac{\omega}{4}\right)\frac{M}{r}\,dB \tag{2.4.28}$$

where m, P and $\tan\left(\dfrac{\pi}{4} \pm \dfrac{\omega}{4}\right)$ in the above formulae are functions of latitude B. For conic projections, $n_0' = m_0' = 0$, letting:

$$m = a_0 + a_2B^2 + a_3B^3 \tag{2.4.29}$$

$$P = b_0 + b_2B^2 + b_3B^3 \tag{2.4.30}$$

$$\tan^2\left(\frac{\pi}{4} \pm \frac{\omega}{4}\right) = c_0 + c_2B^2 + c_3B^3 \tag{2.4.31}$$

By establishing the distribution pattern of distortion along the outer and middle parallels, the above function is determined. Then, by equalizing the absolute values of distortion along the outer and middle parallels, i.e., setting $n_S = n_N$ and $n_N - 1 = 1 - n_m$, we can determine constants (C, α). Applying this to formula (2.4.26), we have:

$$\left.\begin{array}{l} \rho = C - S^{(m)}, \quad \delta = \alpha l \\[2mm] m = a_0 + a_2B^2 + a_3B^3 \\[2mm] C = \dfrac{r_S S_N^{(m)}}{r_S - r_N}, \quad \alpha = \dfrac{2r_m r_N}{r_m(C - S_N^{(m)}) + r_N(C - S_m^{(m)})} \end{array}\right\} \tag{2.4.32}$$

where:

$$S^{(m)} = \int_{B_S}^B mMdB, \quad S_N^{(m)} = \int_{B_S}^{B_N} mMdB, \quad S_m^{(m)} = \int_{B_S}^{B_m} mMdB$$

Applying this to formula (2.4.27), we have:

$$\left.\begin{array}{l} \rho^2 = \dfrac{2}{\alpha}(C - S^{(P)}), \quad \delta = \alpha l \\[2mm] P = P_0 + P_2B^2 + P_3B^3 \\[2mm] C = \dfrac{r_S S_N^{(P)}}{r_S^2 - r_N^2}, \quad \alpha = \dfrac{2r_m^2 r_N^2}{\left(r_m\sqrt{C - S_N^{(P)}} + r_N\sqrt{C - S_m^{(P)}}\right)} \end{array}\right\} \tag{2.4.33}$$

where:

$$S^{(P)} = \int_{B_S}^{B} PMrdB, \quad S_N^{(P)} = \int_{B_S}^{B_N} PMrdB, \quad S_m^{(P)} = \int_{B_S}^{B_m} PMrdB$$

Applied to formula (2.4.28), this provides:

$$\left.\begin{aligned}
\rho &= \frac{C}{U^{(\omega)\alpha}}, \quad \delta = \alpha l \\[2mm]
\tan^2 &\left(\frac{\pi}{4} \pm \frac{\omega}{4}\right) = c_0 + c_2 B^2 + c_3 B^3 \\[2mm]
\alpha &= \frac{\ln r_S - \ln r_N}{S_N^{(\omega)}}, \quad C = \frac{2 r_m r_N U_m^{(\omega)\alpha} U_N^{(\omega)\alpha}}{\alpha(r_m U_m^{(\omega)\alpha} + r_N U_N^{(\omega)\alpha})}
\end{aligned}\right\} \tag{2.4.34}$$

where:

$$S_N^{(\omega)} = \int_{B_S}^{B_N} \tan^2\left(\frac{\pi}{4} \pm \frac{\omega}{4}\right)\frac{M}{r}dB, \quad U^{(\omega)} = \exp(S^{(\omega)}),$$

$$U_N^{(\omega)} = \exp(S_N^{(\omega)}), \quad U_m^{(\omega)} = \exp(S_m^{(\omega)})$$

With formulae (2.4.1) and (2.4.2), coordinates and distortion for an arbitrary conic projection can be calculated. Integration of $S^{(m)}$, $S^{(P)}$, $S^{(\omega)}$ in the above formulae can be accomplished using Romberg numerical integration, and m, p, and $\tan^2\left(\frac{\pi}{4} \pm \frac{\omega}{4}\right)$ can be found using the method of pivot elimination.

Example 1

Given $B_S = 18°$, $B_N = 54°$, $B_m = 36°$, and $P_S = P_N = 1.01$, $P_m = 0.99$.
    Through calculation:

    $P = 1.02636364 - 0.23948643B^2 + 0.23455661B^3$

    $\alpha = 0.58940267, \quad C = 3.280965309 \times 10^{13}$

This is a conic projection with a small amount of areal distortion.

Example 2

Given $B_S = 18°$, $B_N = 54°$, $B_m = 36°$, and $\omega_S = \omega_N = \omega_m = 0°30'$.
    Through derivation:

    $\tan^2\left(\frac{\pi}{4} \pm \frac{\omega}{4}\right) = 0.97703110 + 0.20899465B^2 - 0.20469250B^3$

    $\alpha = 0.59637363, \quad C = 10\ 428\ 714.12$

This is a conic projection with a small amount of angular distortion.

2.5   PSEUDOAZIMUTHAL PROJECTIONS

Pseudoazimuthal projections may be defined in the spherical coordinate system such that almucantars are projected as concentric circles, and verticals are projected as curves which are symmetrical about the central meridian, and which intersect at the center of the circles. Some of the pseudoazimuthal projections by G.A. Ginzburg of Russia are suitable for maps of the Pacific. One of them may be calculated from the following formulae:

$$
\left.\begin{aligned}
&x = \rho \sin\delta, \quad y = \rho\cos\delta \\[4pt]
&\rho = 3R\sin\frac{Z}{3}, \quad \delta = a - C\cdot\frac{Z}{Z_n}\sin 2a \\[4pt]
&C = 0.1, \quad Z_n = \frac{2}{3}\pi \\[4pt]
&\text{centerpoint: } \varphi_0 = +20°, \quad \lambda_0 = -25° \\[4pt]
&\tan\varepsilon = \frac{3C}{Z_n}\tan\frac{Z}{3}\sin 2a \\[4pt]
&\mu_1 = \cos\frac{Z}{3}\sec\varepsilon \\[4pt]
&\mu_2 = 3\sin\frac{Z}{3}\csc Z\left(1 - 2C\cdot\frac{Z}{Z_n}\right)\cos 2a \\[4pt]
&P = \mu_1\mu_2\cos\varepsilon, \quad \tan\frac{\omega}{2} = \frac{1}{2}\sqrt{\frac{\mu_1^2 + \mu_2^2}{P} - 2}
\end{aligned}\right\} \tag{2.5.1}
$$

where ε is the angular distortion shown by the angle between a given meridian and parallel.

Another pseudoazimuthal projection (Liu and Li 1963) is suitable for Chinese maps. It may be calculated from the following formulae:

$$
\left.\begin{aligned}
&x = \rho\sin\delta, \quad y = \rho\cos\delta \\[4pt]
&\rho = RZ, \quad \delta = a + 0.011697143\cdot Z\sin 3(15° + a) \\[4pt]
&R = 6\,370\,892 \text{ m} \\[4pt]
&\text{centerpoint: } \varphi_0 = 35°, \quad \lambda_0 = 150° \\[4pt]
&\tan\varepsilon = 0.011697143\cdot Z\sin 3(15° + a) \\[4pt]
&\mu_1 = \sec\varepsilon \\[4pt]
&\mu_2 = \frac{Z}{\sin Z}[1 + 3\times 0.011697143\cdot Z\cos 3(15° + a)] \\[4pt]
&P = \mu_2, \quad \tan\frac{\omega}{2} = \frac{1}{2}\sqrt{\frac{\mu_1^2 + \mu_2^2}{P} - 2}
\end{aligned}\right\} \tag{2.5.2}
$$

The line of equal distortion for this projection consists of three petal-shaped loops; its shape is appropriate to the shape of the Chinese boundary. Area distortion for most sections of China is below 1.9 percent, and maximum angular distortion is below 1°.

2.6 PSEUDOCYLINDRICAL PROJECTIONS

Now we introduce several pseudocylindrical equivalent projections.

By definition, parallels are projected as parallel straight lines, and meridians are curves that are symmetrical about the central meridian. Snyder (1977) lists more than 80 such projections. Tobler (1973, 1986b) derives several more. The ones in common use are presented here, with several enhancements.

2.6.1 Sanson–Flamsteed projection

Sanson–Flamsteed projection is a pseudocylindrical equivalent projection on which meridians are projected as sine curves and the poles are points. No distortion of length occurs along the central meridian or on any parallel of the projection. The coordinate formulae for the sphere have the form:

$$y = R\varphi, \quad x = R\lambda\cos\varphi \tag{2.6.1}$$

The distortion formulae can be expressed as:

$$\left.\begin{aligned} &\tan\varepsilon = \lambda\sin\varphi, \quad m = \sec\varepsilon \\ &P = 1, \quad n = 1 \\ &\tan\frac{\omega}{2} = \frac{1}{2}\tan\varepsilon = \frac{1}{2}\lambda\sin\varphi \end{aligned}\right\} \tag{2.6.2}$$

2.6.2 Eckert VI projection

This is a pseudocylindrical equivalent projection on which meridians are projected as sine curves and the poles become lines. The coordinate equations for the projection are written in the form:

$$\left.\begin{aligned} &y = \frac{2R}{\sqrt{\pi+2}}\alpha, \quad x = \frac{2R\lambda}{\sqrt{\pi+2}}\cos^2\frac{\alpha}{2} \\ &\sin\alpha + \alpha = \frac{\pi+2}{2}\sin\varphi \end{aligned}\right\} \tag{2.6.3}$$

The distortion equations of the projection are:

$$\left.\begin{aligned} &\tan\varepsilon = \frac{1}{2}\lambda\sin\alpha, \quad p = 1 \\ &n = \frac{2}{\sqrt{\pi+2}}\sec\varphi\cos^2\frac{\alpha}{2} \\ &m = \sqrt{\frac{\pi+2}{2}\cos\varphi\sec^2\frac{\alpha}{2}}\sec\varepsilon \\ &\tan\frac{\omega}{2} = \frac{1}{2}\sqrt{m^2+n^2-2} \end{aligned}\right\} \tag{2.6.4}$$

2.6.3 Kavraisky projection

This is a pseudocylindrical equivalent projection in which meridians are projected as sine curves and the polar point becomes a polar line. Its coordinate equations can be written as:

$$y = R\sqrt[4]{3}\,\alpha, \quad x = \frac{2}{3}R\sqrt[4]{3}\lambda\cos\alpha$$

$$\sin\alpha = \frac{\sqrt{3}}{2}\sin\varphi \qquad\qquad (2.6.5)$$

The distortion equations of the projection are written in the form:

$$\tan\varepsilon = \frac{2}{3}\lambda\sin\alpha, \quad P = 1$$

$$n = \frac{2}{\sqrt[4]{27}}\cos\alpha\sec\varphi$$

$$m = \frac{\sqrt[4]{27}}{2}\sec\alpha\cos\varphi\sec\varepsilon \qquad\qquad (2.6.6)$$

$$\tan\frac{\omega}{2} = \frac{1}{2}\sqrt{m^2 + n^2 - 2}$$

2.6.4 Mollweide projection

This is a pseudocylindrical equivalent projection on which meridians are projected as ellipses and points represent the poles. The meridians at the distance of ±90° difference of longitude from the central meridian will form a circle on the projection and its area equals the area of a terrestrial hemisphere. Its coordinate equations are written as:

$$y = R\sqrt{2}\sin\alpha, \quad x = 2\sqrt{2}\,R\frac{\lambda}{\pi}\cos\alpha$$

$$2\alpha + \sin 2\alpha = \pi\sin\varphi \qquad\qquad (2.6.7)$$

The distortion equations of the projection take the form:

$$\tan\varepsilon = \frac{2\lambda}{\pi}\tan\alpha, \quad P = 1$$

$$n = \frac{2\sqrt{2}}{\pi}\cdot\frac{\cos\alpha}{\cos\varphi}$$

$$m = \frac{\pi}{2\sqrt{2}}\cdot\frac{\cos\varphi}{\cos\alpha}\sec\varepsilon \qquad\qquad (2.6.8)$$

$$\tan\frac{\omega}{2} = \frac{1}{2}\sqrt{m^2 + n^2 - 2}$$

2.6.5 Eckert IV projection

This is a pseudocylindrical equivalent projection on which meridians are projected as ellipses and the pole becomes a polar line. The polar line is half as long as the equator. The meridians at ±180° form two semicircles on the projection. The projection coordinate equations can be written as:

$$\left.\begin{array}{l} y = 2R\sqrt{\dfrac{\pi}{4+\pi}}\sin\alpha, \quad x = \dfrac{2R}{\sqrt{\pi(4+\pi)}}\lambda(1+\cos\alpha) \\[3mm] 2\alpha + \sin2\alpha + 4\sin\alpha = 2(4+\pi)\sin\varphi \end{array}\right\} \tag{2.6.9}$$

The projection distortion equations take the form:

$$\left.\begin{array}{l} \tan\varepsilon = \dfrac{\lambda}{\pi}\tan\alpha, \quad P = 1 \\[3mm] n = \dfrac{2}{\sqrt{\pi(4+\pi)}}\dfrac{1+\cos\alpha}{\cos\varphi} \\[3mm] m = \dfrac{\sqrt{\pi(4+\pi)}}{2}\dfrac{\cos\varphi}{1+\cos\alpha}\sec\varepsilon \\[3mm] \tan\dfrac{\omega}{2} = \dfrac{1}{2}\sqrt{m^2+n^2-2} \end{array}\right\} \tag{2.6.10}$$

2.6.6 Oblique pseudocylindrical projection

Oblique Sanson projection

The coordinate equations take the form:

$$y = R\left(\dfrac{\pi}{2} - Z\right), \quad x = R\sin Z(\pi - a) \tag{2.6.11}$$

The central point of the projection is at 45°N, 90°W.

Oblique Mollweide projection

The coordinate equations can be written as:

$$\left.\begin{array}{l} y = R\sqrt{2}\sin\alpha, \quad x = 2\sqrt{2}R\dfrac{\pi - 2a}{2\pi}\sin\alpha \\[3mm] 2\left(\dfrac{\pi}{2} - \alpha\right) + \sin2\left(\dfrac{\pi}{2} - \alpha\right) = \pi\cos Z \end{array}\right\} \tag{2.6.12}$$

The projection's central points are at: 45°N, 30°W; 30°N, 80°E; 50°N, 20°E; 35°N, 150°E.

2.7 PSEUDOCONICAL PROJECTIONS

By definition, parallels are projected as a group of concentric arcs, and meridians are projected as curves that are symmetrical about the central meridian. The commonly used Bonne projection is an equivalent pseudoconical projection that has no distortion of length along parallels.
    The projection equations are:

$$y = q - \rho\cos\delta, \quad x = \rho\sin\delta \tag{2.7.1}$$

where q is the ordinate of the centre of a circle. It is a constant.

$$\rho = C - S, \quad C = N_0 \cot B_0 + S_0.$$

$$\delta = \frac{r}{\rho} l, \quad S \text{ stands for the length of an arc of the meridian.}$$

B_0 is the latitude of the parallel intersecting all meridians at a right angle.

The projection distortion equations are as follows:

$$\left.\begin{array}{l} \tan \varepsilon = l \left(\sin B - \dfrac{r}{\rho} \right) \\[2mm] m = \sec \varepsilon, \quad n = 1 \\[2mm] P = 1, \quad \tan \dfrac{\omega}{2} = \dfrac{1}{2} \tan \varepsilon \end{array}\right\} \tag{2.7.2}$$

2.8 POLYCONICAL PROJECTIONS

2.8.1 Ordinary polyconic projection

The projection conditions for the ordinary polyconic projection can be given as:

1. The projection of the parallels forms a group of coaxial arcs that have a radius of $\rho = N \cot B$.
2. No distortion of length occurs along the projected parallels.
3. No distortion of length occurs along the central meridian of the projection.

The projection coordinate equations are as follows:

$$y = S + \rho(1 - \cos \delta), \quad x = \rho \sin \delta \tag{2.8.1}$$

where $\rho = N \cot B$, $\delta = l \sin B$.
   Its projection distortion equations are:

$$\left.\begin{array}{l} \tan \varepsilon = \dfrac{\delta - \sin \delta}{\cos \delta - \left(1 + \dfrac{M}{N} \tan^2 B \right)} \\[5mm] P = 1 + 2 \cdot \dfrac{N}{M} \cot^2 B \sin^2 \dfrac{\delta}{2}, \quad n = 1 \\[5mm] m = \left(1 + 2 \cdot \dfrac{N}{M} \cot^2 B \sin^2 \dfrac{\delta}{2} \right) \sec \varepsilon \\[5mm] \tan \dfrac{\omega}{2} = \dfrac{1}{2} \sqrt{\dfrac{m^2 + n^2 - 2mn \cos \varepsilon}{mn \cos \varepsilon}} \end{array}\right\} \tag{2.8.2}$$

2.8.2 Rectangular polyconic projection

The projection conditions for the rectangular polyconic projection are as follows:

1. The projection of the meridians and the parallels intersect at right angles and are curves symmetrical about the central meridian and the equator.

2. The projection parallels are a group of coaxial arcs that have a radius of $\rho = N \cot B$ with the center of the circles located on the meridian.

3. No distortion of length occurs along the projection's central meridian and the equator.

The projection coordinate equations can be written as:

$$y = S + \rho(1 - \cos\delta), \quad x = \rho\sin\delta \tag{2.8.3}$$

where:

$$\rho = N\cot B, \quad \tan\frac{\delta}{2} = \frac{1}{2}l\sin B$$

The projection distortion equations are as follows:

$$\left.\begin{array}{l} n = \dfrac{1}{\sin B} \cdot \dfrac{\partial\delta}{\partial l} \\[2ex] m = 1 + \dfrac{N}{M}\cot^2 B\sin^2\dfrac{\delta}{2} \\[2ex] P = mn, \quad \tan\dfrac{\omega}{2} = \dfrac{m-n}{2\sqrt{P}} \end{array}\right\} \tag{2.8.4}$$

2.8.3 Modified polyconic projection

This is an internationally specified projection used to compile the map on the millionth scale with a 4° latitude span and 6° extent in longitude. Its projection conditions are given as:

1. The parallels on the south and north edge of the projection become an arc of a circle with radius of $\rho = N\cot B$ and the center of the circles is located on the central meridian. The projection of the meridians is as straight lines. The other parallels can be obtained by linear interpolation.

2. The projection of parallels on the south or the north side has no distortion of length. The projection of the meridian 2° distant from the central meridian has no distortion of length.

The projection coordinate equations can be written as:

$$x = x_s + \frac{x_N - x_s}{4}(B - B_s), \quad y = y_s + \frac{y_N - y_s}{4}(B - B_s) \tag{2.8.5}$$

where:

$$y_s = \rho_s(1 - \cos\delta_s), \quad x_s = \rho_s\sin\delta_s, \quad y_N = y_0 + \rho_N(1 - \cos\delta_N), \quad x_N = \rho_N\sin\delta_N$$

$$\rho_s = N_s\cot B_s, \quad \rho_N = N_N\cot B_N, \quad \delta_s = l\sin B_s, \quad \delta_N = l\sin B_N.$$

$$y_0 = \Delta S - 0.271\cos^2\frac{B_s + B_N}{2}\,(\text{mm})$$

ΔS is the arc length of a longitude between the southern parallel and the northern parallel. Its length on the map has been modified and its scale is now millimeters.

N_s, N_N are multiplied by 10^{-6} and take mm as their unit. The unit of B, B_s is degrees.

2.8.4 Lagrange's projection

This is a conformal projection on which the projected parallels and meridians become groups of coaxial circles. The axes passing through the centers of the circles are perpendicular to each other.

The projection coordinate equations are:

$$y = \frac{K \sin\delta}{1 + \cos\delta \cos(\alpha l)}, \quad x = \frac{K \cos\delta \sin(\alpha l)}{1 + \cos\delta \cos(\alpha l)} \tag{2.8.6}$$

where:

$$K = \frac{n_0 \, r_0}{\alpha}(1 + \sec\delta_0), \quad n_0 \text{ is the ratio of length at the central point,}$$

$$\alpha = \sqrt{1 + \frac{1 - n^2}{1 + n^2}\cos^2 B_0}, \quad n = \frac{b}{a} \text{ is the ratio of the projected region,}$$

$$\tan\frac{\delta_0}{2} = \frac{\sin B_0}{\alpha}, \quad \beta = \tan\left(45° + \frac{\delta_0}{2}\right)U_0^{-\alpha}, \quad \tan\left(45° + \frac{\delta}{2}\right) = \beta U^\alpha.$$

Its projection distortion equations are:

$$\mu = \frac{\alpha K \cos\delta}{r(1 + \cos\delta \cos\alpha l)}, \quad P = \mu^2, \quad \omega = 0 \tag{2.8.7}$$

2.8.5 Polyconic projection with tangent difference parallel (Fang 1983)

This is a kind of polyconic projection on which meridional intervals on the same parallel decrease away from central meridian by equal increments. It has been made especially for the 1:14 000 000 world map by the Chinese Map Publishing House. The coordinate equations are as follows:

$$\left.\begin{aligned}
y_0 &= (\varphi + 0.06683225\varphi^4) \cdot 100\mu_0 R \\
y_n &= y_0 + 0.20984\varphi \cdot 100\mu_0 R \\
x_n &= \sqrt{112^2 - y_n^2} + 20 \quad (\text{cm}) \\
\rho &= \frac{x_n^2 + (y_n - y_0)^2}{2(y_n - y_0)}, \quad \sin\delta_{\varphi_n} = \frac{x_n}{\rho} \\
\delta_{\varphi_i} &= \frac{\delta_{\varphi_n}}{\lambda_n}\left(1.1 - 0.11106126 tg\frac{\lambda_i}{5}\right)\lambda_i \\
y &= y_0 + \rho(1 - \cos\delta_{\varphi_i}), \quad x = \rho \sin\delta_{\varphi_i}
\end{aligned}\right\} \tag{2.8.8}$$

On the equator the values are:

$$y = 0, \quad x_i = \frac{x_n}{\lambda_n}\left(1.1 - 0.11106126 tg\frac{\lambda_i}{5}\right)\lambda_i \tag{2.8.9}$$

where the units of φ, λ are radians; $R = 6\ 371\ 116$ m, $\mu_0 = 1{:}14\ 000\ 000$.

2.8.6   A family of polyconic projections (Yang 1983e)

The projection condition for a family of polyconic projections can be summed up as:

1. The projection parallels and meridians are curves symmetrical about the central meridian and the equator.

2. The projection parallels form a series of non-concentric arcs of circles with radius $\rho = N \cot B$ and whose center is located on the central meridian.

3. The proportion of length on the central meridian is $m_0 = f(B)$.

4. The proportion of length on the parallels is $n = F(B, l)$.

Now we introduce three kinds of polyconic projection systems.
     The first is the polyconic projection system with:

$$m_0 = 1 - q_1 \cos^2 KB, \quad n = m_0\left(1 + \frac{1}{2R^2}S_n^2\right)$$

The projection equations can be written as:

$$y = q - \rho\cos\delta, \quad x = \rho\sin\delta \tag{2.8.10}$$

where:

$$q = y_0 + \rho, \quad y_0 = \int_0^a m_0 M dB, \quad \rho = N\cot B$$

$$\delta = m_0\left\{l\sin B + \frac{l^3}{6}\sin B\cos^2 B(1 + \eta^2)\right\}, \quad m_0 = 1 - q_1\cos^2 KB.$$

The projection distortion equations can be written as:

$$\left.\begin{array}{l}
\tan\varepsilon = -\dfrac{(1 + \eta^2)\cot^2 B\dfrac{\partial\delta}{\partial B} + [(m_0 - 1) - (1 + \eta^2)\cot^2 B]\sin\delta}{1 + (1 + \eta^2)\cot^2 B + [(m_0 - 1) - (1 + \eta^2)\cot^2 B]\cos\delta} \\[4mm]
n = m_0\left[1 + \dfrac{l^2}{2}\cos^2 B(1 + \eta^2)\right] \\[4mm]
P = n\{1 + (1 + \eta^2)\cot^2 B + [(m_0 - 1) - (1 + \eta^2)\cot^2 B]\cos\delta\} \\[4mm]
m = \dfrac{P}{n}\sec\varepsilon, \quad \tan\dfrac{\omega}{2} = \dfrac{1}{2}\sqrt{\dfrac{m^2 + n^2}{P} - 2}
\end{array}\right\} \tag{2.8.11}$$

where:

$$m_0 = 1 - q_1\cos^2 KB, \quad \delta = m_0\left[l\sin B + \frac{l^3}{6}\sin B\cos^2 B(1 + \eta^2)\right]$$

$$\frac{\partial\delta}{\partial B} = \frac{Kq_1\sin 2KB}{m_0}\cdot\delta + \cot B\left[\delta - \frac{l^3}{6}m_0\sin^3 B(1 + 2\eta^2)\right]$$

In this projection system, if $q_1 = 0$, that is,

$$m_0 = 1, \quad n = 1 + \frac{1}{2R^2} S_n^2 \tag{2.8.12}$$

Thus it is a polyconic projection similar to the Gauss–Krüger projection.

The second is a polyconic projection system with:

$$m_0 = \sum_{i=0}^{n} a_{2i} B^{2i}, \quad n = m_0 \left(1 + \frac{1}{2R^2} S_n^2 \right) \tag{2.8.13}$$

The third is the polyconic projection system with:

$$m_0 = 1 - q_1 \sin^2 KB, \quad n = m_0 \left(1 + \frac{1}{2R^2} S_n^2 \right) \tag{2.8.14}$$

2.8.7 Hammer–Aitoff projection

This is a type of equivalent projection developed for the presentation of the entire world. The coordinates of the latitude–longitude grid are obtained by multiplying the abscissa of the transverse azimuthal equivalent projection by 2. Its meridians are relabeled to make the meridians of both sides of the central point stand for $180°$ instead of the original $90°$. For this reason, the entire latitude–longitude grid of the Hammer–Aitoff projection is packed into an ellipse with $a = 2\sqrt{2}R$ as the semimajor axis and $b = \sqrt{2}R$ the semiminor axis. The projection coordinate equations are:

$$y = \frac{\sqrt{2}R \sin\varphi}{\sqrt{1 + \cos\varphi \cos\frac{\lambda}{2}}}, \quad x = \frac{2\sqrt{2}R \cos\varphi \sin\frac{\lambda}{2}}{\sqrt{1 + \cos\varphi \cos\frac{\lambda}{2}}} \tag{2.8.15}$$

And the projection distortion equations are:

$$\left. \begin{array}{l} m = \dfrac{\sqrt{E}}{R}, \quad n = \dfrac{\sqrt{G}}{r} \\[2mm] P = 1, \quad \tan\dfrac{\omega}{2} = \dfrac{1}{2}\sqrt{m^2 + n^2 - 2} \end{array} \right\} \tag{2.8.16}$$

where:

$$E = \left(\frac{\partial x}{\partial \varphi} \right)^2 + \left(\frac{\partial y}{\partial \varphi} \right)^2, \quad G = \left(\frac{\partial x}{\partial \lambda} \right)^2 + \left(\frac{\partial y}{\partial \lambda} \right)^2$$

The oblique Hammer–Aitoff projection is a commonly used one.

2.9   OTHER TYPES OF PROJECTIONS

2.9.1   Conformal projection in which the projected parallel is an ellipse

This projection is also called the Littrow projection. Its coordinate equation is:

$$y + ix = R\sinh(q + i\lambda) \tag{2.9.1}$$

i.e.:

$$y + ix = R(\sinh q \cosh i\lambda + \cosh q \sinh i\lambda)$$
$$= R(\sinh q \cos \lambda + i \cosh q \sin \lambda)$$
$$\therefore \quad y = R \sinh q \cos \lambda, \quad x = R \cosh q \sin \lambda \tag{2.9.2}$$

Noticing that $q = \ln \tan \left(45° + \dfrac{\varphi}{2} \right)$, then we have:

$$\sinh q = \tan\varphi, \quad \cosh q = \sec\varphi$$

Hence equation (2.9.2) can be rewritten as:

$$y = R \tan\varphi \cos\lambda, \quad x = R \sec\varphi \sin\lambda \tag{2.9.3}$$

According to (2.9.3) we can get the equation of a parallel:

$$\frac{y^2}{R^2 \tan^2\varphi} + \frac{x^2}{R^2 \sec^2\varphi} = 1 \tag{2.9.4}$$

According to (2.9.2) we can also get the equation of a meridian:

$$\frac{x^2}{R^2 \sin^2\lambda} - \frac{y^2}{R^2 \cos^2\lambda} = 1 \tag{2.9.5}$$

It can be seen that in this type of projection the projected parallel forms an ellipse and the projected meridian becomes a hyperbola. The projection distortion equations are as follows:

$$\mu = \sec\varphi \sqrt{\tan^2\varphi + \cos^2\lambda}, \quad P = \mu^2, \quad \omega = 0 \tag{2.9.6}$$

2.9.2 Conformal projection in which the projected parallel is a hyperbola

The coordinate equation of this type of projection is:

$$y + ix = R \sin(q + i\lambda) \tag{2.9.7}$$

i.e.:

$$y + ix = R(\sin q \cos i\lambda + \cos q \sin i\lambda)$$
$$= R(\sin q \cosh\lambda - i \cos q \sinh\lambda)$$
$$\therefore \quad y = R \sin q \cosh\lambda, \quad x = -R \cos q \sinh\lambda \tag{2.9.8}$$

From the above equation we can obtain the equation of a parallel:

$$\frac{y^2}{R^2 \sin^2 q} - \frac{x^2}{R^2 \cos^2 q} = 1 \tag{2.9.9}$$

And the equation of a meridian:

$$\frac{y^2}{R^2 \cosh^2\lambda} + \frac{x^2}{R^2 \sinh^2\lambda} = 1 \tag{2.9.10}$$

It is obvious that in this type of projection the projected parallel is a hyperbola and the meridian is an ellipse. Its projection distortion equations can be written as:

$$\mu = \cosh q \sqrt{\cos^2 q + \sinh^2\lambda}, \quad P = \mu^2, \quad \omega = 0 \tag{2.9.11}$$

2.9.3 Oblated equal-area projection (Snyder 1988)

This is a type of equivalent projection with approximate distortion isograms as ellipses or rectangles. The projection is modified from the azimuthal equivalent projection with the distortion isograms made to fit the outline of a mapped region.

The coordinate equations for this type of projection are:

$$y = nR\sin(2N/n), \quad x = mR\sin(2M/m)\cos N/\cos(2N/n) \tag{2.9.12}$$

where:

$$M = \arcsin(y'/2), \quad N = \arcsin[(x'/2)\cos M/\cos(2M/m)]$$

$$y' = 2\sin\frac{Z}{2}\cos a, \quad x' = 2\sin\frac{Z}{2}\sin a$$

The shape of distortion isograms is determined by parameters m, n in the above equations and varies with the size of these as well as the ratio of n to m.

For example, given the coordinates for a projection centered at a point in the Atlantic Ocean at $\varphi_0 = 0°$, $\lambda_0 = 30°W$, with the axis rotation angle $a_0 = 15°$ (since the α in the above equation is replaced by direction of the angle α, then $\alpha = a - a_0$), $m = 1.45$, $n = 3.0$.

The projection distortion equations are given as:

$$\mu_1 = \frac{1}{R}\left[\left(\frac{\partial x}{\partial z}\right)^2 + \left(\frac{\partial y}{\partial z}\right)^2\right]^{1/2}$$

$$\mu_2 = \frac{1}{R\sin z}\left[\left(\frac{\partial x}{\partial a}\right)^2 + \left(\frac{\partial y}{\partial a}\right)^2\right]^{1/2} \tag{2.9.13}$$

$$P = 1, \quad \tan\frac{\omega}{2} = \frac{1}{2}[\mu_1^2 + \mu_2^2 - 2]^{1/2}$$

2.9.4 $m = n^k$ rectangular projection (Zhou 1957, Li 1987)

The $m = n^k$ rectangular projection will be the equivalent projection when $k = -1$, the equidistant projection when $k = 0$, and the conformal projection when $k = 1$. When $-1 < k < 0$ the projection has little area distortion; when $0 < k < 1$ the projection has little angular distortion.

m = n^k *cylindrical projection*

The coordinate equations of the projection are:

$$y = r_0^k\int_0^B \frac{M}{r^k}dB = r_0^k A, \quad x = r_0 l \tag{2.9.14}$$

The projection distortion equations can be given as:

$$n = \frac{r_0}{r}, \quad m = n^k, \quad p = n^{k+1}, \quad \sin\frac{\omega}{2} = \left|\frac{m-n}{m+n}\right| \tag{2.9.15}$$

$m = n^k$ *conical projection*

The coordinate equations of the projection are:

$$\left.\begin{aligned}
\rho &= \{\alpha^k(1-k)(C-A)\}\frac{1}{1-k}, \quad \delta = \alpha l \\
y &= \rho_S - \rho\cos\delta, \quad x = \rho\sin\delta
\end{aligned}\right\}$$

(2.9.16)

Its projection distortion equations can be given as:

$$n = \frac{\alpha\rho}{r}, \quad m = n^k, \quad p = n^{k+1}, \quad \sin\frac{\omega}{2} = \left|\frac{m-n}{m+n}\right|$$

(2.9.17)

where:

$$A = \int_0^B \frac{M}{r^k} dB \quad (k \neq 1)$$

After the value of k is fixed, the parameters α, c can be determined in many ways. For example, if the absolute value of length distortion of the edge parallel and the central parallel are equal then the expressions for α, c can be written as follows:

$$\left.\begin{aligned}
C &= \frac{r_s^{1-k}A_N - r_N^{1-k}A_s}{r_s^{1-k} - r_N^{1-k}} \\
\alpha &= \frac{2^{1-k}r_m^{1-k}r_N^{1-k}}{(1-k)[r_m(C-A_m)^{1/(1-k)} + r_N(C-A_m)^{1/(1-k)}]^{1-k}}
\end{aligned}\right\}$$

(2.9.18)

The integral A in the above equation can be obtained by Romberg's numerical integral method.

2.9.5 Three kinds of modified projections (Yang 1983d)

Modified conic equidistant projection

The coordinate equations of the projection are:

$$y = \rho_s - \rho\cos\delta, \quad x = \rho\sin\delta$$

(2.9.19)

where $\rho = C - S$, $\delta = \alpha l - q\sin Kl$
  If $Kl = 90°$, then we have:

$$n_1 = n_2, \quad C = \frac{n_1 S_2 - n_2 S_1}{n_1 - n_2}, \quad \alpha = \frac{n_1 r_1}{\rho_1} = \frac{n_2 r_2}{\rho_2}$$

The projection distortion equations are given as:

$$\left.\begin{aligned}
m &= 1, \quad n = \frac{\rho}{r}(\alpha - Kq\cos Kl) \\
P &= n, \quad \sin\frac{\omega}{2} = \left|\frac{m-n}{m+n}\right|
\end{aligned}\right\}$$

(2.9.20)

Modified cylindrical equidistant projection

The coordinate equations for the projection are:

$$y = s, \quad x = Cl - q \sin Kl \tag{2.9.21}$$

When $Kl = 90°$, $C = n_1 r_1$, $q = \dfrac{C - n_2 r_2}{K}$

The projection distortion equations are given as:

$$
\left.
\begin{aligned}
m &= 1, \quad n = \frac{\rho}{r}(C - Kq \cos Kl) \\
P &= n, \quad \sin\frac{\omega}{2} = \left|\frac{m - n}{m + n}\right|
\end{aligned}
\right\} \tag{2.9.22}
$$

Modified azimuthal equidistant projection

The coordinate equations of the projection are:

$$y = \rho \cos \delta, \quad x = \rho \sin \delta \tag{2.9.23}$$

where $\rho = C - S$, $\delta = l - q \sin Kl$
When $Kl = 90°$, $C = S_1 + n_1 r_1$; when $l = 0°$,

$$q = \frac{\rho_2 - n_2 r_2}{K \rho_2}$$

The projection distortion equations are given as:

$$
\left.
\begin{aligned}
m &= 1, \quad n = \frac{\rho}{r}(1 - Kq \cos Kl) \\
P &= n, \quad \sin\frac{\omega}{2} = \left|\frac{m - n}{m + n}\right|
\end{aligned}
\right\} \tag{2.9.24}
$$

When this projection is applied to the map of all of China, the calculating equations are as follows:

$$
\left.
\begin{aligned}
\varphi_0 &= 30°, \quad \lambda_0 = 105° \\
\rho &= RZ, \quad \delta = a + 0.00385 \sin 3(a + 15°)
\end{aligned}
\right\} \tag{2.9.25}
$$

The distorting effect of applying the modified azimuthal equidistant projection to the Chinese nation-wide map is the same as the pseudoazimuthal projection. The merit of the modified azimuthal equidistant projection lies in its nature of equidistance and the simple calculation.

2.9.6 Ordinary perspective azimuthal projection

The coordinate equations of this projection are written in the form:

$$
\left.
\begin{aligned}
y &= \rho \cos a, \quad x = \rho \sin a \\
\rho &= \frac{(D + R)R \sin Z}{D + R \cos Z}
\end{aligned}
\right\} \tag{2.9.26}
$$

where D is the distance from viewing point to center of the sphere.

The projection distortion equations are:

$$\mu_1 = \frac{(R+D)(R+D\cos Z)}{(R\cos Z + D)^2}, \quad \mu_2 = \frac{R+D}{R\cos Z + D}$$

$$P = \mu_1\mu_2, \quad \sin\frac{\omega}{2} = \left|\frac{\mu_1 - \mu_2}{\mu_1 + \mu_2}\right|$$

$$(2.9.27)$$

Equation (2.9.26) is the same for many of the perspective azimuthal projections including the gnomonic projection ($D = 0$), the stereographic projection ($D = R$) and the orthographic projection ($D = \infty$).

The coordinate equations for the perspective azimuthal projection presented in geographical coordinates (φ, λ) can be written as follows:

$$y = \frac{(R+D)R[\sin\varphi\cos\varphi_0 - \cos\varphi\sin\varphi_0\cos(\lambda - \lambda_0)]}{D + R[\sin\varphi\sin\varphi_0 - \cos\varphi\cos\varphi_0\cos(\lambda - \lambda_0)]}$$

$$x = \frac{(R+D)R\cos\varphi\sin(\lambda - \lambda_0)}{D + R[\sin\varphi\sin\varphi_0 - \cos\varphi\cos\varphi_0\cos(\lambda - \lambda_0)]}$$

$$(2.9.28)$$

We give the meridian's equation in the perspective azimuthal projection:

$$x^2[D^2\cos^2\varphi_0 + D^2\sin^2\varphi_0\cos^2(\lambda - \lambda_0) - R^2\cos^2(\lambda - \lambda_0)]$$
$$+ y^2[D^2\sin^2(\lambda - \lambda_0) - R^2\sin^2\varphi_0\sin^2(\lambda - \lambda_0)] + 2xy\sin\varphi_0\sin(\lambda - \lambda_0)\cdot$$
$$\cos(\lambda - \lambda_0)(D^2 - R^2) + 2R^2(R+D)x\cos\varphi_0\sin(\lambda - \lambda_0)\cos(\lambda - \lambda_0)$$
$$+ 2R^2(R+D)y\sin\varphi_0\cos\varphi_0\sin^2(\lambda - \lambda_0) = (R+D)^2R^2\cos^2\varphi_0\sin^2(\lambda - \lambda_0)$$

$$(2.9.29)$$

We can also give the parallel's equation in the perpective azimuthal projection:

$$x^2(D\sin\varphi_0 + R\sin\varphi)^2 + y^2[(D^2 - R^2)\cos^2\varphi_0 + (D\sin\varphi_0 + R\sin\varphi)^2]$$
$$- 2(R+D)R\cos\varphi_0(D\sin\varphi + R\sin\varphi_0)y = (R+D)^2R^2(\sin^2\varphi_0 - \sin^2\varphi)$$

$$(2.9.30)$$

2.9.7 The perspective azimuthal projection with variable view point (Yang 1987c)

The coordinate equations for the perspective azimuthal projection with a variable viewing point can be given as:

$$y = \rho\cos a, \quad x = \rho\sin a$$

$$\rho = \frac{R\sin Z(R + D(Z))}{R\cos Z + D(Z)}$$

$$(2.9.31)$$

Its projection distortion equations are:

$$\mu_1 = \frac{(R+D)(R+D\cos Z) + D'\sin Z(R\cos Z - R)}{(R\cos Z + D)^2}$$

$$\mu_2 = \frac{R+D}{R\cos z + D}, \quad P = \mu_1\mu_2, \quad \sin\frac{\omega}{2} = \left|\frac{\mu_1 - \mu_2}{\mu_1 + \mu_2}\right|$$

$$(2.9.32)$$

where:

$$D' = \frac{dD}{dZ}, \quad D = D(Z)$$

The above equation is just the ordinary perspective azimuthal projection when D is a constant. Sometimes it is also referred to as perspective azimuthal projection with single view point.

The relation between a variable view point $D(Z)$ and $\rho = f(Z)$ can be given as follows:

$$D = \frac{R(\rho \cos Z - R \sin Z)}{R \sin Z - \rho} \tag{2.9.33}$$

Example 1: Given the azimuthal equidistant projection $\rho = RZ$, from (2.9.33) we can obtain the following value:

$$D = \frac{R(\sin Z - Z \cos Z)}{Z - \sin Z} \tag{2.9.34}$$

Example 2: Given the azimuthal equivalent projection $\rho = 2R \sin \frac{Z}{2}$, according to (2.9.33) we can obtain the following result:

$$D = R + 2R \cos \frac{Z}{2} \tag{2.9.35}$$

It can be seen that the perspective azimuthal projection with a variable view point is a system which includes the perspective azimuthal projections and non-perspective azimuthal projection.

Now we give some examples to illustrate the application of the perspective azimuthal projection with a variable viewing point.

1. Letting $R = 1$, for the ordinary perspective azimuthal projection in which the value of Z is in $0°–90°$ we can get the alternate azimuthal projections using different viewing locations. They are: $D = 1$ (conformal); $1 < D < 1.57$ (little angular distortion); $1.75 < D < 2$ (approximately equidistant); $2 < D < 2.4$ (little area distortion); $2.4 < D < 3$ (approximately equal-area).

2. We can explore an azimuthal projection having arbitrary characteristics of our own choosing and be flexible through the use of the perspective azimuthal projection with a variable view point. For example, for the approximate perspective azimuthal equivalent projection covering a fixed region and equal-area, $Z_k = 30°$, $D = 2.931851653$. In another way, we can get azimuthal projections with different characteristics by creating a variable view point function $D(Z)$. In the literature, Yang (1987c) notes that we can get an azimuthal projection having little angular distortion by letting $D = \cosh \frac{Z}{2}$.

2.9.8   A generalized equation for azimuthal projections (Yang 1987b)

The coordinate equations are:

$$\left. \begin{array}{l} y = \rho \cos a, \quad x = \rho \sin a \\ \rho = \left(\dfrac{2}{1 + \cos Z} \right)^k R \sin Z \end{array} \right\} \tag{2.9.36}$$

And its projection distortion equations are:

$$\mu_1 = \left(\frac{2}{1 + \cos Z}\right)^k \left(\frac{k \sin^2 Z}{1 + \cos Z} + \cos Z\right)$$

$$\mu_2 = \left(\frac{2}{1 + \cos Z}\right)^k, \quad \rho = \mu_1 \mu_2, \quad \sin\frac{\omega}{2} = \left|\frac{\mu_1 - \mu_2}{\mu_1 + \mu_2}\right|$$

(2.9.37)

Equation (2.9.36) includes the following well-known azimuthal projections: $k = 1$, conformal projection; $k = \frac{3}{4}$, Breusing's projection; $k = \frac{2}{3}$, approximately equidistant projection; $k = \frac{1}{2}$, equal-area projection; $k = 0$, orthographic projection.

The coordinate equations given by geographical coordinates (φ, λ) are as follows:

$$y = R\frac{2^k(\cos\varphi_0 \sin\varphi - \sin\varphi_0 \cos\varphi \cos(\lambda - \lambda_0))}{(1 + \sin\varphi_0 \sin\varphi + \cos\varphi_0 \cos\varphi \cos(\lambda - \lambda_0))^k}$$

$$x = R\frac{2^k \cos\varphi \sin(\lambda - \lambda_0)}{(1 + \sin\varphi_0 \sin\varphi + \cos\varphi_0 \cos\varphi \cos(\lambda - \lambda_0))^k}$$

(2.9.38)

2.10   PROJECTION FROM THE ELLIPSOID TO THE SPHERE

When double projection is used to realize a projection from the ellipsoid to the plane, we first project from the ellipsoid to the sphere according to some condition, and then from the sphere to the plane.

It is generally required that a parallel projected from the ellipsoid to a sphere is still a parallel and a meridian projected from ellipsoid to sphere is still a meridian. The angle included between two meridians is in direct proportion to their corresponding difference of longitude on the ellipsoid.

2.10.1   Conformal projection from ellipsoid to sphere

The coordinate equation of this projection is written in the form:

$$\tan\left(45° + \frac{\varphi}{2}\right) = \beta U^\alpha, \quad \Delta\lambda = \alpha l$$

(2.10.1)

where:

$$\tan\varphi_0 = \frac{R}{N_0}\tan B_0, \quad \alpha = \frac{\sin B_0}{\sin\varphi_0}, \quad \beta = \tan\left(45° + \frac{\varphi_0}{2}\right)U_0^{-\alpha}$$

When $R = \sqrt{M_0 N_0}$ and $m_0 = n_0 = 1$, then:

$$\tan\varphi_0 = \sqrt{\frac{M_0}{N_0}}\tan B_0, \quad \alpha = \sqrt{1 + e'^2 \cos^4 B_0}$$

(2.10.2)

The distortion equations of the projection are:

$$\mu = \frac{\alpha R \cos\varphi}{N \cos B}, \quad P = \mu^2, \quad \omega = 0$$

(2.10.3)

2.10.2 Equivalent projection from ellipsoid to sphere

The coordinate equations of this projection are written in the form:

$$\Delta\lambda = l, \quad \sin\varphi = \frac{F}{R^2} \tag{2.10.4}$$

where:

$$F = \int_0^B MrdB, \quad R_2 = \int_0^{90°} MrdB$$

The projection distortion is:

$$n = \frac{R\cos\varphi}{N\cos B}, \quad m = \frac{1}{n}, \quad P = 1, \quad \sin\frac{\omega}{2} = \left|\frac{m-n}{m+n}\right| \tag{2.10.5}$$

2.10.3 Equidistant projection from ellipsoid to sphere

The coordinate equations of this projection are written in the form:

$$\Delta\lambda = l, \quad \varphi = \frac{S}{R} \tag{2.10.6}$$

where:

$$S = \int_0^B MdB, \quad R = \frac{2}{\pi}\int_0^{90°} MdB$$

The projection distortion is:

$$m = 1, \quad n = \frac{R\cos\varphi}{N\cos B}, \quad P = 1, \quad \sin\frac{\omega}{2} = \left|\frac{m-n}{m+n}\right| \tag{2.10.7}$$

2.11   SPACE MAP PROJECTION

2.11.1 Space oblique Mercator projection

The launching of an Earth-sensing satellite by the National Aeronautics and Space Administration (NASA) in 1972 led to a new era of continuing mapping from space. This satellite, first called ERTS-1 and then renamed Landsat 1 in 1975, was followed by two others, all of which circled the Earth in a nearly circular orbit inclined about 99° to the Equator and scanning a swath about 185 km (officially 100 nautical miles) from an altitude of about 919 km. The fourth and fifth Landsat satellites (launched in 1982 and 1984, respectively) involved circular orbits inclined about 98° and scanning from an altitude of about 705 km (Snyder 1987a, pp. 214–215).

In 1974 Alden P. Colvocoresses of the US Geological Survey conceived geometrically a new projection, which he called the space oblique Mercator projection, as a time–space related projection for future satellites similar to Landsat, and which would maintain a ground track true-to-scale and be basically conformal in the region of the swath. The second author of the current revision of this book developed the mathematics and published in the USA in several locations (Snyder 1978, 1981, 1982, 1987a); the latter

(1987a) is the most easily followed. It is also available in the USGS software, General Cartographic Transformation Package (GCTP) and its private derivatives. Since the details and lengthy formulae of this projection have been made available in the literature they are not repeated here, but they have also been published in some detail in several Chinese sources, such as Huang Guoshou (1988), Wu and Yang (1989, pp. 348–356, 444), and Yang (1990a, pp. 470–482).

However, Cheng Yang, working under Yang Qihe in Zhengzhou for his master's degree, referenced the SOM and Snyder's (1978) formulae in Cheng (1990a), and Snyder's (1981) formulae in Cheng (1991), and used them for much of his inspiration to develop the space conformal projection (below).

2.11.2 Space conformal projection

Although Cheng Yang's space conformal projection was initially published in several Chinese journals (Cheng 1990a, 1990b, 1991), he moved to the USA in 1992, obtained his doctorate from the University of South Carolina at Columbia, SC, and is currently employed at the Oak Ridge National Laboratory in Oak Ridge, Tennessee, with his name reversed in the American tradition to Yang Cheng. His formulae for the above projection have appeared in US publications (Cheng 1992, 1996). They are more detailed than those for the space oblique Mercator projection and therefore do not seem to merit repetition here, since the reader can refer to English publications for them as well as for the formulae of the space oblique Mercator projection.

We have briefly introduced the coordinate equations and the projection distortion equations of some commonly used, well known map projections as well as a few Chinese innovations. Readers should read the relevant publications for more detail.

Some important map projections such as the poly-focus projection (by N. Kadmon and Wang Qiao), composite projections (by Yang Qihe), as well as the further exploration for new projections from the affine transformation of equal-area projections and the linear transformation of the gnomonic projection (by Yang Qihe) will be presented in Chapters 10 and 11.

2.12   CHINESE TOPOGRAPHIC MAPS

2.12.1 Projection and coordinate systems

Projection

Chinese basic scale topographic maps (1:500 000; 1:250 000; 1:100 000; 1:50 000; 1:25 000; 1:10 000; 1:5 000) adopt the Gauss–Krüger projection (transverse Mercator projection). Of these the 1:500 000–1:25 000 scales use the Gauss–Krüger projection with 6° zones; the 1:10 000 and larger scales use the Gauss–Krüger projection with 3° zones.

The projection for the 1:1 000 000 topographic map is the conformal conic projection (Lambert projection, Lambert 1772). The projection condition is that the absolute values of the scale error at the edge parallels and on the central parallel should be equal. The rule for map sheet division is in concordance with the IMW projection, defined by the IGU and uniformly used throughout the world.

Coordinate systems

The Gauss–Krüger projection uses a plane rectangular coordinate system taking the projected central meridian as ordinate (y) and the projected equator as abscissa (x), for which the origin is at the intersecting point of the ordinate and the abscissa. The y-coordinate takes a positive sign when going north from the equator and takes a negative sign when towards south; the x-coordinate takes a positive sign when going east from the central meridian and takes a negative sign when towards west. It is defined so that the ordinate of each zone is shifted 500 kilometers to the west in order to refrain from using negative values. That is, the coordinates for the origin are (500 000, 0). Hence the shifted ordinate is $X = x + 500\ 000$ m.

Since the zone boundaries occur along meridians, there are identical projections for each zone. For the sake of noting in which zone a certain point belongs, it is defined so that there should be a sign denoting the projection zone ahead of the X value. For example, $n \times 1\ 000\ 000 + X$. The resulting coordinates are called universal coordinates.

2.12.2 Map sheet division and numbering systems

A new national standard for the sheet division and numbering for the national basic scale topographic maps was made in 1991. Since then, the sheet division and numbering of newly edited and revised maps adhere to the new standard.

Map sheet division

Based on the 1:1 000 000 sheet, the map sheet division for Chinese basic scale topographic maps is made according to the listed differences of longitude Δl and of latitude ΔB. The differences of longitude and the differences of latitude for each scale are shown in Table 2.2.

Table 2.2 Sizes of topographic maps sheet

Nominal map scale	Sheet margin dimensions	
	Along meridian	Along parallel
1:1 000 000	6°00′	4°00′
1:500 000	3°00′	2°00′
1:250 000	1°30′	1°00′
1:100 000	0°30′	0°20′
1:50 000	15′00″	10′00″
1:25 000	7′30″	5′00″
1:10 000	3′45″	2′30″
1:5 000	1′52.5″	1′15″

Numbering system

Numbering of the 1:1 000 000 topographic maps   The sheet numbering of the 1:1 000 000 map is composed of a row number (a character code) and a column number (a numerical code). As an example, the sheet number of the Beijing 1:1 000 000 topographic map is J50.

Numbering of the 1:500 000–1:5 000 topographic maps   The 1:500 000–1:5 000 sheet numbering system is based on the sheet numbering of the 1:1 000 000 map. It also uses the method of row and column numbering. That is, the 1:1 000 000 sheet is divided into rows and columns according to the difference in longitude and the difference in latitude for a specific scale of topographic map. The rows are numbered sequentially from the top down with Arabic numerals and the columns sequentially from left to right. Formed of a three-digit number, zero should precede a row or column code if it has a less than three digits. As a sequence, the row code is followed by a column code and they are placed after the sheet number of the 1:1 000 000 map.

In order to deal with different scales, different characters are used as distinct codes for each scale (see Table 2.3).

Table 2.3   Codes for each scale of topographic map

Map scale	1:500 000	1:250 000	1:100 000	1:50 000	1:25 000	1:10 000	1:5 000
Code	B	C	D	E	F	G	H

Thus the number of a 1:500 000–1:5 000 topographic map is composed of five elements, a 10-digit code.

Examples: numbering of a map at a scale of:

- 1:500 000 is J50B001002;
- 1:250 000 is J50C003003;
- 1:100 000 is J50D012012;
- 1:50 000 is J50E024024;
- 1:25 000 is J50F048048;
- 1:10 000 is J50G096096;
- 1:5 000 is J50H192192.

CHAPTER THREE

General Theory of Map Projection Transformation

3.1 RESEARCH OBJECTIVES AND BASIC METHODS

3.1.1 Research objectives

Map projection transformation is a recent development as a field of research on map projections. It involves exploring the theory and methods of transformation of the coordinates of a point from one type of projection to another.

In geodetic and topographic surveying, coordinate transformation between different systems, i.e. computation of coordinate zone transformations, is always needed. For a long time, many computational methods have been presented in China and in other countries. In conventional cartography, in order to combine all the basic characteristics onto the geographic graticule when compiling a new map, map projection transformation is implemented using photography in the transferring of coordinate data and rectification. By using these methods the coordinates of only a few points need to be rigorously transformed. For most points the subsequent transformation is approximate. This is characterized by the fact that we need not create a rigorous mathematical relationship for transformation of all points between two types of projections. In mapping, the rigorous mathematical relations to compute a zone transformation for a projection involved have been described; this is usually limited to a transformation between different coordinate systems for the same type of projection. For this reason, it belongs to a special case of map projection transformation, and is one of the fields of research in map projection transformation. The methods described in this chapter can also be applied to study maps whose projection is not given, as on ancient maps (Tobler 1965, 1994) or the geometry implicit in 'mental maps' (Tobler 1976). By studying map projection transformation we can also develop some new map projections.

With the development of computer-assisted cartography, it has become more important and rather urgent to delve into the theories and methods of map projection transformation; in 'auto-carto' or automated cartography we must previously have developed the mathematical relationship (i.e. mathematical model) in order to complete the coordinate transformation from one type of projection to another. There is no such general relationship or program that we can write for computer-assisted transformations. As a result, map projection transformation has become a component of computer-assisted cartography.

Economic activities of mankind cannot take place without geographical space. All specialized data should involve terrain-based data related to spatial location. Therefore we should study data manipulation for digitized maps and the orientation and transformation of spatial information in map databases to gear it to requirements for constructing all kinds of specialized information systems.

In applications for the presentation and orientation of telemetered images, we must develop coordinate transformation methods for calculating the position circle and target point on the projection plane. In the field of remote sensing from space, we should delve into the theory and methods for the coordinate transformation of dynamic spatial projections. A topological map is an important method of presentation frequently used in modern maps and atlases. A study to retain topological properties after transformation of the plane features is one of the aspects of research in map projection transformation.

From the above we conclude that the requirements of computer-assisted mapping, construction of geographic information systems, and the presentation and orientation of telemetered images, as well as remote sensing from space, and topological transformations, etc., cause map projection transformation to become increasingly a branch of the science of mathematical cartography. Thus it involves the study of theory and methods, and their application to the spatial information registration model, spatial digital processing, and transformation between spatial and planar points.

In a broad sense, map projection transformation can be treated as a study of the theory and methods of spatial data registration, processing, and transformation, and their applications. It can be described by the following relationships:

$$\begin{array}{c} \{x_i', y_i', z_i'\} \\ (\{x_i', y_i', z_i', t_i\}) \end{array} \Leftrightarrow \{\varphi_i, \lambda_i\} \Leftrightarrow \{x_i, y_i\} \Leftrightarrow \{X_i, Y_i\}$$

3.1.2   Basic methods

In a narrow sense, map projection transformation can also be regarded as creating the corresponding relationships of a function of a point on two different planes. Suppose the coordinates of a point on one plane are (x, y) and its coordinates on another plane are (X, Y), then its map projection transformation relationship takes the form:

$$X = F_1(x, y), \quad Y = F_2(x, y) \tag{3.1.1}$$

There are many different ways to complete the coordinate transformation of a point from one plane to another and find their relationship. Here we introduce three kinds that are frequently used (Wu and Yang 1981).

1.   Analytical transformation methods

Usually these kinds of methods aim at finding the analytical calculating formulae of a coordinate transformation between two projection planes. They are also classified as follows, according to the calculation methods used.

Inverse or indirect transformation   This solution involves inverse computation of the geographical coordinates (φ, λ) of the original map projection point by transitional methods, then obtaining its coordinates by substituting it into the equations for the new projection, i.e.,

$$[x, y] \rightarrow [\varphi, \lambda] \rightarrow [X, Y]$$

If the projection equation takes the form of polar coordinates, as in conical, pseudoconical, polyconic, azimuthal, or pseudoazimuthal projections, etc., we need to convert the plane rectangular coordinates (x, y) of the original projection points into plane polar coordinates (ρ, δ) and find their geographical coordinates (φ, λ). Finally, substituting (φ, λ) into the equations for the new projection yields (X, Y), i.e.,

$$[x, y] \rightarrow [\rho, \delta] \rightarrow [\varphi, \lambda] \rightarrow [X, Y]$$

In the case of oblique projections, it is necessary to convert the polar coordinates (ρ, δ) into spherical polar coordinates (Z, a) and further to convert to spherical geographical coordinates (φ', λ'), then temporarily to ellipsoidal geographical coordinates (φ, λ). The final transformation gives (X, Y) by substituting (φ, λ) into the equations for the new projection, i.e.,

$$[x, y] \rightarrow [\rho, \delta] \rightarrow [Z, a] \rightarrow [\varphi', \lambda'] \rightarrow [\varphi, \lambda] \rightarrow [X, Y]$$

Direct transformation   This method need not inversely compute the geographical coordinates of the original map projection points. On the contrary it directly finds the rectangular coordinate relationship of two projection points. For example, we can obtain the relationship of coordinate transformation between two conformal projections in terms of the theory of complex variables:

$$Y + iX = f(y + ix) \tag{3.1.2}$$

that is,

$$[x, y] \rightarrow [X, Y]$$

Combined transformation   This is a method of combining two transformations into one. Generally it is used to find latitude φ from the x-coordinate of the original projection point, then to find coordinates (X, Y) of the new projection point according to (φ, y), i.e.,

$$[x, y \rightarrow \varphi, y] \rightarrow [X, Y]$$

2.   Numerical transformation

If the analytical equations for the original projection point are unknown, or it is difficult to determine the direct relationship between two projection points, we may use the method of polynomial approximation to build a transformation relationship for two types of projections. For example, a bivariate polynomial of the third degree or order is as follows:

$$\left.\begin{aligned}
X &= a_{00} + a_{10}x + a_{01}y + a_{20}x^2 + a_{11}xy + a_{02}y^2 \\
&\quad + a_{30}x^3 + a_{21}x^2y + a_{12}xy^2 + a_{03}y^3 \\
Y &= b_{00} + b_{10}x + b_{01}y + b_{20}x^2 + b_{11}xy + b_{02}y^2 \\
&\quad + b_{30}x^3 + b_{21}x^2y + b_{12}xy^2 + b_{03}y^3
\end{aligned}\right\} \tag{3.1.3}$$

By selecting plane rectangular coordinates (x_i, y_i) and (X_i, Y_i) of ten points common to the two projections and forming a system of linear equations, we can obtain the values of (a_{ij}, b_{ij}). This is a transformation method by the direct solution of a polynomial.

In order to obtain a least-squares approximation for points in the region of transformation of the two projections, it would be better to select more than ten points. According to the theory of the method of least-squares, our objective is to minimize the sum of the squares of the differences between the calculated values of the transformation to the new projection and the true coordinate values, that is minimize:

$$\varepsilon = \sum_{i=1}^{n}(X_i - X_i')^2, \quad \varepsilon' = \sum_{i=1}^{n}(Y_i - Y_i')^2 \tag{3.1.4}$$

According to the principle of minimization, the first derivative of ε and ε' with respect to a_{ij} and b_{ij}, respectively, should be made equal to zero. Thus we obtain two sets of linear equations and can obtain the values of a_{ij} and b_{ij} by solving them. This method belongs to direct transformation that uses the least-squares method to determine the polynomial coefficients.

3. Numerical analytical transformation

If the equations for the new projection are given but the original ones are unknown, we can use a polynomial such as equations (3.1.3), replace (X, Y) with (φ, λ), find the geographical coordinates (φ, λ) of the original projection points, and finally achieve transformation between the two projections by substituting their values into the equations for the new projection.

The three methods described above are now commonly used methods in the research of map projection transformation, but they are not the only ones. We should explore the use of other effective transformation methods in accordance with a specific objective, especially in the case of computer-assisted mapping. Map projection transformation is now widely used in surveying, cartography, remote sensing from space, and in other space technology.

3.2   EQUATIONS FOR MAP PROJECTION TRANSFORMATION

3.2.1   Equations for transformation

The equations for the original map projection may be given as follows:

$$x = f_1(\varphi, \lambda), \quad y = f_2(\varphi, \lambda) \tag{3.2.1}$$

The equations for the new map projection take the form:

$$X = \psi_1(\varphi, \lambda), \quad Y = \psi_2(\varphi, \lambda) \tag{3.2.2}$$

The inverse form of equations (3.2.1) is:

$$\varphi = \varphi(x, y), \quad \lambda = \lambda(x, y) \tag{3.2.3}$$

Substituting equation (3.2.3) into (3.2.2) gives:

$$X = F_1(x, y), \quad Y = F_2(x, y) \tag{3.2.4}$$

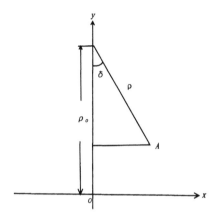

Figure 3.1 The relationship between the plane polar coordinates and the plane rectangular coordinates

If the equations for polar coordinates for the projection are given as:

$$\rho = \phi_1(\varphi, \lambda), \quad \delta = \phi_2(\varphi, \lambda) \tag{3.2.5}$$

then we can obtain the relationship between the plane polar coordinates and the plane rectangular coordinates from Figure 3.1:

$$\left.\begin{array}{l} \delta = \arctan\dfrac{x}{\rho_0 - y} \\[2mm] \rho = \sqrt{(\rho_0 - y)^2 + x^2} \end{array}\right\} \tag{3.2.6}$$

The three equations (3.2.3), (3.2.4), and (3.2.6) are called the equations of map projection transformation.

Now let us give some examples to illustrate these equations.

Example 1: Transformation from a conformal cylindrical projection to an equivalent (equal-area) cylindrical projection

The equations for a conformal cylindrical projection on the sphere take the form:

$$x = r_0\lambda, \quad y = r_0\ln\tan\left(45° + \frac{\varphi}{2}\right) \tag{3.2.7}$$

and the equations for an equivalent cylindrical projection on the sphere take the form:

$$X = r_0\lambda, \quad Y = \frac{R^2}{r_0}\sin\varphi \tag{3.2.8}$$

but the inverse equations for (3.2.7) are:

$$\varphi = 2\arctan(e_1^{y/r_0}) - \frac{\pi}{2}, \quad \lambda = \frac{x}{r_0} \tag{3.2.9}$$

where $e_1 = 2.718281828.\ldots$

Then we can obtain the following equation for the direct transformation from a conformal cylindrical to an equivalent cylindrical projection by substituting equation (3.2.9) into (3.2.8):

$$X = x; \quad Y = \frac{R^2}{r_0} \cdot \frac{e_1^{2y/r_0} - 1}{e_1^{2y/r_0} + 1} \tag{3.2.10}$$

Example 2: Transformation from a conformal conical to a conformal cylindrical projection

Since the equations for a conformal conical projection are as follows:

$$x = \rho \sin \delta, \quad y = \rho_s - \rho \cos \delta \qquad (3.2.11)$$

where

$$\rho = \frac{C}{U^\alpha}, \quad \delta = \alpha l$$

and the equations for a conformal cylindrical projection are as follows:

$$X = r_0 l, \quad Y = r_0 \ln U \qquad (3.2.12)$$

from equations (3.2.11) we have:

$$\rho = \sqrt{(\rho_s - y)^2 + x^2}, \quad \delta = \arctan\frac{x}{\rho_s - y} \qquad (3.2.13)$$

But

$$\ln U = \frac{1}{\alpha}(\ln C - \ln\rho) \qquad (3.2.14)$$

Hence we get the following equations for transformation of coordinates:

$$X = \frac{r_0}{\alpha} \arctan\frac{x}{\rho_s - y}, \quad Y = \frac{r_0}{\alpha}(\ln C - \ln\rho), \qquad (3.2.15)$$

where:

$$\rho = \sqrt{(\rho_s - y)^2 + x^2}$$

Example 3: Transformation from the Sanson–Flamsteed pseudocylindrical equivalent ('sinusoidal') projection to the Kavraisky sinusoidal pseudocylindrical equivalent projection

Since the equations for the Sanson projection have the form:

$$x = R\lambda \cos\varphi, \quad y = R\varphi \qquad (3.2.16)$$

but the equations for the Kavraisky projection are:

$$\left. \begin{aligned} X &= \frac{2}{3}R\sqrt[4]{3}\lambda \cos\alpha, \quad Y = R\sqrt[4]{3}\alpha \\ \sin\alpha &= \frac{\sqrt{3}}{2}\sin\varphi \end{aligned} \right\} \qquad (3.2.17)$$

From equations (3.2.16) we can obtain the inverse equations for the Sanson projection as follows:

$$\varphi = \frac{y}{R}, \quad \lambda = \frac{x}{R\cos\dfrac{y}{R}} \qquad (3.2.18)$$

Incorporating equations (3.2.18) with (3.2.17) we have:

$$
\left.
\begin{aligned}
X &= \frac{2}{3}\sqrt[4]{3}\,\frac{x}{\cos\dfrac{y}{R}}\cos\left[\arcsin\left(\frac{\sqrt{3}}{2}\sin\frac{y}{R}\right)\right] \\[2em]
Y &= R\sqrt[4]{3}\arcsin\left(\frac{\sqrt{3}}{2}\sin\frac{y}{R}\right)
\end{aligned}
\right\}
\tag{3.2.19}
$$

These are the equations for direct transformation from the Sanson projection to the Kavraisky projection.

3.2.2 Differential equations for inverse transformation (Urmayev 1962)

Inverse transformation is a commonly used method in map projection transformation. Assuming that the coordinates of the original projection are given, we can find their geographic coordinates (φ, λ) from the following relationships:

$$
\varphi = \varphi(x, y), \quad \lambda = \lambda(x, y)
\tag{3.2.20}
$$

Differentiating equations (3.2.20) yields:

$$
d\varphi = \varphi_x dx + \varphi_y dy, \quad d\lambda = \lambda_x dx + \lambda_y dy
\tag{3.2.21}
$$

Introducing the following determinant,

$$
h' = \varphi_x \lambda_y - \varphi_y \lambda_x
\tag{3.2.22}
$$

Solving equations (3.2.21) gives:

$$
\left.
\begin{aligned}
dx &= \frac{\lambda_y}{h'}d\varphi - \frac{\varphi_y}{h'}d\lambda \\[1em]
dy &= -\frac{\lambda_x}{h'}d\varphi + \frac{\varphi_x}{h'}d\lambda
\end{aligned}
\right\}
\tag{3.2.23}
$$

Introducing equations (3.2.1) we have:

$$
dx = x_\varphi d\varphi + x_\lambda d\lambda, \quad dy = y_\varphi d\varphi + y_\lambda d\lambda
\tag{3.2.24}
$$

Comparing equations (3.2.23) and (3.2.24) we get:

$$
\left.
\begin{aligned}
x_\varphi &= +\frac{\lambda_y}{h'}, \quad x_\lambda = -\frac{\varphi_y}{h'} \\[1em]
y_\varphi &= -\frac{\lambda_x}{h'}, \quad y_\lambda = +\frac{\varphi_x}{h'}
\end{aligned}
\right\}
\tag{3.2.25}
$$

Hence:

$$
h = x_\varphi y_\lambda - y_\varphi x_\lambda = \frac{\varphi_x \lambda_y - \varphi_y \lambda_x}{h'^2} = \frac{1}{h'}
\tag{3.2.26}
$$

Thus equations (3.2.25) can be written as:

$$x_\varphi = h\lambda_y, \quad x_\lambda = -h\varphi_y \atop y_\varphi = -h\lambda_x, \quad y_\lambda = h\varphi_x \Big\}$$

(3.2.27)

and:

$$\left. \begin{array}{l} e = h^2(\lambda_x^2 + \lambda_y^2) \\ f = -h^2(\varphi_x\lambda_x + \varphi_y\lambda_y) \\ g = h^2(\varphi_x^2 + \varphi_y^2) \end{array} \right\}$$

(3.2.28)

From the theory of map projection distortion (Tissot 1881) we have:

the scale factor along meridians: $m = \dfrac{\sqrt{e}}{M}$

the scale factor along parallels: $n = \dfrac{\sqrt{g}}{r}$

(3.2.29)

the scale factor for area: $P = \dfrac{h}{Mr}$

the angular distortion of the intersection between meridians and parallels: $\tan\varepsilon = -\dfrac{f}{h}$

From equations (3.2.22), (3.2.26), (3.2.28), and (3.2.29) we have:

$$PMr = h$$

(3.2.30)

$$m = Pr\sqrt{\lambda_x^2 + \lambda_y^2}$$

(3.2.31)

$$n = PM\sqrt{\varphi_x^2 + \varphi_y^2}$$

(3.2.32)

Then:

$$\frac{1}{PMr} = \frac{1}{h} = h' = \varphi_x\lambda_y - \varphi_y\lambda_x$$

(3.2.33)

$$\tan\varepsilon = \frac{\varphi_x\lambda_x + \varphi_y\lambda_y}{PMr} = \frac{\varphi_x\lambda_x + \varphi_y\lambda_y}{\varphi_x\lambda_y - \varphi_y\lambda_x}$$

(3.2.34)

Noting that:

$$d\varphi = \frac{dS}{M}$$

(3.2.35)

equation (3.2.32) can be written as:

$$S_x^2 + S_y^2 = \frac{n^2}{P^2}$$

(3.2.36)

Introducing

$$dF = Mr\,d\varphi$$

(3.2.37)

equation (3.2.33) may be rewritten as:

$$F_x\lambda_y - F_y\lambda_x = \frac{1}{P}$$

(3.2.38)

Since

$$dq = \frac{M}{r} d\varphi \tag{3.2.39}$$

equation (3.2.32) can take the following form:

$$n = Pr\sqrt{q_x^2 + q_y^2} \tag{3.2.40}$$

If the relationships in equations (3.2.20) have been determined, we can calculate the proportions of length (linear scale factors) and angular distortion of the map projection using the above equations. If the values of (n/PM), P, and $\tan \varepsilon$ are known, we can obtain

$$\varphi_x^2 + \varphi_y^2 = \frac{n^2}{P^2 M^2}$$

from equation (3.2.32). This is a non-linear partial differential equation of the first order. Its integral equation takes a form like $\varphi = \varphi(x, y)$. From that we can obtain an equation for the latitude. If the equations for latitude and the area scale factor are determined, then we can obtain an equation for longitude from (3.2.33) and (3.2.38), since equations (3.2.33) and (3.2.38) are linear partial differential equations of the first order.

3.3   PLANE FEATURE TRANSFORMATION

The meaning of a plane feature varies widely. Figures such as a plane map, plane nomograph, and plane design drawing, as well as the appearance of views of the natural world and society are generally called plane features. The concepts of a plane feature transformation also vary widely. For example, projective geometry transformations, nomographic transformations, and map projection transformations are natural topics of research into plane feature transformation. Here we explore the general concepts and methods of plane feature transformation using the methods of map projection.

3.3.1   General theory

Plane feature transformation can be regarded as the transformation of coordinates of points on planes, that is, transformation from coordinates (x, y) of a point on one plane to the coordinates (x', y') of the same point on another plane. As a result, the object of plane feature transformation is to find the corresponding relationships of a point on two different planes, i.e.,

$$x' = f_1(x, y), \quad y' = f_2(x, y) \tag{3.3.1}$$

There are many ways for determining the above equations. In this book we emphasize a transformation method that introduces intermediate variables (u, v).

Suppose the relationship between coordinates (x, y) of the point on the first plane and variables (u, v) is as follows:

$$x = \psi_1(u, v), \quad y = \psi_2(u, v) \tag{3.3.2}$$

and that the relationship between the coordinates (x', y') of the point on the second plane and variables (u, v) is:

$$x' = \psi_3(u, v), \quad y' = \psi_4(u, v) \tag{3.3.3}$$

Equation (3.3.2) can be inversely written as:

$$u = u(x, y), \quad v = v(x, y) \tag{3.3.4}$$

Substituting the above equations (3.3.4) into (3.3.3) we obtain:

$$\left. \begin{array}{l} x' = \psi_3(u(x, y), v(x, y)) \\ y' = \psi_4(u(x, y), v(x, y)) \end{array} \right\} \tag{3.3.5}$$

or:

$$x' = f_1(x, y), \quad y' = f_2(x, y) \tag{3.3.6}$$

This is the relationship for plane feature transformation. The variables (u, v) can be fixed at random. Thus there are many ways to determine equations (3.3.1).

In map projection, studies of transformation of the coordinates of a point between the surface of a sphere and a plane become even more complicated. As a result, we introduce spherical polar coordinates (z, a) as intermediate variables to achieve a transformation between two kinds of planes. These coordinates can be explained with a diagram, as shown in Figure 3.2.

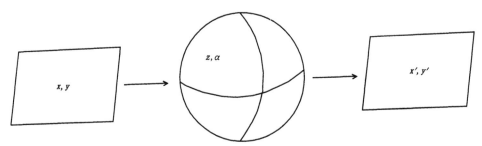

Figure 3.2   Transformation between two kinds of planes taking the sphere as intermediate

In general, there is a one-to-one correspondence of points on a plane to points on a spherical surface. Hence, we can regard features on a plane as the mapping from those on a spherical surface and vice versa. Therefore, the concept of a plane feature is broad.

Now let us substitute variables (u, v) in the above equation for variables (z, a) and discuss distortion theory for plane transformations.

Proportion of length

The ratio of a differential segment ds' at a point on the field of the second plane to the corresponding ds on the field of the first plane is called the proportion of length. Noting that:

$$\mu = \frac{ds'}{ds} \tag{3.3.7}$$

in the case of transformation using variables (z, a), we can obtain the following equation in accordance with equations (3.3.3):

$$ds'^2 = E'dz^2 + 2F'dz\,da + G'da^2$$

where:

$$E' = \left(\frac{\partial x'}{\partial z}\right)^2 + \left(\frac{\partial y'}{\partial z}\right)^2, \quad F' = \frac{\partial x' \partial x'}{\partial z \partial a} + \frac{\partial y' \partial y'}{\partial z \partial a}, \quad G' = \left(\frac{\partial x'}{\partial a}\right)^2 + \left(\frac{\partial y'}{\partial a}\right)^2$$

From equation (3.3.2) we have:

$$ds^2 = Edz^2 + 2Fdzda + Gda^2$$

Hence,

$$\mu^2 = \frac{E'dz^2 + 2F'dzda + G'da^2}{Edz^2 + 2Fdzda + Gda^2} \tag{3.3.8}$$

Then we can obtain the equation for proportion of length by using equation (3.3.1):

$$\mu^2 = \frac{edx^2 + 2fdxdy + gdy^2}{dx^2 + dy^2} \tag{3.3.9}$$

Taking into account the equation for the calculation of coordinate azimuths on the first plane:

$$\tan a = \frac{dx}{dy} \tag{3.3.10}$$

we have

$$\mu^2 = e\cos^2 a + f\sin 2a + g\sin^2 a \tag{3.3.11}$$

where:

$$e = \left(\frac{\partial x'}{\partial x}\right)^2 + \left(\frac{\partial y'}{\partial x}\right)^2, \quad f = \frac{\partial x' \partial x'}{\partial x \partial y} + \frac{\partial y' \partial y'}{\partial x \partial y}, \quad g = \left(\frac{\partial x'}{\partial y}\right)^2 + \left(\frac{\partial y'}{\partial y}\right)^2$$

If $a = 0°$, then $\mu_y = \sqrt{e}$ $\tag{3.3.12}$

If $a = 90°$, then $\mu_x = \sqrt{g}$ $\tag{3.3.13}$

Since the element of length for intermediate variables (z, a) on the spherical surface is $d\sigma$, we can obtain the following in terms of the theory of map projection:

$$\mu' = \frac{ds'}{d\sigma}, \quad \bar{\mu} = \frac{ds}{d\sigma}$$

Finally we obtain:

$$\mu = \frac{\mu'}{\bar{\mu}} \tag{3.3.14}$$

Proportion of area

The ratio of differential area dF' on the second plane to the corresponding dF on the first one is called the proportion of area. It can be written as:

$$p = \frac{dF'}{dF} = \mu_x \mu_y \sin\theta \tag{3.3.15}$$

where:

$$\tan\theta = \frac{h}{f}, \quad h = \sqrt{eg - f^2}$$

and θ is the angle between abscissa and ordinate of the projection.

Taking equations (3.3.12) and (3.3.13) into consideration we have:

$$p = \sqrt{eg}\sin\theta = h = \frac{\partial x'}{\partial x} \cdot \frac{\partial y'}{\partial y} - \frac{\partial y'}{\partial x} \cdot \frac{\partial x'}{\partial y} \qquad (3.3.16)$$

Ellipse of distortion

Generally two orthogonal directions passing through an arbitrary point on the first plane are not orthogonal when they project on the second plane, but there is at least one pair of orthogonal directions that retain their orthogonality after projection. We call this pair the principal directions (Tissot 1881).

A differential circle of unit radius will normally become a differential ellipse when it projects from the first plane to the second one. We call it the ellipse of distortion. Symbols (a, b) denote respectively the semimajor axis and semiminor axis of the ellipse of distortion. We obtain the following equation using analytical geometry:

$$a^2 + b^2 = \mu_x^2 + \mu_y^2, \quad ab = \mu_x\mu_y\sin\theta \qquad (3.3.17)$$

Maximum angular distortion

The equation for the maximum angular distortion is:

$$\sin\frac{\omega}{2} = \frac{a - b}{a + b} \qquad (3.3.18)$$

The equation for the distortion of plane feature transformation from the first plane to the second appears the same as the corresponding equation for distortion in the theory of map projection transformation, but some of the concepts are different.

3.3.2   Basic methods

The basic methods for plane feature transformation include analytical transformation and numerical transformation. In this section we delve into the general methods of analytical transformation.

We can describe any plane feature by using rectangular or polar coordinates. The plane rectangular coordinate grid or polar coordinate grid can be taken as the control network for plane features. As a result, to study the transformation of rectangular or polar coordinate grids on the first plane is to study the transformation of plane features.

Transformation of rectangular coordinate grids using map projection methods

A grid of rectangular coordinates can be treated as the mapping of latitude and longitude lines for a normal cylindrical projection, i.e.,

$$x = C\lambda, \quad y = f(\varphi) \qquad (3.3.19)$$

This shows that the mapping of a plane coordinate grid from the first plane to the second one can be taken as the transformation from a cylindrical projection to others. In other words, we can take the shape of latitude and longitude lines for any type of projection as the mapping of rectangular coordinates on the second plane, i.e.,

$$x' = f_1(\varphi, \lambda), \quad y' = f_2(\varphi, \lambda) \qquad (3.3.20)$$

In this case, we can derive the equations for the calculation of distortion of the plane feature transformations.

From equation (3.3.19), we have:

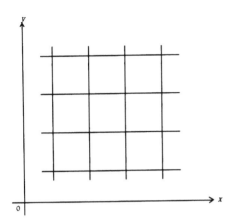

Figure 3.3 The rectangular coordinate system

$$\bar{m} = \frac{dy}{Rd\varphi}, \quad \bar{n} = \frac{dx}{rd\lambda} = \frac{C}{R\cos\varphi}, \quad \bar{p} = \bar{m}\bar{n}$$

From equation (3.3.20):

$$m' = \frac{\sqrt{E'}}{R}, \quad n' = \frac{\sqrt{G'}}{R\cos\varphi}, \quad p' = m'n'\sin\theta, \quad \tan\theta = \frac{H'}{F'}$$

Incorporating them we obtain:

$$\left.\begin{array}{l} \mu_x = \dfrac{n'}{\bar{n}}, \quad \mu_y = \dfrac{m'}{\bar{m}} \\[2mm] p = \dfrac{p'}{\bar{p}}, \quad \sin\dfrac{\omega}{2} = \dfrac{a-b}{a+b} \end{array}\right\} \qquad (3.3.21)$$

Transformation of polar coordinate grids using projection methods

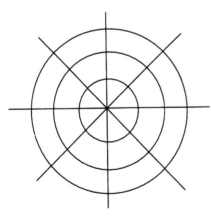

Figure 3.4 The polar coordinate system

A grid of polar coordinates (Figure 3.4) can be taken as the mapping of latitude and longitude lines for a normal azimuthal projection, i.e.,

$$\rho = f(\varphi), \quad \delta = \lambda \qquad (3.3.22)$$

This proves that the mapping of a grid of polar coordinates from the first plane to the second one can be taken as the transformation from a normal azimuthal projection to others. In other words, we can take the shape of latitude and longitude lines for any type of projection as the mapping of a polar coordinate grid onto the second plane, i.e.,

$$x' = f_1(\varphi, \lambda), \quad y' = f_2(\varphi, \lambda) \qquad (3.3.20)$$

From equation (3.3.22) we have:

$$\overline{m} = -\frac{d\rho}{R\,d\varphi}, \quad \overline{n} = \frac{\rho}{R\cos\varphi} \tag{3.3.23}$$

The other equations for calculation are the same as above.

Indirect method for the transformation of a rectangular coordinate grid

We can take the grid of rectangular coordinates on the first plane as the mapping of a family of curves for any projection. To maintain the generality, suppose the relationship for the projection is as follows:

$$x = \psi_1(z, a), \quad y = \psi_2(z, a) \tag{3.3.24}$$

and its inverse equations are:

$$z = z(x, y), \quad a = a(x, y) \tag{3.3.25}$$

The equations for parameters of the ordinates corresponding to curves on the surface of a sphere are as follows:

$$z = z(x, y_0), \quad a = a(x, y_0) \tag{3.3.26}$$

and the equations for parameters of the abscissa corresponding to curves on the surface of a sphere are as follows:

$$z = z(x_0, y), \quad a = a(x_0, y) \tag{3.3.27}$$

We can project the ordinate and abscissa onto the second plane by using any type of projection if equations (3.3.26) and (3.3.27) are determined. Hence we have:

$$x' = \psi_3(z, a), \quad y' = \psi_4(z, a) \tag{3.3.28}$$

In this case we can use the related equations described above to compute the distortion.

Direct method for the transformation of a rectangular coordinate grid

Substituting equation (3.3.25) into (3.3.28), we obtain the relationship for the direct transformation of a grid of rectangular coordinates from the first plane to the second one:

$$x' = f_1(x, y), \quad y' = f_2(x, y) \tag{3.3.29}$$

In this case we can also use the related equations described above to compute the distortion.

3.3.3 Examples

Example 1

Assuming that the grid of rectangular coordinates on the first plane is a mapping of latitude and longitude lines according to the cylindrical equidistant projection (as shown in Figure 3.5) and that it is projected onto the second plane with a transverse or equatorial orthographic projection, let us present a graph of the transformation.

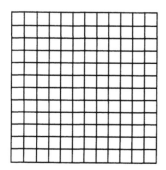

Figure 3.5 The grid of the equidistant projection

Since the equations for the cylindrical equidistant projection are as follows:

$$x = R\Delta\lambda, \quad y = R\varphi \tag{3.3.30}$$

we obtain:

$$\varphi = \frac{y}{R}, \quad \Delta\lambda = \frac{x}{R} \tag{3.3.31}$$

We now transform the grid of rectangular coordinates from the first plane to the second according to the transverse orthographic projection. Its equations are as follows:

$$x' = R\sin z \sin a, \quad y' = R\sin z \cos a \tag{3.3.32}$$

or:

$$x' = R\cos\varphi \sin\Delta\lambda, \quad y' = R\sin\varphi \tag{3.3.33}$$

Substituting (3.3.31) into the above we obtain the direct transformation equation for a plane feature:

$$x' = R\sin\frac{x}{R}\cos\frac{y}{R}, \quad y' = R\sin\frac{y}{R} \tag{3.3.34}$$

Figure 3.6 shows the transformation graph as drawn by a pocket calculator.

Suppose $y = 2$ cm when $\varphi = 60°$; then we get:

$$R = \frac{y}{\varphi} = \frac{2}{\frac{\pi}{3}} = \frac{6}{\pi} \text{ cm}$$

Figure 3.6 The grid of transverse orthographic projection

From equations (3.3.34) we have:

$$
\left.
\begin{aligned}
e &= \cos^2\frac{y}{R} + \sin^2\frac{y}{R}\sin^2\frac{x}{R} \\[2mm]
g &= \cos^2\frac{x}{R}\cos^2\frac{y}{R} \\[2mm]
f &= -\sin\frac{x}{R}\sin\frac{y}{R}\cos\frac{x}{R}\cos\frac{y}{R} \\[2mm]
h &= \cos\frac{x}{R}\cos^2\frac{y}{R}
\end{aligned}
\right\} \tag{3.3.35}
$$

Thus the equations for distortion of the transformed feature are as follows:

$$
\left.
\begin{aligned}
\mu_y &= \sqrt{e} = \sqrt{\cos^2\frac{y}{R} + \sin^2\frac{y}{R}\sin^2\frac{x}{R}} \\[2mm]
\mu_x &= \sqrt{g} = \cos\frac{x}{R}\cos\frac{y}{R} \\[2mm]
\tan\varepsilon &= -\frac{f}{h} = \sin\frac{x}{R}\tan\frac{y}{R} \\[2mm]
p &= h = \cos\frac{x}{R}\cos^2\frac{y}{R}
\end{aligned}
\right\} \tag{3.3.36}
$$

The map projection method for a grid of polar coordinates is identical with this example.

It must be pointed out that this method of projection transformation is different from the map projection transformation in concept. In this case the surface of the sphere is an interim supplementary one but not the true surface of the earth.

Example 2

Assuming that the first plane is a transverse cylindrical conformal (transverse Mercator) projection plane, but that the second is a normal Mercator projection plane, we will try to find the mapping of the rectangular coordinate grid from the first plane to the second.

Coming from the theory of map projection we obtain the coordinate equations for the transverse Mercator projection:

$$x = R \operatorname{arctanh}(\cos\varphi \sin\Delta\lambda), \quad y = R \arctan(\tan\varphi \sec\Delta\lambda) \tag{3.3.37}$$

By a simplified solution of the above we obtain the inverse transformation equation as follows:

$$\left. \begin{aligned} \varphi &= \arcsin\left(\sin\frac{y}{R}\operatorname{sech}\frac{x}{R}\right) \\ \Delta\lambda &= \arctan\left(\sec\frac{y}{R}\sinh\frac{x}{R}\right) \end{aligned} \right\} \tag{3.3.38}$$

Since the coordinate equations for the normal Mercator projection are:

$$x' = R\Delta\lambda, \quad y' = R \ln\tan\left(\frac{\pi}{4} + \frac{\varphi}{2}\right) \tag{3.3.39}$$

substituting (3.3.38) into the above, we can find the transformation of the rectangular coordinate grid from the first plane to the second. From (3.3.39) we have:

$$\frac{x'}{R} = \Delta\lambda, \quad \tanh\frac{y'}{R} = \sin\varphi \tag{3.3.40}$$

Substituting (3.3.38) into (3.3.40), we obtain the direct transformation equation from the transverse Mercator projection to the normal Mercator projection:

$$\left. \begin{aligned} \tan\frac{x'}{R} &= \sec\frac{y}{R}\sinh\frac{x}{R} \\ \tanh\frac{y'}{R} &= \sin\frac{y}{R}\operatorname{sech}\frac{x}{R} \end{aligned} \right\} \tag{3.3.41}$$

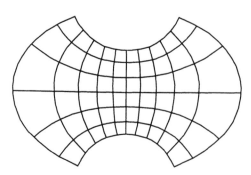

Figure 3.7 The mapping of a rectangular coordinate grid projected from the transverse conformal cylindrical projection to the Mercator projection

Figure 3.7 shows the mapping of a rectangular coordinate grid projected from the transverse conformal cylindrical projection to the Mercator projection. This is an example of the application of plane feature transformation to a map projection transformation.

Now we discuss the application of plane feature transformation using a nomograph,

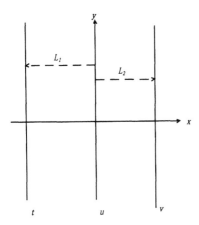

Figure 3.8 The tri-parallel rulers of a collinear diagram

It can also be written as:

$$\begin{vmatrix} -ac & a[F(t)+d] & 1 \\ 0 & \dfrac{-ab}{a+b}[F(u)+e] & 1 \\ bc & b[F(v)-d-e] & 1 \end{vmatrix} = 0 \qquad (3.3.44)$$

Then we obtain each scale as follows:

$$\left. \begin{aligned} x_t &= -ac, & y_t &= a[F(t)+d] \\ x_u &= 0, & y_u &= \dfrac{-ab}{a+b}[F(u)+e] \\ x_v &= bc, & y_v &= b[F(v)-d-e] \end{aligned} \right\} \qquad (3.3.45)$$

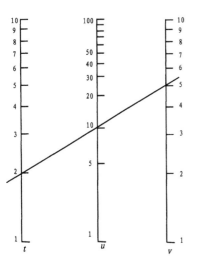

Figure 3.9 Graduations of the logarithmic scale nomograph

which was a popular calculating device prior to the advent of pocket calculators and computers.

Example 3

Suppose that the collinear determinant for the tri-parallel rulers of a collinear diagram as shown in Figure 3.8 is as follows:

$$\begin{vmatrix} -L_1 & f(t) & 1 \\ 0 & f(u) & 1 \\ L_2 & f(v) & 1 \end{vmatrix} = 0 \qquad (3.3.42)$$

Then the determinant expansion may be expressed as:

$$F(t) + F(u) + F(v) = 0 \qquad (3.3.43)$$

We can freely choose the values of the coefficients a, b, c, d, and e so that the diagram is drawn to an appropriate scale. As an example, let us draw the $u = tv$ collinear diagram. It can also be written as:

$$\log t - \log u + \log v = 0 \qquad (3.3.46)$$

Assume that the diagram is 30 cm high and 20 cm wide, with a scale range for u from 1 to 100 and for t and v from 1 to 10. From equations (3.3.45) we can obtain the following equations for calculating each scale ($a = b = 1$, $c = 10$, $d = e = 0$):

$$\left. \begin{aligned} x_t &= -10, & y_t &= 30(\log t) \\ x_u &= 0, & y_u &= 15(\log u) \\ x_v &= 0, & y_v &= 30(\log v) \end{aligned} \right\} \qquad (3.3.47)$$

Figure 3.9 is approximately a 1/5-scale drawing. As shown, graduations of the logarithmic scale nomograph are large at the beginning, but the intervals become shorter and shorter with an increase in scale values. In order to improve the precision of the logarithmic scale, we should make the graduations shorter at the beginning and larger at the top by transforming the tri-parallel scale.

To retain the collinearity of the nomograph, we may use the gnomonic projection transformation method. As the first step, transform the tri-parallel scale on the first plane using the gnomonic projection method. Its transformation equations are as follows:

$$\left. \begin{array}{l} x = R \cot\varphi \, \sin\Delta\lambda \\ y = y_0 - R \cot\varphi \, \cos\Delta\lambda \end{array} \right\} \qquad (3.3.48)$$

where the coefficients y_0, R are determined by the following: for the scale u, $x = 0$, $\Delta\lambda = 0$; for $y = 0$, let us set $\varphi_0 = 25°$, and for $y = 30$ cm, set $\varphi = 90°$. Hence we get $y_0 = 30$ cm, and $R = y_0 \tan\varphi_0 = 13.989$ cm.

The inverse transformation equations for (3.3.48) are as follows:

$$\left. \begin{array}{l} \varphi = \arctan \dfrac{R}{\sqrt{x^2 + (y_0 - y)^2}} \\[3mm] \Delta\lambda = \arctan \dfrac{x}{y_0 - y} \end{array} \right\} \qquad (3.3.49)$$

The inverse transformation equations for each scale are as follows:

$$\left. \begin{array}{ll} u - \text{scale:} & \varphi = \arctan \dfrac{R}{30 - 15(\log u)}, \quad \Delta\lambda = 0 \\[4mm] t - \text{scale:} & \varphi = \arctan \dfrac{R}{\sqrt{100 + 30(1 - \log t)^2}} \\[4mm] & \Delta\lambda = \arctan \dfrac{-10}{30(1 - \log t)} \\[4mm] v - \text{scale:} & \varphi = \arctan \dfrac{R}{\sqrt{100 + 30(1 - \log v)^2}} \\[4mm] & \Delta\lambda = \arctan \dfrac{10}{30(1 - \log v)} \end{array} \right\} \qquad (3.3.50)$$

Finally we obtain a transformation equation to transform the nomograph from the first plane to the second:

$$x' = R \tan z \sin a, \quad y' = R \tan z \cos a \qquad (3.3.51)$$

$$\left. \begin{array}{l} \cos z = \sin\varphi \sin\varphi_0 + \cos\varphi \, \cos\varphi_0 \, \cos\Delta\lambda \\ \sin z \sin a = \cos\varphi \sin\Delta\lambda \\ \sin z \cos a = \sin\varphi \cos\varphi_0 - \cos\varphi \sin\varphi_0 \, \cos\Delta\lambda \\ \varphi_0 = 25°, \quad R = y_0 \tan\varphi_0 \end{array} \right\} \qquad (3.3.52)$$

Figure 3.10 shows the transformed nomograph.

Comparing Figures 3.9 and 3.10 we can conclude that the transformed nomograph has improved intervals between the numerical values. In other words, the precision of the nomograph for measurements has substantially improved. This transformation method is flexible for adjusting the scale and interval to meet all kinds of requirements.

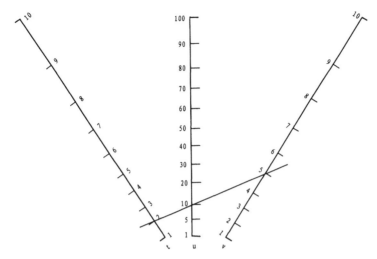

Figure 3.10   The logarithmic scale nomograph after a gnomonic projection transformation

3.4   COORDINATE SYSTEMS ON THE SURFACES OF THE ELLIPSOID AND SPHERE

3.4.1   Coordinate system on the surface of the ellipsoid

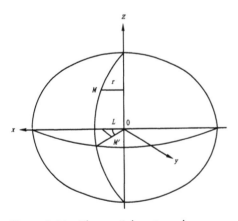

Figure 3.11   The spatial rectangular coordinate system of the surface of an ellipsoid of revolution

The surface of the earth ellipsoid is created by rotating an ellipse about the earth's axis. We call it the surface of an ellipsoid of revolution. Figure 3.11 shows the spatial rectangular coordinate system. The system has its origin at the center of rotational ellipsoid with the z-axis coinciding with the rotation axis and the xOy plane coinciding with the equatorial plane.

Symbols a, b, respectively, denote the semimajor axis and semiminor axis of the ellipsoid of revolution. The equation for the spatial rectangular coordinate system is then given by:

$$\frac{x^2}{a^2} + \frac{y^2}{a^2} + \frac{z^2}{b^2} = 1 \qquad (3.4.1)$$

A common method of defining the position on the surface of an ellipsoid is with curvilinear coordinates. Let us draw a sphere that has the same center as that of the ellipsoid of revolution and that has semimajor axis $a = OR$ as its radius, with the sphere tangent to the ellipsoid of revolution along the equatorial circle. As is shown in Figure 3.12, by extending $M'M$ to the sphere, we have an intersection point M''. The reduced latitude u of point M is the angle formed by the intersection of OM'' with the equatorial plane. The geocentric latitude ϕ of point M is the angle formed by the intersection between OM and the equatorial plane. B is the geographic latitude of point M, i.e. the angle of intersection between a normal (perpendicular) line at point M' and the equatorial plane. By definition,

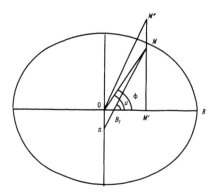

Figure 3.12 Three latitudes on the surface of an ellipsoid

the values of u, ϕ, and B are zero at the equator, positive in a northward direction, and negative southerly.

Furthermore, suppose that the angle of intersection between the meridian passing through point M and the prime meridian plane zOx is L. We call L the geographic longitude.

It is apparent that the curves for a constant B (or u or ϕ) form a family of latitudes, and the curves of constant L form a family of longitudes. The curvilinear grid or graticule formed by these two sets of orthogonal curves is called the geographical coordinate system. As a result, the geographic latitude B and longitude L are called geographic coordinates, and geocentric latitude ϕ and geographic longitude L are called geocentric coordinates.

Equations for the earth ellipsoid surface with respect to parameters u *and* L *as well as the inverse transformation*

From Figures 3.11 and 3.12 we find:

$$x = OM' \cos L, \quad y = OM' \sin L, \quad OM' = a \cos u$$

or:

$$x = a \cos u \cos L, \quad y = a \cos u \sin L$$

Substituting the above into (3.4.1),

$$z = b \sin u$$

Then the equations for the earth ellipsoid with parameters u and L are:

$$x = a \cos u \cos L, \quad y = a \cos u \sin L, \quad z = b \sin u \tag{3.4.2}$$

From (3.4.2) we obtain the corresponding inverse transformation equations:

$$u = \arcsin\left(\frac{z}{b}\right), \quad L = \arctan\left(\frac{y}{x}\right) \tag{3.4.3}$$

Equation for the earth ellipsoid surface with respect to parameters ϕ *and* L *as well as its inverse transformation*

From (3.4.2) we obtain the radius of the latitude circle:

$$r = \sqrt{x^2 + y^2} = a \cos u \tag{3.4.4}$$

The equations

$$r = a \cos u, \quad z = b \sin u \tag{3.4.5}$$

are the parametric expressions for the meridian ellipse

$$\frac{r^2}{a^2} + \frac{z^2}{b^2} = 1 \tag{3.4.6}$$

Relating to geocentric latitude ϕ and geocentric radius ρ

$$z = \rho \sin\phi, \quad r = \rho \cos\phi \tag{3.4.7}$$

we obtain:

$$\tan\phi = \frac{z}{r}$$

Substituting (3.4.5) into the above gives:

$$\tan\phi = \frac{b}{a}\tan u = \sqrt{1 - e^2}\,\tan u \tag{3.4.8}$$

or

$$\tan u = \frac{1}{\sqrt{1 - e^2}}\tan\phi \tag{3.4.9}$$

Since:

$$\sin u = \frac{\tan u}{\sqrt{1 + \tan^2 u}}, \quad \text{and} \quad \cos u = \frac{1}{\sqrt{1 + \tan^2 u}}$$

replacing with (3.4.9) and rearranging,

$$\sin u = \frac{\sin\phi}{\sqrt{1 - e^2\cos^2\phi}}, \quad \cos u = \frac{\sqrt{1 - e^2}\,\cos\phi}{\sqrt{1 - e^2\cos^2\phi}} \tag{3.4.10}$$

Substituting into equations (3.4.2) gives the equations for the surface of the earth ellipsoid with respect to the parameters ϕ and L:

$$\left.\begin{array}{l} x = \dfrac{a\sqrt{1 - e^2}}{\sqrt{1 - e^2\cos^2\phi}}\cos\phi\,\cos L \\[3mm] y = \dfrac{a\sqrt{1 - e^2}}{\sqrt{1 - e^2\cos^2\phi}}\cos\phi\,\sin L \\[3mm] z = \dfrac{a\sqrt{1 - e^2}}{\sqrt{1 - e^2\cos^2\phi}}\sin\phi \end{array}\right\} \tag{3.4.11}$$

From the first and second equations of (3.4.11) we have:

$$\tan L = \frac{y}{x} \tag{3.4.12}$$

$$\sqrt{x^2 + y^2} = \frac{a\sqrt{1 - e^2}}{\sqrt{1 - e^2\cos^2\phi}}\cos\phi$$

Combining the above with (3.4.11) we obtain:

$$\tan\phi = \frac{z}{\sqrt{x^2 + y^2}} \tag{3.4.13}$$

Finally we obtain the inverse transformation equations of (3.4.11):

$$\left.\begin{array}{l} \phi = \arctan\left(\dfrac{z}{\sqrt{x^2 + y^2}}\right) \\[4mm] L = \arctan\left(\dfrac{y}{x}\right) \end{array}\right\}$$

(3.4.14)

Equation for the earth ellipsoid surface with respect to parameters B and L

For convenience, we shall consider the meridian plane $L = 0$. From (3.4.2) we can get the parametric equations of the meridian:

$$x = a\cos u, \quad z = b\sin u$$

(3.4.15)

In Figure 3.11 the slope of a line tangent to a point along the prime meridian and in the plane of the meridian is as follows:

$$\tan(90° + B) = -\cot B = \frac{dz}{dx}$$

Differentiating (3.4.15) gives:

$$\cot B = \frac{b}{a}\cot u$$

Thus we have:

$$\tan u = \sqrt{1 - e^2}\,\tan B$$

(3.4.16)

and:

$$\sin u = \frac{\sqrt{1 - e^2}\,\sin B}{\sqrt{1 - e^2\sin^2 B}}, \quad \cos u = \frac{\cos B}{\sqrt{1 - e^2\sin^2 B}}$$

(3.4.17)

Substituting into (3.4.2), we obtain the equations for the surface of the ellipsoid using the geographical coordinates parameters:

$$x = N\cos B\cos L, \quad y = N\cos B\sin L, \quad z = N(1 - e^2)\sin B$$

(3.4.18)

where $N = a/(1 - e^2\sin^2 B)^{1/2}$.

By using the above we can derive its inverse transformation equation:

$$\left.\begin{array}{l} B = \arctan\left(\dfrac{z}{(1 - e^2)\sqrt{x^2 + y^2}}\right) \\[4mm] L = \arctan\left(\dfrac{y}{x}\right) \end{array}\right\}$$

(3.4.19)

*Rectangular coordinate equations for spatial points and
their inverse transformation*

As illustrated in Figure 3.13, when point P is situated in space instead of on the surface of the ellipsoid, then $PP_0 = H$ is the ellipsoidal height. Obviously,

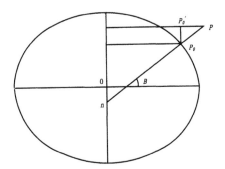

Figure 3.13 Rectangular coordinate system for spatial points

$$P_0'P = H\cos B, \quad P_0P_0' = H\sin B$$

With regard to the rectangular coordinate equations (3.4.18) for point P_0, we can write the rectangular coordinate equations for point P as follows:

$$\begin{bmatrix} x \\ y \\ z \end{bmatrix} = \begin{bmatrix} (N+H)\cos B\cos L \\ (N+H)\cos B\sin L \\ [N(1-e^2)+H]\sin B \end{bmatrix} \quad (3.4.20)$$

Now we find the inverse transformation equation of (3.4.20).

Using iteration From (3.4.20) we have:

$$\sqrt{x^2+y^2} = (N+H)\cos B \quad (3.4.21)$$

and

$$\tan L = \frac{y}{x} \quad (3.4.22)$$

Substituting (3.4.21) into the third equation of (3.4.20) yields (omitting the final $+H\,e^2\sin B$ term in the iteration):

$$z = \sqrt{x^2+y^2}\,\tan B - e^2 r\tan B \quad (3.4.23)$$

Using the Newton iteration method to find the value of latitude B:

$$\left.\begin{array}{l} B_{i+1} = B_i - \dfrac{f(B_i)}{f'(B_i)}, \quad \left|B_{i+1} - B_i\right| < 10^{-8} \\[2mm] B_0 = \arctan\left(z/(x^2+y^2)^{1/2}\right) \quad (i = 0, 1, 2, \ldots) \end{array}\right\} \quad (3.4.24)$$

where:

$$f(B) = \sqrt{x^2+y^2}\,\tan B - e^2 r\tan B - z$$

and:

$$f'(B) = \sqrt{x^2+y^2}\,(1+\tan^2 B) - e^2 r(1+\tan^2 B) + e^2 M\sin B\tan B$$

From (3.4.21) we have:

$$H = \frac{\sqrt{x^2+y^2}}{\cos B} - N \quad (3.4.25)$$

Equations (3.4.22), (3.4.24), and (3.4.25) are the inverse transformation equations using iteration for (3.4.20).

Example: Denoting i as the number of iterations, X', Y', Z' are check values computed by substituting the iterated results of B, L, H into (3.4.20). The following results were obtained using a pocket calculator:

Given $X = 4\,524\,737.356$, $Y = 0$, $Z = 4\,494\,498.711$

From two iterations $(i = 2)$: $B = 45°$, $L = 0$, $H = 10\,000$

Check values: $X' = 4\,524\,737.356$, $Y' = 0$, $Z' = 4\,494\,498.711$

Using a direct method For a ellipsoidal height of $H < 10\ 000$ m, equations for direct solution that will keep the precision of H within 0.001 m and the precision of B within 0.00001^2 are as follows (Zhu 1986):

$$
\left.
\begin{aligned}
&r = (x^2 + y^2 + z^2)^{1/2} \\[4pt]
&\sin^2\varphi = \frac{z^2}{r^2} \\[4pt]
&\sin^2 2\varphi = 4(\sin^2\varphi - \sin^4\varphi) \\[4pt]
&p = a(1 + e'^2\sin^2\varphi)^{-1/2}, \quad q = \frac{e^2}{2 - e^2} \\[4pt]
&H = (r - p)\left(1 - \frac{1}{2}q^2\sin^2 2\varphi\right) \\[4pt]
&N = a[1 - e^2(\sin^2\varphi + q^2\sin^2 2\varphi)]^{-1/2} \\[6pt]
&B = \arctan\left[\frac{z}{\sqrt{x^2 + y^2}}\left(1 + \frac{e^2}{1 - e^2 + \dfrac{H}{N}}\right)\right] \\[6pt]
&L = \arctan\frac{y}{x}
\end{aligned}
\right\}
\tag{3.4.26}
$$

Example: For the Krasovskiy ellipsoid, the parameters are $a = 6\ 378\ 245$ m, $e^2 = 0.006693421623$, and $e'^2 = 0.00678525415$.

Given $x = 4\ 524\ 737.356$ m, $y = 0$, and $z = 4\ 494\ 498.711$ m, we can obtain B, L, H by using (3.4.26). The following results were computed with a pocket calculator:

$$B = 45°, \quad L = 0, \quad H = 9\ 999.999623 \text{ m}$$

They can also be computed by using the following direct solution equation (Butekovich 1964):

$$
\left.
\begin{aligned}
&\tan B = \frac{z + e'^2 a\sin^3\theta}{p - e^2 a\cos^3\theta} \\[6pt]
&\tan L = \frac{y}{x} \\[6pt]
&H = \rho\sec B - N = a\csc B - (1 - e^2)N
\end{aligned}
\right\}
\tag{3.4.27}
$$

where

$$\tan\theta = \frac{z}{\rho\sqrt{1 - e^2}} = \frac{za}{\rho b}, \quad \rho = (x^2 + y^2)^{1/2}$$

With equation (3.4.27), when $0 < H < 700$ km, the error in latitude B is below $0.0001''$. When $700 < H < 3\ 700$ km, the error is below 0.001^2, but the error of H is very large. However, the value of H can be calculated with the following (Butekovich 1964):

$$H = (\rho' - a)a/N \tag{3.4.28}$$

where:

$$\rho' = \sqrt{x^2 + y^2 + z^2(1 + e'^2)}$$

Here are results computed with a PC-1500 pocket calculator:

Given: $x = 4\ 524\ 737.356$ m, $y = 0$, $z = 4\ 494\ 498.711$ m

Computed result: $B = 45.00011592°$, $L = 0$, $H = 10\ 035.84146$ m, $H' = 9\ 999.99914$ m
where H' has just been computed with equation (3.4.28).

3.4.2   Coordinate system on the surface of the sphere

Transformation between geographical and spherical polar coordinates

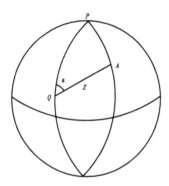

Figure 3.14 Spherical polar coordinates system

As illustrated in Figure 3.14, a new polar point $Q(\varphi_0, \lambda_0)$ is on the surface of a sphere, the geographical coordinates of point A are (φ, λ), and its corresponding polar coordinates are (Z, a).

Transformation from geographical coordinates to spherical polar coordinates   Consider the spherical triangle PQA. According to the Law of Cosines we have:

$$\cos Z = \cos(90° - \varphi_0)\cos(90° - \varphi)$$
$$+ \sin(90° - \varphi_0)\sin(90° - \varphi)\cos(\lambda - \lambda_0)$$

i.e. $\cos Z = \sin\varphi_0\sin\varphi + \cos\varphi_0\cos\varphi\cos(\lambda - \lambda_0)$

(3.4.29)

From the first sine–cosine formula we have:

$$\sin Z\cos a = \sin(90° - \varphi_0)\cos(90° - \varphi) - \cos(90° - \varphi_0)\sin(90° - \varphi)\cos(\lambda - \lambda_0)$$

i.e. $\sin Z\cos a = \cos\varphi_0\sin\varphi + \sin\varphi_0\cos\varphi\cos(\lambda - \lambda_0)$ (3.4.30)

From the Law of Sines we also have:

$$\frac{\sin Z}{\sin(\lambda - \lambda_0)} = \frac{\sin(90° - \varphi)}{\sin a}$$

i.e. $\sin Z\sin a = \cos\varphi\sin(\lambda - \lambda_0)$ (3.4.31)

Hence, we can obtain the following results:

$$\left. \begin{aligned} \cos Z &= \sin\varphi\sin\varphi_0 + \cos\varphi\cos\varphi_0\cos(\lambda - \lambda_0) \\ \tan a &= \frac{\cos\varphi\sin(\lambda - \lambda_0)}{\sin\varphi\cos\varphi_0 - \cos\varphi\sin\varphi_0\cos(\lambda - \lambda_0)} \end{aligned} \right\}$$

(3.4.32)

Transformation from spherical polar coordinates into geographical coordinates   In the same way, from the spherical triangle PQA we have the following equation according to the Law of Cosines:

$$\cos(90° - \varphi) = \cos(90° - \varphi_0)\cos Z + \sin(90° - \varphi_0)\sin Z\cos a$$

i.e. $\sin\varphi = \sin\varphi_0\cos Z + \cos\varphi_0\sin Z\cos a$ (3.4.33)

The first sine–cosine formula gives:

$$\sin(90° - \varphi_0)\cos(\lambda - \lambda_0) = \sin(90° - \varphi_0)\cos Z - \cos(90° - \varphi_0)\sin Z \cos a$$

i.e. $\cos\varphi\cos(\lambda - \lambda_0) = \cos\varphi_0\cos Z - \sin\varphi_0\sin Z \cos a$ (3.4.34)

The Law of Sines gives:

$$\frac{\sin(90° - \varphi)}{\sin a} = \frac{\sin Z}{\sin(\lambda - \lambda_0)}$$

i.e. $\cos\varphi\sin(\lambda - \lambda_0) = \sin Z \cos a$ (3.4.35)

By combining them we can obtain the following results:

$$\left. \begin{aligned} \sin\varphi &= \sin\varphi_0\cos Z + \cos\varphi_0\sin Z \cos a \\[2mm] \tan(\lambda - \lambda_0) &= \frac{\sin Z \sin a}{\cos\varphi_0\cos Z - \sin\varphi_0\sin Z \cos a} \end{aligned} \right\} \qquad (3.4.36)$$

The calculation of azimuth   The second equation of (3.4.32) may be rewritten as follows:

$$\tan a = \frac{\cos\varphi\sin(\lambda - \lambda_0)}{\cos\varphi_0\sin\varphi - \sin\varphi_0\cos\varphi\cos(\lambda - \lambda_0)} = \frac{C}{D}$$

Then we can calculate the value of azimuth a according to the following cases:

$$\Delta\lambda = \lambda - \lambda_0 = 0 \begin{cases} \varphi - \varphi_0 > 0, & a = 0° \\ \varphi - \varphi_0 = 0, & a = \text{variable} \\ \varphi - \varphi_0 < 0, & a = 180° \end{cases}$$

$$\Delta\lambda = \lambda - \lambda_0 > 0 \begin{cases} D > 0, & 0° < a < 90° \\ D = 0, & a = 90° \\ D < 0, & 90° < a < 180° \end{cases}$$

$$\Delta\lambda = \lambda - \lambda_0 < 0 \begin{cases} D > 0, & 90° < a < 180° \\ D = 0, & a = 270° \\ D < 0, & 270° < a < 360° \end{cases}$$

Transformation between geographical coordinates and spherical rectangular coordinates

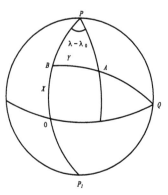

From geographical coordinates into spherical rectangular coordinates   As shown in Figure 3.15, POP_1 is the central longitude or central meridian, for which the value is λ_0. The new polar point Q is situated on the equator, and its longitude equals $(\lambda_0 + 90°)$.

The geographical coordinates of point A on the spherical surface are (φ, λ). Drawing a perpendicular circle QAB which intersects the central longitude at point B, let $BA = X$ and $BO = Y$. Thus the spherical rectangular coordinates of point A are (X, Y). Relating to spherical rectangular triangle PAB, we have:

Figure 3.15   Spherical rectangular coordinate system

$$\cos(\lambda - \lambda_0) = \cot(90° - \varphi)\tan(90° - Y)$$
$$\sin X = \sin(90° - \varphi)\sin(\lambda - \lambda_0)$$

Then we obtain the equation for transformation from geographical coordinates into spherical rectangular coordinates:

$$\left.\begin{array}{l} \tan Y = \tan\varphi\sec(\lambda - \lambda_0) \\ \sin X = \cos\varphi\,\sin(\lambda - \lambda_0) \end{array}\right\} \tag{3.4.37}$$

Transformation from spherical rectangular coordinates into geographical coordinates From spherical rectangular triangle PAB we have:

$$\cos(90° - \varphi) = \cos(90° - Y)\cos X$$
$$\sin(90° - Y) = \tan X\cot(\lambda - \lambda_0)$$

Then we obtain:

$$\left.\begin{array}{l} \sin\varphi = \cos X\,\sin Y \\ \tan(\lambda - \lambda_0) = \tan X\,\sec Y \end{array}\right\} \tag{3.4.38}$$

3.5 THREE KINDS OF LATITUDE FUNCTIONS q, S, F ON THE SURFACE OF THE ELLIPSOID

3.5.1 Conformal latitude and conformal latitude function q

In terms of the theory of map projection, the equation for conformal projection from an ellipsoidal surface into a spherical surface is as follows:

$$\tan\left(\frac{\pi}{4} + \frac{\varphi}{2}\right) = \tan\left(\frac{\pi}{4} + \frac{B}{2}\right)\left(\frac{1 - e\sin B}{1 + e\sin B}\right)^{e/2} \tag{3.5.1}$$

We usually let

$$q = \ln\left(\tan\left(\frac{\pi}{4} + \frac{B}{2}\right)\left(\frac{1 - e\sin B}{1 + e\sin B}\right)^{e/2}\right) \tag{3.5.2}$$

where q is the conformal latitude function or the conformal isometric latitude. Thus we have:

$$q = \ln\tan\left(\frac{\pi}{4} + \frac{\varphi}{2}\right) \tag{3.5.3}$$

Here q is called the conformal latitude.
The equation for calculating conformal latitude is:

$$\varphi = 2\arctan(e_1^q) - \frac{\pi}{2} \tag{3.5.4}$$

where $e_1 = 2.718281828.\ldots$.
From the above we can easily obtain the value of φ if q is given.
Now we derive the relation between φ and B. Applying equation (3.5.3) we have:

$$\varphi = 2\arctan\left\{\left(\frac{\pi}{4} + \frac{B}{2}\right)\left(\frac{1 - e\sin B}{1 + e\sin B}\right)^{e/2}\right\} - \frac{\pi}{2} \tag{3.5.5}$$

Let:

$$F(B, e) = \arctan\left\{\left(\frac{\pi}{4} + \frac{B}{2}\right)\left(\frac{1 - e\sin B}{1 + e\sin B}\right)^{e/2}\right\}$$

Then equation (3.5.5) can be rewritten:

$$\varphi = 2F(B, e) - \frac{\pi}{2} \tag{3.5.6}$$

Since e is very small, $F(B, e)$ can be expanded into a power series of e at $e = 0$. That is:

$$F(B, e) = F(B, e)\Big|_{e=0} + e\left[\frac{\partial F(B, e)}{\partial e}\right]_{e=0} + \frac{e^2}{2}\left[\frac{\partial^2 F(B, e)}{\partial e^2}\right]_{e=0}$$

$$+ \frac{e^3}{6}\left[\frac{\partial^3 F(B, e)}{\partial e^3}\right]_{e=0} + \frac{e^4}{24}\left[\frac{\partial^4 F(B, e)}{\partial e^4}\right]_{e=0} + \cdots \tag{3.5.7}$$

$$+ \frac{e^8}{8!}\left[\frac{\partial^8 F(B, e)}{\partial e^8}\right]_{e=0}$$

When $e = 0$, the values of all stages of the partial differential coefficients with respect to e are as follows:

$$\frac{\partial F(B, e)}{\partial e}\Big|_{e=0} = 0, \quad \frac{\partial^3 F(B, e)}{\partial e^3}\Big|_{e=0} = 0$$

$$\frac{\partial^5 F(B, e)}{\partial e^5}\Big|_{e=0} = 0, \quad \frac{\partial^7 F(B, e)}{\partial e^7}\Big|_{e=0} = 0$$

$$\frac{\partial^2 F(B, e)}{\partial e^2}\Big|_{e=0} = -\frac{1}{2}\sin 2B$$

$$\frac{\partial^4 F(B, e)}{\partial e^4}\Big|_{e=0} = -\frac{5}{2}\sin 2B + \frac{5}{4}\sin 4B$$

$$\frac{\partial^6 F(B, e)}{\partial e^6}\Big|_{e=0} = -\frac{135}{4}\sin 2B + \frac{63}{2}\sin 4B - \frac{39}{4}\sin 6B$$

$$\frac{\partial^8 F(B, e)}{\partial e^8}\Big|_{e=0} = -\frac{4\,197}{8}\sin 2B + \frac{6\,201}{8}\sin 4B - \frac{4\,809}{8}\sin 6B - \frac{2\,031}{16}\sin 8B$$

First substituting the above into (3.5.7) and rearranging, and then into (3.5.5), we obtain:

$$\varphi = B - \left(\frac{1}{2}e^2 + \frac{5}{24}e^4 + \frac{3}{32}e^6 + \frac{1\,399}{53\,760}e^8\right)\sin 2B$$

$$+ \left(\frac{5}{48}e^4 + \frac{7}{80}e^6 + \frac{689}{17\,920}e^8\right)\sin 4B \tag{3.5.8}$$

$$- \left(\frac{13}{480}e^6 + \frac{1\,363}{53\,760}e^8\right)\sin 6B + \frac{677}{107\,520}e^8\sin 8B$$

It can also be rewritten as:

$$B = \varphi + A_2 \sin 2B + A_4 \sin 4B + A_6 \sin 6B + A_8 \sin 8B \tag{3.5.9}$$

where:

$$\varphi = 2\arctan(e_1^q) - \frac{\pi}{2}$$

$$A_2 = \frac{1}{2}e^2 + \frac{5}{24}e^4 + \frac{3}{32}e^6 + \frac{1\,399}{53\,760}e^8$$

$$A_4 = -\left(\frac{5}{48}e^4 + \frac{7}{80}e^6 + \frac{689}{17\,920}e^8\right)$$

$$A_6 = \frac{13}{480}e^6 + \frac{1\,363}{17\,920}e^8$$

$$A_8 = -\frac{677}{107\,520}e^8$$

Formula (3.5.9) is the trigonometric series relation of conformal latitude φ expressed in terms of latitude B.

3.5.2 Equidistant latitude ψ' and equidistant latitude function S

According to the theory of map projections, the equation for an equidistant projection from an ellipsoidal surface to a spherical surface takes the form:

$$S = R\varphi \tag{3.5.10}$$

where:

$$R = a(1 - e^2)\left(1 + \frac{3}{4}e^2 + \frac{45}{64}e^4 + \frac{175}{256}e^6 + \frac{11\,025}{16\,384}e^8 + \ldots\right) \tag{3.5.11}$$

φ is called the equidistant latitude. S (the arc length of the meridian) is called the equidistant latitude function.

The equation for calculating the equidistant latitude φ is:

$$\varphi = \frac{S}{R} \tag{3.5.12}$$

Arc length of the meridian S

On the surface of the ellipsoid, the equation for calculating the arc length of a meridian from the equator to latitude B takes the form:

$$S = \int_0^B M\,dB \tag{3.5.13}$$

where:

$$M = \frac{a(1 - e^2)}{(1 - e^2\sin^2 B)^{3/2}}$$

is the radius of curvature of the meridian, a is semimajor axis of the ellipsoid, and e is the first eccentricity.

According to the binomial theorem,

$$(1 - e^2 \sin^2 B)^{-3/2}$$

can be expanded as the power series:

$$(1 - e^2 \sin^2 B)^{-3/2} = 1 + \frac{3}{2} e^2 \sin^2 B + \frac{15}{8} e^4 \sin^4 B$$

$$+ \frac{35}{16} e^6 \sin^6 B + \frac{315}{128} e^8 \sin^8 B + \ldots$$

(3.5.14)

Using a de Moivre formula,

$$\sin^{2n} B = \frac{1}{2^{2n-1}} \left[\frac{1}{2} \cdot \frac{2n \ldots (n+1)}{n!} - \frac{2n \ldots n}{(n+1)!} \cos 2B + \frac{2n \ldots (n-1)}{(n+2)!} \cos 4B - \ldots \right]$$

(3.5.15)

$$\cos^{2n} B = \frac{1}{2^{2n-1}} \left[\frac{1}{2} \cdot \frac{2n \ldots (n+1)}{n!} + \frac{2n \ldots n}{(n+1)!} \cos 2B + \frac{2n \ldots (n-1)}{(n+2)!} \cos 4B + \ldots \right]$$

(3.5.16)

$$\cos^{2n+1} B = \frac{1}{2^{2n}} \left[\frac{(2n+1) \ldots (n+1)}{(n+1)!} \cos B + \frac{(2n+1) \ldots n}{(n+2)!} \cos 3B + \ldots \right]$$

(3.5.17)

$$\sin^{2n+1} B = \frac{1}{2^{2n}} \left[\frac{(2n+1) \ldots (n+1)}{(n+1)!} \sin B - \frac{(2n+1) \ldots n}{(n+2)!} \sin 3B + \ldots \right]$$

(3.5.18)

From (3.5.15) we obtain:

$$\left. \begin{array}{l} \sin^2 B = \dfrac{1}{2} - \dfrac{1}{2}\cos 2B \\[2mm] \sin^4 B = \dfrac{3}{8} - \dfrac{1}{2}\cos 2B + \dfrac{1}{8}\cos 4B \\[2mm] \sin^6 B = \dfrac{5}{16} - \dfrac{15}{32}\cos 2B + \dfrac{3}{16}\cos 4B - \dfrac{1}{32}\cos 6B \\[2mm] \sin^8 B = \dfrac{35}{128} - \dfrac{7}{16}\cos 2B + \dfrac{7}{32}\cos 4B - \dfrac{1}{16}\cos 6B + \dfrac{1}{128}\cos 8B \end{array} \right\}$$

(3.5.19)

Substituting (3.5.19) into (3.5.14) gives:

$$(1 - e^2 \sin^2 B)^{-3/2} = \left(1 + \frac{3}{4} e^2 + \frac{45}{64} e^4 + \frac{175}{256} e^6 + \frac{11\,025}{16\,384} e^8 + \ldots \right)$$

$$- \left(\frac{3}{4} e^2 + \frac{15}{16} e^4 + \frac{525}{512} e^6 + \frac{2\,205}{2\,048} e^8 \right) \cos 2B$$

$$+\left(\frac{15}{64}e^4 + \frac{105}{256}e^6 + \frac{2\,205}{4\,096}e^8 + \ldots\right)\cos 4B$$

$$-\left(\frac{35}{512}e^6 + \frac{315}{2\,048}e^8 + \ldots\right)\cos 6B + \left(\frac{315}{16\,384}e^8 + \ldots\right)\cos 8B - \ldots$$

Let:

$$\left.\begin{aligned}
A' &= 1 + \frac{3}{4}e^2 + \frac{45}{64}e^4 + \frac{175}{256}e^6 + \frac{11\,025}{16\,384}e^8 + \ldots\\[6pt]
B' &= \frac{3}{4}e^2 + \frac{15}{16}e^4 + \frac{525}{512}e^6 + \frac{2\,205}{2\,048}e^8 + \ldots\\[6pt]
C' &= \frac{15}{64}e^4 + \frac{105}{256}e^6 + \frac{2\,205}{4\,096}e^8 + \ldots\\[6pt]
D' &= \frac{35}{512}e^6 + \frac{315}{2\,048}e^8 + \ldots\\[6pt]
E' &= \frac{315}{16\,384}e^8 + \ldots
\end{aligned}\right\} \qquad (3.5.20)$$

Then we obtain:

$$(1 - e^2\sin^2 B)^{-3/2} = A' - B'\cos 2B + C'\cos 4B - D'\cos 6B + E'\cos 8B - \ldots \qquad (3.5.21)$$

Substituting the above equation into (3.5.13) and differentiating,

$$S = a(1 - e^2)\left(A' \cdot B - \frac{B'}{2}\sin 2B + \frac{C'}{4}\sin 4B - \frac{D'}{6}\sin 6B + \frac{E'}{8}\sin 8B - \ldots\right) \qquad (3.5.22)$$

This is the equation for calculating the arc length of a meridian from the equator to latitude B. The symbol ψ represents equidistant latitude. It can also be rewritten as:

$$\psi = B - A_2\sin 2B - A_4\sin 4B - A_6\sin 6B - A_8\sin 8B \qquad (3.5.23)$$

where:

$$\psi = \frac{S}{R} = \frac{S}{a(1 - e^2)A'}$$

$$A_2 = \frac{B'}{2A'}, \quad A_4 = -\frac{C'}{4A'}$$

$$A_6 = \frac{D'}{6A'}, \quad A_8 = -\frac{E'}{8A'}$$

Formula (3.5.23) is the trigonometric series relation of the equidistant latitude ψ expressed in terms of latitude B.

3.5.3 Equivalent latitude τ and equivalent latitude function F

In terms of the theory of map projection, the relation for an equivalent (equal-area) projection from the surface of an ellipsoid onto the surface of a sphere takes the form:

$$R^2 \sin \varphi = F \tag{3.5.24}$$

where:

$$R^2 = a^2(1 - e^2)\left[\frac{1}{2(1 - e^2)} + \frac{1}{4e} \ln \frac{1 + e}{1 - e}\right] \tag{3.5.25}$$

φ is here called the equivalent latitude, and F is called the equivalent latitude function. The equation for computing φ takes the form:

$$\varphi = \arcsin\left(\frac{F}{R^2}\right) \tag{3.5.26}$$

Equivalent latitude function F

According to map projection theory, the trapezoidal area enclosed with a 1 arc difference of longitude on the surface of the ellipsoid and the curve from the equator to latitude B takes the form:

$$F = \int_0^B Mr\,dB \tag{3.5.27}$$

Observe that:

$$Mr = \frac{a^2(1 - e^2)\cos B}{(1 - e^2 \sin^2 B)^2}$$

and:

$$(1 - e^2 \sin^2 B)^{-2} = 1 + 2e^2 \sin^2 B + 3e^4 \sin^4 B + 4e^6 \sin^6 B + 5e^8 \sin^8 B + \ldots$$

Substituting into (3.5.27) and differentiating term by term, furthermore taking into account the condition that if $B = 0$, $F = 0$, we have:

$$F = a^2(1 - e^2)\left(\sin B + \frac{2}{3}e^2 \sin^3 B + \frac{3}{5}e^4 \sin^5 B + \frac{4}{7}e^6 \sin^7 B + \frac{5}{9}e^8 \sin^9 B + \ldots\right)$$

$$\tag{3.5.28}$$

Considering (3.5.24) and (3.5.28), we obtain:

$$R^2 = F_{90^\circ} = a^2(1 - e^2)\left(1 + \frac{2}{3}e^2 + \frac{3}{5}e^4 + \frac{4}{7}e^6 + \frac{5}{9}e^8 + \ldots\right) \tag{3.5.29}$$

From (3.5.26),

$$\sin\varphi = \frac{\sin B + \frac{2}{3}e^2 \sin^3 B + \frac{3}{5}e^4 \sin^5 B + \frac{4}{7}e^6 \sin^7 B + \frac{5}{9}e^8 \sin^9 B + \ldots}{1 + \frac{2}{3}e^2 + \frac{3}{5}e^4 + \frac{4}{7}e^6 + \frac{5}{9}e^8 + \ldots}$$

Expanding

$$\left(1 + \frac{2}{3}e^2 + \frac{3}{5}e^4 + \frac{4}{7}e^6 + \frac{5}{9}e^8 + \ldots\right)^{-1}$$

as a series and neglecting higher terms, after rearranging we get the following:

$$\sin\varphi = \sin B + \sin B \cos^2 B \left(-\frac{2}{3}e^2 - \frac{34}{45}e^4 - \frac{766}{945}e^6 - \frac{12\,014}{14\,175}e^8 \right)$$

$$+ \sin B \cos^4 B \left(\frac{3}{5}e^4 + \frac{46}{35}e^6 + \frac{367}{175}e^8 \right) \tag{3.5.30}$$

$$+ \sin B \cos^6 B \left(-\frac{4}{7}e^6 - \frac{116}{63}e^8 \right) + \frac{5}{9}e^8 \sin B \cos^8 B$$

Since $(\varphi - B)$ is very small, by developing $\sin\varphi$ into a Taylor series at B and using terms up to the fourth order $(\varphi - B)^4$ we have:

$$\sin\varphi = \sin B + (\varphi - B)\cos B - \frac{1}{2}(\varphi - B)^2 \sin B - \frac{1}{6}(\varphi - B)^3 \cos B + \frac{1}{24}(\varphi - B)^4 \sin B$$

$$\tag{3.5.31}$$

From the above we obtain:

$$\varphi - B = \frac{\sin\varphi - \sin B}{\cos B} + \frac{1}{2}\frac{(\varphi - B)^3 \sin B}{\cos B} + \frac{1}{6}(\varphi - B)^3 - \frac{1}{24}\frac{(\varphi - B)^4 \sin B}{\cos B} \tag{3.5.32}$$

In equation (3.5.32), the $(\varphi - B)$ terms on the left can be approximately replaced with $\dfrac{\sin\varphi - \sin B}{\cos B}$; the value of $(\sin\varphi - \sin B)$ can be determined from (3.5.30). After rearranging we can get:

$$\varphi - B = -\left(\frac{1}{3}e^2 + \frac{31}{180}e^4 + \frac{1243}{15\,120}e^6 + \frac{18\,563}{604\,800}e^8 \right)\sin 2B$$

$$+ \left(\frac{17}{360}e^4 + \frac{113}{3\,780}e^6 + \frac{47\,963}{60\,480}e^8 \right)\sin 4B \tag{3.5.33}$$

$$- \left(\frac{173}{43\,560}e^6 - \frac{4\,207}{777\,600}e^8 \right)\sin 6B - \frac{2\,671}{10\,886\,400}e^8 \sin 8B$$

Substituting into the right side of (3.5.32) and iterating again, with the symbol τ representing equivalent latitude, we finally obtain:

$$B = \varphi + A_2 \sin 2B + A_4 \sin 4B + A_6 \sin 6B + A_8 \sin 8B \tag{3.5.34}$$

where:

$$\varphi = \arcsin\left(\frac{F}{R^2} \right)$$

$$A_2 = \frac{1}{3}e^2 + \frac{31}{180}e^4 + \frac{59}{560}e^6 + \frac{126\,853}{518\,400}e^8$$

$$A_4 = -\left(\frac{17}{360}e^4 + \frac{61}{1\,260}e^6 + \frac{3\,622\,447}{94\,089\,600}e^8 \right)$$

$$A_6 = \frac{383}{45\,360}e^6 + \frac{6\,688\,039}{658\,627\,200}e^8$$

$$A_8 = -\frac{27\,787}{23\,522\,400}e^8$$

Formula (3.5.34) is the trigonometric series relation of the equivalent latitude τ expressed in terms of latitude B.

The three latitude functions q, S, and F discussed above are a function of geographical latitude B. In surveying, the computation of the footpoint latitude B_f merely proceeds through the inverse solution of the equidistant isometric latitude function S to find latitude B. In map projection transformation we often deal with the practical problem of finding the inverse solution of these three kinds of latitude functions, that is, finding latitude B. For this reason, the inverse transformation of these three kinds of functions has been studied by many surveyors and cartographers. Many algorithms have been presented. They can be summed up as consisting of two types of methods – the first, iteration, and the second, direct. Now we shall discuss each type.

3.6   INVERSE TRANSFORMATION OF q, S, F – ITERATION METHOD

3.6.1   Inverse transformation of the conformal latitude function q

The iteration method for the inverse solution of conformal latitude function q is as follows:

$$\left.\begin{array}{l} B_{i+1} = B_i - \dfrac{F(B_i)}{F'(B_i)} \quad (i = 0, 1, \ldots, n) \\[4mm] B_0 = 2\arctan(e_1^q) - \dfrac{\pi}{2}, \quad |B_{i+1} - B_i| < \varepsilon \end{array}\right\} \tag{3.6.1}$$

where:

$$F(B) = \ln\left[\tan\left(\frac{\pi}{4} + \frac{B}{2}\right)\left(\frac{1 - e\sin B}{1 + e\sin B}\right)^{e/2}\right] - q_k \tag{3.6.2}$$

$$F'(B) = \frac{M}{r} = \frac{1}{(1 + \eta^2)\cos B} \tag{3.6.3}$$

ε depends on the required precision.

3.6.2   Inverse transformation of the equidistant latitude function S

Newton's iterative method for the inverse solution of the equidistant latitude function S is as follows:

$$B_{i+1} = B_i - \frac{F(B_i)}{F'(B_i)}, \quad |B_{i+1} - B_i| < 10^{-8}$$

$$B_0 = \frac{S}{A'(1 - e^2)} \quad (i = 0, 1, 2, \ldots, n)$$

(3.6.4)

where:

$$F(B) = a(1 - e^2)\left(A' \cdot B - \frac{B'}{2}\sin 2B + \frac{C'}{4}\sin 4B \right.$$

(3.6.5)

$$\left. - \frac{D'}{6}\sin 6B + \frac{E'}{8}\sin 8B \right) - S_k$$

$$F'(B) = a(1 - e^2)(A' - B'\cos 2B + C'\cos 4B - D'\cos 6B + E'\cos 8B)$$

(3.6.6)

By using the above iteration equations we can achieve a precision for the footpoint latitude as large as 10^{-8} radians.

3.6.3 Inverse transformation of the equivalent latitude function *F*

Newton's iteration method applied to the inverse solution of the equivalent latitude function F is as follows:

$$B_{i+1} = B_i - \frac{f(B_i)}{f'(B_i)} \quad (i = 0, 1, \ldots, n)$$

$$B_0 = \arcsin\left(\frac{F}{R^2}\right), \quad |B_{i+1} - B_i| < \varepsilon$$

(3.6.7)

where:

$$f(B) = \int_0^B MrdB - F_k, \quad f'(B) = Mr$$

$$\int_0^B MrdB = a^2(1 - e^2)\left[\frac{\sin B}{2(1 - e^2\sin^2 B)} + \frac{1}{4e}\ln\frac{1 + e\sin B}{1 - e\sin B} \right]$$

(3.6.8)

ε depends on the required precision.

3.7 INVERSE TRANSFORMATION OF *q, S, F* — TRIGONOMETRIC SERIES METHOD WITH CONSTANT COEFFICIENTS

3.7.1 Lagrange series

Taking account of a special formula such as:

$$y = a + x\varphi(y)$$

(3.7.1)

the function $\varphi(y)$ is an analytical function at the point $y = a$; if x is not large, then y is a function of x. It is an analytical function at point $x = 0$. If there is a point at which $x = 0$, at that point $y = a$.

For more general cases, let us consider a function $u(u = f(y))$ of y. Suppose it is an analytical function at $y = a$. Because x is small and y is a function of x, u is also a function of x. Then it can be developed into a power series expansion of x at point $y = a$:

$$u = u_0 + x\left(\frac{\partial u}{\partial x}\right)_0 + \frac{x^2}{2!}\left(\frac{\partial^2 u}{\partial x^2}\right)_0 + \ldots + \frac{x^n}{n!}\left(\frac{\partial^n u}{\partial x^n}\right)_0 + \ldots \tag{3.7.2}$$

where the subscript 0 (zero) denotes the values of the function and its derivatives when $x = 0$. Since there is an $x = 0$, there is a point $y = a$. Thus we obtain $u_0 = f(a)$.

Here we give the expressions of coefficients in (3.7.2). From (3.7.1) we know that y is a function of two variables x and a. Thus u is also a function of x and a. Differentiating (3.7.1) with respect to x and a yields:

$$[1 - x\varphi'(y)]\frac{\partial y}{\partial x} = \varphi(y)$$

$$[1 - x\varphi'(y)]\frac{\partial y}{\partial a} = 1$$

Hence we have:

$$\frac{\partial y}{\partial x} = \varphi(y)\frac{\partial y}{\partial a} \tag{3.7.3}$$

Generally when $u = f(y)$ we have the same as the above, that is:

$$\frac{\partial u}{\partial x} = \varphi(y)\frac{\partial u}{\partial a} \tag{3.7.4}$$

In addition, no matter what form function $F(y)$ takes, if its derivative with respect to y exists, the following must be true:

$$\frac{\partial}{\partial x}\left[F(y)\frac{\partial u}{\partial a}\right] = \frac{\partial}{\partial a}\left[F(y)\frac{\partial u}{\partial x}\right] \tag{3.7.5}$$

By directly differentiating (3.7.5) and substituting into (3.7.3) and (3.7.4), the validity of (3.7.5) can be verified.

From the above equations we can prove that the following equation is correct, by using the inductive method:

$$\frac{\partial^n u}{\partial x^n} = \frac{\partial^{n-1}}{\partial x^{n-1}}\left(\varphi^n(y)\cdot\frac{\partial u}{\partial a}\right) \tag{3.7.6}$$

When $x = 0$, then we must have the condition $y = a$ and $\frac{\partial u}{\partial a} = f'(a)$. As a result, (3.7.6) may be rewritten as:

$$\left(\frac{\partial^n u}{\partial x^n}\right)_0 = \frac{d^{n-1}}{da^{n-1}}[\varphi^n(a)f'(a)] \tag{3.7.7}$$

Substituting into the series expansion (3.7.2),

$$f(y) = f(a) + x\varphi(a)f'(a) + \frac{x^2}{2!}\frac{d}{da}[\varphi^2(a)f'(a)] + \ldots$$

$$+ \frac{x^n}{n!}\frac{d^{n-1}}{da^{n-1}}[\varphi^n(a)\cdot f'(a)] + \ldots \tag{3.7.8}$$

This is a Lagrange series. If $f(y) \equiv y$ exists, then we have:

$$y = a + x\varphi(a) + \frac{x^2}{2!}\frac{d}{da}[\varphi^2(a)] + \ldots + \frac{x^n}{n!}\frac{d^{n-1}}{da^{n-1}}[\varphi^n(a)] + \ldots \qquad (3.7.9)$$

In view of the above, for the many equations of the form (3.7.1) we can easily find the equations for their inverse algorithms by using (3.7.9).

Now we discuss the solution of the direct equations for three kinds of latitude functions. Rewriting (3.5.9) in the following form:

$$B = \varphi + \varphi(B) \qquad (3.7.10)$$

where:

$$\varphi(B) = A_2 \sin 2B + A_4 \sin 4B + A_6 \sin 6B + A_8 \sin 8B$$

we can use the Lagrange series to find an equation for the inverse solution of (3.7.10); e.g.,

$$B = \varphi + \varphi(\varphi) + \frac{1}{2}\frac{d}{d\varphi}[\varphi(\varphi)]^2 + \frac{1}{6}\frac{d^2}{d\varphi^2}[\varphi(\varphi)]^3 + \frac{1}{24}\frac{d^3}{d\varphi^3}[\varphi(\varphi)]^4 \qquad (3.7.11)$$

From (3.7.10) we have:

$$\varphi(\varphi) = A_2 \sin 2\varphi + A_4 \sin 4\varphi + A_6 \sin 6\varphi + A_8 \sin 8\varphi \qquad (3.7.12)$$

If we differentiate (3.7.12) in accordance with the requirements of (3.7.11) and delete the high-degree terms, after rearranging we have:

$$\frac{1}{2}\frac{d}{d\varphi}[\varphi(\varphi)]^2 = (-A_2 A_4 - A_4 A_6 - A_6 A_8)\sin 2\varphi + (A_2^2 - 2A_2 A_6 - 2A_4 A_8)\sin 4\varphi \qquad (3.7.13)$$
$$+ (3A_2 A_4 - 3A_2 A_8)\sin 6\varphi + (2A_4^2 + 4A_2 A_6)\sin 8\varphi$$

$$\frac{1}{6}\frac{d^2}{d\varphi^2}[\varphi(\varphi)]^3 = \left(-\frac{1}{2}A_2^3 - A_2 A_4^2 + \frac{1}{2}A_2^2 A_6 - \frac{1}{2}A_4^2 A_6 + A_2 A_4 A_8\right)\sin 2\varphi$$
$$+ (-4A_2^2 A_4 - 2A_4^3 - 4A_2 A_4 A_6 + 2A_2^2 A_8 - A_2 A_4 A_6)\sin 4\varphi$$
$$+ \left(\frac{3}{2}A_2^3 - \frac{9}{2}A_2 A_4^2 - 9A_2^2 A_6 - 9A_4^2 A_6 - 9A_2 A_4 A_8\right)\sin 6\varphi \qquad (3.7.14)$$
$$+ (8A_2^2 A_4 - 19A_2 A_4 A_6 - 16A_2^2 A_8 - 16A_4^2 A_8$$
$$- 25A_2 A_4 A_6)\sin 8\varphi$$

$$\frac{1}{24}\frac{d^3}{d\varphi^3}[\varphi(\varphi)]^4 = (-8.2A_2^3 A_8 - 18.3A_2^3 A_4)\sin 2\varphi + (-1.3A_2^4$$
$$- 6.2A_2^2 A_4^2 + 120A_2^3 A_6)\sin 4\varphi + (21.5A_2^3 A_8 \qquad (3.7.15)$$
$$- 12.5A_2^3 A_4)\sin 6\varphi + (2.7A_2^4 - 66A_2^3 A_6)\sin 8\varphi$$

Substituting the above into (3.7.11) and rearranging gives:

$$B = \varphi + B_2 \sin 2\varphi + B_4 \sin 4\varphi + B_6 \sin 6\varphi + B_8 \sin 8\varphi \qquad (3.7.16)$$

where:

$$B_2 = A_2 - A_2A_4 - A_4A_6 - \frac{1}{2}A_2^3 - A_2A_4^2 + \frac{1}{2}A_2^2A_6 - 18.3A_2^3A_4$$

$$B_4 = A_4 + A_2^2 - 2A_2A_6 - 4A_2^2A_4 - 1.3A_2^4$$

$$B_6 = A_6 + 3A_2A_4 - 3A_2A_8 + \frac{3}{2}A_2^3 - \frac{9}{2}A_2A_4^2 - 9A_2^2A_6 - 12.5A_2^3A_4 \qquad (3.7.17)$$

$$B_8 = A_8 + 2A_4^2 + 4A_2A_6 + 8A_2^2A_4 + 2.7A_2^4$$

That is, (3.7.16) and (3.7.17) are the equations for the inverse solution of (3.7.10).

3.7.2 Inverse transformation for the conformal latitude function q

For the Krasovskiy ellipsoid, the values of the elements of the ellipsoid are as follows:

$$a = 6\,378\,245 \text{ m}, \quad e^2 = 0.006693421623$$

From equations (3.5.9), (3.7.16), and (3.7.17) we can obtain the formula for computing directly the inverse solution for the conformal latitude function (or isometric latitude) q.

$$B = \varphi + B_2\sin 2\varphi + B_4\sin 4\varphi + B_6\sin 6\varphi + B_8\sin 8\varphi \qquad (3.7.18)$$

where:

$$\varphi = 2\arctan(e_1^q) - \frac{\pi}{2}, \quad e_1 = 2.718281828\ldots$$

$$B_2 = 0.33560695588 \times 10^{-2}$$
$$B_4 = 0.65700353 \times 10^{-5}$$
$$B_6 = 0.176221 \times 10^{-7}$$
$$B_8 = 0.608 \times 10^{-10}$$

Table 3.1 lists values of the conformal latitude q computed with the above equations, with checks of the recomputed B against the initial values. The accuracy is within 0.0001″.

Table 3.1 Inverse calculation with constant coefficients for conformal latitude functions

Isometric latitude q	B, recomputed	B, initial
0.0867940561	5°00′00″.00003	5°00′00″.0000
0.3540886221	20°00′00″.00003	20°00′00″.0000
0.8766353329	45°00′00″.00004	45°00′00″.0000
1.3111514950	60°00′00″.00003	60°00′00″.0000
2.0211105644	75°00′00″.00003	75°00′00″.0000
4.7346413800	89°00′00″.00003	89°00′00″.0000

For the International Ellipsoid 1975 (recommended by the IUGG in 1975), also called the New Ellipsoid for short, the elements of the ellipsoid are as follows:

$$a = 6\,378\,140 \text{ m}, \quad e^2 = 0.0066943849996$$

The coefficients in (3.7.18) are as follows:

$$B_2 = 0.33565539449 \times 10^{-2}$$
$$B_4 = 0.6571932 \times 10^{-5}$$
$$B_6 = 0.176297 \times 10^{-7}$$
$$B_8 = 0.608 \times 10^{-10}$$

(3.7.19)

3.7.3 Inverse transformation for the equidistant latitude function S

Now we delve into the formula for computing the direct solution of the equidistant isometric latitude function S (useful for finding the footpoint latitude). From equations (3.5.20), (3.5.23) and (3.7.17), we can obtain the formula as follows:

$$B = \psi + B_2\sin 2\psi + B_4\sin 4\psi + B_6\sin 6\psi + B_8\sin 8\psi$$ (3.7.20)

where:

$$\psi = S \times 0.1570460641219 \times 10^{-6}$$
$$B_2 = 0.25184647783 \times 10^{-2}$$
$$B_4 = 0.36998873 \times 10^{-5}$$
$$B_6 = 0.74449 \times 10^{-8}$$
$$B_8 = 0.1828 \times 10^{-10}$$

The unit of S is m, and ψ is in radians. Table 3.2 shows values of the arc length of a meridian or the equidistant latitude function for the Krasovskiy ellipsoid computed with the above equations, with checks against the initial values.

Table 3.2 Inverse calculation with constant coefficient for equidistant latitude functions

Equidistant latitude function S (m)	B, recomputed	B, initial
552 895.344	4°59′59″.99998	5°00′00″.0000
2 212 405.724	19°59′59″.99998	20°00′00″.0000
4 985 032.290	44°59′59″.99996	45°00′00″.0000
6 654 189.092	59°59′59″.99996	60°00′00″.0000
8 327 081.746	74°59′59″.99997	75°00′00″.0000
9 890 441.195	88°59′59″.99995	89°00′00″.0000

The given accuracy of the arc length of meridian is within 0.001 m, or an accuracy of 0.0001″ in the footpoint latitude. The maximum computing error is 0.00006″. For this reason, the accuracy of (3.7.20) is 0.0001″.

For the New Ellipsoid, the coefficients in the equation for computing the footpoint latitude are as follows:

$$\psi = S \times 0.1570486874728 \times 10^{-6}$$
$$B_2 = 0.25188284763 \times 10^{-2}$$
$$B_4 = 0.37009560 \times 10^{-5}$$
$$B_6 = 0.74481 \times 10^{-8}$$
$$B_8 = 0.1829 \times 10^{-10}$$

(3.7.21)

3.7.4 Inverse transformation for the equivalent latitude function F

From equations (3.5.34), (3.7.16), and (3.7.17) we can obtain the formula for computing the direct solution of the equivalent (equal-area) latitude function for the Krasovskiy ellipsoid as follows:

$$B = \tau + B_2 \sin 2\tau + B_4 \sin 4\tau + B_6 \sin 6\tau + B_8 \sin 8\tau \qquad (3.7.22)$$

where:

$$\tau = \arcsin\left(\frac{F}{R^2}\right), \quad R^2 = 40\ 591\ 120.141234 \text{ km}^2$$

$$B_2 = 0.22388876665 \times 10^{-2}$$
$$B_4 = 0.28823802 \times 10^{-5}$$
$$B_6 = 0.507852 \times 10^{-8}$$
$$B_8 = 0.1198 \times 10^{-10}$$

The computed values of F as well as a comparison of the recomputed with the initial values of B are illustrated in Table 3.3.

Table 3.3 Inverse calculation with constant coefficient for equivalent latitude functions

Equivalent latitude function F (km^2)	B, recomputed	B, initial
3 522 057.4488	5°00′00″.00000	5°00′00″.00000
13 828 153.3670	20°00′00″.00000	20°00′00″.00000
28 637 923.3464	45°00′00″.00000	45°00′00″.00000
35 113 485.8311	60°00′00″.00000	60°00′00″.00000
39 196 206.6016	75°00′00″.00000	75°00′00″.00000
40 584 882.3346	89°00′00″.00000	89°00′00″.00000

The given accuracy of F is within 0.0001 km^2, equivalent to a latitude error of 0.000003″. Thus the accuracy of (3.7.22) reaches 0.0001″.

For the New Ellipsoid, the coefficients in equation (3.7.22) for computing the equivalent isometric latitude function F are as follows:

$$\tau = \arcsin\left(\frac{F}{R^2}\right), \quad R^2 = 40\ 589\ 770.614524 \text{ km}^2$$

$$B_2 = 0.22392110268 \times 10^{-2}$$
$$B_4 = 0.28832129 \times 10^{-5}$$
$$B_6 = 0.508072 \times 10^{-8}$$
$$B_8 = 0.1199 \times 10^{-10}$$

$$(3.7.23)$$

3.8 INVERSE TRANSFORMATION OF q, S, F — TAYLOR SERIES METHOD WITH VARIED COEFFICIENTS

3.8.1 Inverse transformation of the conformal latitude function q

Here we explore the various series algorithms for the inverse transformation of the conformal latitude function (or isometric latitude) q (Yang 1985b). According to (3.5.2) the equation for the isometric latitude is as follows:

$$q = \int_0^B \frac{M}{r} dB = \ln\left[\tan\left(\frac{\pi}{4} + \frac{B}{2}\right)\left(\frac{1 - e\sin B}{1 + e\sin B}\right)^{e/2}\right]$$

From the above it is clear that $B = f(q)$.

Expanding the above into a power series expansion of $(q - q_0)$ at B_0 yields:

$$B = B_0 + \left(\frac{dB}{dq}\right)_0 (q - q_0) + \frac{1}{2}\left(\frac{d^2 B}{dq^2}\right)_0 (q - q_0)^2 + \frac{1}{6}\left(\frac{d^3 B}{dq^3}\right)_0 (q - q_0)^3$$

$$+ \frac{1}{24}\left(\frac{d^4 B}{dq^4}\right)_0 (q - q_0)^4 + \cdots$$

After further derivation we obtain the derivatives for each order:

$$\left.\begin{array}{l}
\dfrac{dB}{dq} = (1 + \eta^2)\cos B \\[2mm]
\dfrac{d^2 B}{dq^2} = t(-1 - 4\eta^2 - 3\eta^4)\cos^2 B \\[2mm]
\dfrac{d^3 B}{dq^3} = (-1 + t^2 - 5\eta^2 + 13t^2\eta^2 - 7\eta^4 + 27t^2\eta^4)\cos^3 B \\[2mm]
\dfrac{d^4 B}{dq^4} = t(5 - t^2 + 56\eta^2 - 40t^2\eta^2)\cos^4 B
\end{array}\right\} \qquad (3.8.2)$$

where:

$\quad t = \tan B, \quad \eta = e\cos^2 B$, e is the second eccentricity.

Hence the power series expansion for the inverse transformation of the isometric latitude q takes the form:

$$B = B_0 + (1 + \eta_0^2)\cos B_0 \, \Delta q + \frac{1}{2}t_0(-1 - 4\eta_0^2 - 3\eta_0^4)\cos^2 B_0 \, \Delta q^2$$

$$+ \frac{1}{6}(-1 + t_0^2 - 5\eta_0^2 + 13t_0^2\eta_0^2 - 7\eta_0^4 + 27t_0^2\eta_0^4)\cos^3 B_0 \, \Delta q^3 \qquad (3.8.3)$$

$$+ \frac{1}{24}t_0(5 - t_0^2 + 56\eta_0^2 - 40t_0^2\eta_0^2)\cos^4 B_0 \, \Delta q^4$$

Equation (3.8.3) has constant coefficients, which fixes its point of expansion at B_0. If it is used to find B inversely for a value of isometric latitude q far from B_0, and a highly accurate result is expected, we must calculate its terms up to a rather high order. For this reason, the method is not suitable.

If B_0 varies with isometric latitude q, that is, B_0 is a function of q (or $B_0 = f_1(q)$), then the error of latitude B for the inverse solution for q and B_0 will be reduced. In this way we can ensure the required high precision by calculating only the terms of low order. This method is called the variable-coefficient algorithm for the inverse transformation of isometric latitude.

It is critical to derive the equation for calculating $B_0 = f_1(q)$. Considering conformal projection from the surface of the ellipsoid to the surface of the sphere, we have:

$$\ln\tan\left(\frac{\pi}{4} + \frac{\varphi}{2}\right) = q \tag{3.8.4}$$

For the Krasovskiy ellipsoid, we can obtain the following equation from (3.8.4):

$$B - \varphi = 692.23'' \sin 2B - 0.96'' \sin 4B \tag{3.8.5}$$

With the above equation we find that $(B - \varphi)$ has a maximum value of $11'32''$ near $B = 45°$. Hence we may take φ in (3.8.4) as the latitude B_0 for the point of expansion for the variable coefficients. The calculating formula for $B_0 = f_1(q)$ can be written:

$$B_0 = 2\arctan(e_1^q) - \frac{\pi}{2} \tag{3.8.6}$$

To check the precision of terms of low order in (3.8.3) when calculating B_0 with the above equation, we separately calculate the errors of omission of the third- and fourth-order terms with respect to q at intervals of $5°$. The maximum error of omission of the third-order terms is found to be $-0.0023''$ when $B_0 = 70°$ or $B_0 = 69°52'34''$. The maximum error of omission of the fourth-order terms will be $0.0000043''$ when $B_0 = 45°$ or $B_0 = 45°11'32''$.

Hence we obtain the following two equations:

- The equation for the inverse solution of B from the isometric latitude q with $0.002''$ of calculating accuracy:

$$B = B_0 + (1 + \eta_0^2)\cos B_0(q - q_0) + \frac{1}{2}t_0(-1 - 4\eta_0^2 - 3\eta_0^4)\cos^2 B_0(q - q_0)^2 \tag{3.8.7}$$

where B_0 is calculated with (3.8.6); B_0 and B are in radians.

- The equation for the inverse solution of B from isometric latitude with $0.00001''$ of calculating accuracy:

$$B = B_0 + (1 + \eta_0^2)\cos B_0(q - q_0) + \frac{1}{2}t_0(-1 - 4\eta_0^2 - 3\eta_0^4)$$

$$\times \cos^2 B_0(q - q_0)^2 + \frac{1}{6}(-1 + t_0^2 - 5\eta_0^2 + 13t_0^2\eta_0^2 \tag{3.8.8}$$

$$- 7\eta_0^4 + 27t_0^2\eta_0^4)\cos^3 B_0(q - q_0)^3$$

where B_0 is calculated with (3.8.6); B_0 and B are in radians.

In the following we will derive the formula for inverse calculation of B from the conformal isometric latitude by linear interpolation with variable coefficients.

From (3.8.4) we obtain:

$$\ln\tan\left(\frac{\pi}{4} + \frac{\varphi}{2}\right) = \ln\left[\tan\left(\frac{\pi}{4} + \frac{B}{2}\right)\left(\frac{1 - e\sin B}{1 + e\sin B}\right)^{e/2}\right] = q \tag{3.8.9}$$

The zero-order approximation expression of (3.8.9) takes the form:

$$\varphi_0 = 2\arctan(e_1^q) - \frac{\pi}{2}$$

and its first-order approximation expression is:

$$\ln\tan\left(\frac{\pi}{4} + \frac{B_0}{2}\right) = q - \ln\left(\frac{1 - e\sin\varphi_0}{1 + e\sin\varphi_0}\right)^{e/2}$$

Let:

$$q_1 = q - \frac{e}{2}\ln\left(\frac{1 - e\sin\varphi_0}{1 + e\sin\varphi_0}\right) \tag{3.8.10}$$

Thus we obtain:

$$B_0 = 2\arctan(e_1^{q_1}) - \frac{\pi}{2} \tag{3.8.11}$$

Finally we obtain the interpolation formula as follows:

$$B = B_0 + (1 + \eta_0^2)\cos B_0(q - q_0) \tag{3.8.12}$$

where B_0 is calculated with (3.8.9), (3.8.10), (3.8.11), and

$$q_0 = \ln\left[\tan\left(\frac{\pi}{4} + \frac{B_0}{2}\right)\left(\frac{1 - e\sin B_0}{1 + e\sin B_0}\right)^{e/2}\right] \tag{3.8.13}$$

The results achieved by computer are presented in Table 3.4. It can be seen that the accuracy of inverse transformation is 0.0001″.

Table 3.4 Inverse calculation with variable coefficients for conformal latitude functions

Isometric latitude q	B, recomputed	B, initial
0.08679405586	4°59′59″.99996	5°00′00″.0000
0.3540886221	20°00′00″.0000	20°00′00″.0000
0.8766353324	44°59′59″.9999	45°00′00″.0000
1.311151494	59°59′59″.9999	60°00′00″.0000
2.021110564	75°00′00″.0000	75°00′00″.0000
4.734641355	88°59′59″.9999	89°00′00″.0000

3.8.2 Inverse transformation for the equidistant latitude function S

Let us discuss the variable-coefficient method for the inverse solution of the equidistant latitude function S, that is, to find the footpoint latitude. The equation for calculating the equidistant latitude function is the arc length of the meridian and has been given in (3.5.22) and (3.5.20). For the Krasovskiy ellipsoid we have:

$$S = 6\,367\,558.497\,\text{arc}\,B - 16\,036.4803\sin 2B + 16.8281\sin 4B$$
$$- 0.0219753\sin 6B + 0.0000311311\sin 8B \tag{3.8.14}$$

For the New Ellipsoid (introduced by IUGG in 1975) we have:

$$S = 6\,367\,452.128 \operatorname{arc} B - 16\,038.5282 \sin 2B + 16.8326 \sin 4B$$
$$- 0.0219845 \sin 6B + 0.0000311485 \sin 8B \tag{3.8.15}$$

In accordance with (3.8.14) the inverse algorithm for solving the equidistant isometric latitude function takes the form $B = f(S)$. Expanding it into a power series expression of $(S - S_0)$ at B_0 yields:

$$B = B_0 + \left(\frac{dB}{dS}\right)_0 (S - S_0) + \frac{1}{2}\left(\frac{d^2B}{dS^2}\right)_0 (S - S_0)^2 + \frac{1}{6}\left(\frac{d^3B}{dS^3}\right)_0 (S - S_0)^3 + \dots \tag{3.8.16}$$

Derivatives of each order are as follows:

$$\left.\begin{aligned}
\frac{dB}{dS} &= \frac{1}{N}(1 + \eta^2) \\[2mm]
\frac{d^2B}{dS^2} &= \frac{3}{N^2}t(-\eta^2 - \eta^4) \\[2mm]
\frac{d^3B}{dS^3} &= \frac{3}{N^3}(-\eta^2 + t^2\eta^2 - 2\eta^4 + 6t^2\eta^4)
\end{aligned}\right\} \tag{3.8.17}$$

where:

$$t = \tan B, \quad \eta^2 = e'^2 \cos^2 B, \quad N = a(1 - e^2 \sin^2 B)^{-1/2}$$

From (3.8.17) we obtain the power series expansion for the inverse solution of the equidistant latitude function S:

$$B_f = B_0 + \frac{1}{N_0}(1 + \eta_0^2)\Delta S + \frac{3}{2N_0^2}t_0(-\eta_0^2 - \eta_0^4)\Delta S^2$$
$$+ \frac{1}{2N_0^3}(-\eta_0^2 + t_0^2\eta_0^2 - 2\eta_0^4 + 6t_0^2\eta_0^4)\Delta S^3 + \dots \tag{3.8.18}$$

Equation (3.8.18) is a constant-coefficient expression having an expansion point at B_0, which mainly would be used for finding the footpoint latitude. But if it is used to find a footpoint latitude far from B_0, and a result of high accuracy is expected, we must calculate its terms to a rather high order. For this reason, the method is not ideal.

If B_0 varies with the footpoint, that is, B_0 is a function of S, or $B_0 = f_1(S)$, then we can minimize the error of finding the footpoint latitude B_f and B_0. As a result, only terms of low order need be calculated to meet the required accuracy. This is a so-called variable-coefficient algorithm.

The key to the question is in deriving the formula for calculating $B_0 = f_1(S)$. According to the equidistant projection of the surface of the ellipsoid onto the surface of the sphere, we have:

$$\varphi = \frac{S}{R}, \quad R = a(1 - e^2)A' \tag{3.8.19}$$

with the maximum value of $(B - \varphi_0)$ being $8'39''$ in the neighborhood of $45°$.

Hence we can calculate B_0 with (3.8.18), i.e.,

$$B_0 = \frac{S}{R}, \quad R = a(1 - e^2)A' \tag{3.8.20}$$

For the Krasovskiy ellipsoid, $B_f = 45°08'39''$, $B_0 = 45°$, $R = 6\,367\,558.487$ m. Therefore, the value of the second-order term

$$\frac{3}{2N_0^2}t_0(-\eta_0^2 - \eta_0^4)\Delta S^2$$

is $-6.6'' \times 10^{-3}$ and the value of the third-order term

$$\frac{1}{2N_0^3}(-\eta_0^2 + t_0^2\eta_0^2 - 2\eta_0^4 + 6t_0^2\eta_0^4)\Delta S^3$$

is $7''.4 \times 10^{-8}$.

From the above we obtain the equation for calculating the footpoint latitude B_f with an accuracy of up to 0.01:

$$B_f = B_0 + (1 + \eta_0^2)(S - S_0)/N_0 \tag{3.8.21}$$

For the Krasovskiy ellipsoid

$$B_0 = S \times 1.570460641 \times 10^{-7} \tag{3.8.22}$$

where the units of B_f and B_0 are radians, and S is in meters.

For the New Ellipsoid,

$$B_0 = S \times 1.570486875 \times 10^{-7} \tag{3.8.23}$$

The equation for calculating footpoint latitude B_f with an accuracy of up to $0''.0001$ is

$$B_f = B_0 + \frac{1}{N_0}(1 + \eta_0^2)(S - S_0) + \frac{3}{2N_0^2}t_0(-\eta_0^2 - \eta_0^4)(S - S_0)^2 \tag{3.8.24}$$

The equation for calculating B_0 is the same as (3.8.22) and (3.8.23).

If latitude B_0 of the expansion point is calculated from a first-order approximation of (3.8.14),

$$B_0 = (S + 16\,036.4803 \sin 2B)/6\,367\,558.497 \tag{3.8.25}$$

$$B = S/6\,367\,558.497$$

Finally we obtain the formula for calculating the footpoint latitude by linear interpolation with variable coefficients:

$$B_f = B_0 + \frac{1}{N_0}(1 + \eta_0^2)(S - S_0) \tag{3.8.26}$$

The results achieved from (3.8.26), (3.8.25) and (3.8.14) using a computer are shown in Table 3.5. It can be seen that the accuracy of inverse transformation is $0.0001''$.

Table 3.5 Inverse calculation with variable coefficients for equidistant latitude functions

Equidistant latitude function S (m)	B, recomputed	B, initial
552 895.3443	4°59'59".99999	5°00'00".0000
2 212 405.724	20°00'00".0000	20°00'00".0000
4 985 032.291	45°00'00".0000	45°00'00".0000
6 654 189.091	59°59'59".9999	60°00'00".0000
8 327 081.745	75°00'00".0000	75°00'00".0000
9 890 441.794	88°59'59".9999	89°00'00".0000

3.8.3   Inverse transformation for the equivalent latitude function F

Now let us discuss the variable-coefficient method for the inverse solution of the equivalent latitude function F. Equations (3.5.27) and (3.6.8) give the expression for the equivalent latitude function as:

$$F = \int_0^B MrdB = a^2(1 - e^2)\left[\frac{\sin B}{2(1 - e^2\sin^2 B)} + \frac{1}{4e}\ln\frac{1 + e\sin B}{1 - e\sin B}\right] \tag{3.8.27}$$

From the above equation we have:

$$B = f(F) \tag{3.8.28}$$

Expanding it into a power series expression of $(F - F_m)$ at point B_m yields:

$$B = B_m + \frac{1}{M_m r_m}(F - F_m) + \dots \tag{3.8.29}$$

Here B_m is a function of F, i.e., $B_m = f_1(F)$. For this reason, the above equation is a linear-interpolation equation with variable coefficients for the inverse solution of the equivalent isometric latitude function. The B_m will be determined by the following procedure.

From (3.5.24) and (3.5.25) we know that the equation for the equivalent projection from the surface of the ellipsoid to the surface of the sphere takes the form:

$$\begin{aligned} R^2\sin\varphi &= F \\ R^2 &= a^2(1 - e^2)\left[\frac{1}{2(1 - e^2)} + \frac{1}{4e}\ln\frac{1 + e}{1 - e}\right] \end{aligned} \tag{3.8.30}$$

Considering (3.8.27) and (3.8.30) we have:

$$\begin{aligned} B_0 &= \arcsin(F/R^2) \\ B_1 &= \arcsin\left[2(1 - e^2\sin^2 B_0)\left(\frac{F}{a^2(1 - e^2)} - \frac{1}{4e}\ln\frac{1 + e\sin B_0}{1 - e\sin B_0}\right)\right] \end{aligned} \tag{3.8.31}$$

and:

$$B_m = \frac{B_0 + B_1}{2} \tag{3.8.32}$$

Equations (3.8.29), (3.8.32), and (3.8.31) are the linear interpolation expressions with variable coefficients for the inverse solution of the equivalent isometric latitude function F.

The results achieved from the above equation using a computer are presented in Table 3.6. It can be seen that the accuracy of inverse transformation is 0.0001″.

Table 3.6   Inverse calculation with variable coefficient for equivalent latitude functions

Equivalent latitude function F (km²)	B, recomputed	B, initial
3 522 057.449	5°00′00″.0000	5°00′00″.0000
13 828 153.37	19°59′59″.9999	20°00′00″.0000
28 637 923.36	45°00′00″.0000	45°00′00″.0000
35 113 485.83	59°59′59″.9999	60°00′00″.0000
39 196 206.6	75°00′00″.0000	75°00′00″.0000
40 435 277.61	85°00′00″.0001	85°00′00″.0000

3.9   INVERSE TRANSFORMATION OF q, S, F — NUMERICAL METHODS

In the foregoing we introduced the analytical method to derive the equations for calculating the footpoint latitude and inversely solving for the isometric latitude and equivalent latitude function. Now we will discuss a numerical method to derive the equations for the inverse solution of conformal isometric latitude, equidistant isometric latitude function, and the equivalent isometric latitude function.

By a 'numerical method' we mean using an equation of a trigonometric series or power series to approximate the inverse solution of conformal isometric latitude, equidistant isometric latitude, and equivalent isometric latitude; in other words, using the given values of a small number of nodes to determine the value of each coefficient in the equation for calculation.

3.9.1   Approximate method using a formula with a trigonometric series

Taking into account (3.7.18), (3.7.20), and (3.7.22), we can write the general equation for the inverse solution as:

$$B = X + K_2 \sin 2X + K_4 \sin 4X + K_6 \sin 6X + K_8 \sin 8X \tag{3.9.1}$$

where K_2, K_4, K_6, K_8 are undetermined coefficients. The units of X and B are radians.

The equation for the inverse solution of the conformal isometric latitude q takes the form:

$$X = 2\arctan(e_1^q) - \frac{\pi}{2} \tag{3.9.2}$$

The equation for the inverse solution of the equidistant isometric latitude function S takes the form:

$$X = \frac{S}{a(1 - e^2)A'} \tag{3.9.3}$$

The equation for the inverse solution of the equivalent isometric latitude function F takes the form:

$$X = \arcsin(F/R^2) \tag{3.9.4}$$

By using the corresponding direct equation to calculate the values of S_j (or q_j or F_j) of several points in the interval [0, 90], and developing a system of linear equations, we obtain:

$$K_2 \sin 2X_i + K_4 \sin 4X_i + K_6 \sin 6X_i + K_8 \sin 8X_i = B_i - X_i \tag{3.9.5}$$

where $i = 1, 2, 3, 4$.

Solving the above system of linear equations by elimination of the primary element, we can determine the value of each coefficient K_i. This is illustrated as follows.

Example 1: The direct formula for finding the footpoint latitude based on four points

The formula for fitting the footpoint latitude to the four points 15°, 30°, 45°, and 60° takes the form:

$$B = X + K_2 \sin 2X + K_4 \sin 4X + K_6 \sin 6X + K_8 \sin 8X \tag{3.9.6}$$

where:

$X = S_m \cdot 10^{-6} \times 0.15704606$

$K_2 = 0.25184647 \times 10^{-2}, \quad K_4 = 0.36999040 \times 10^{-5},$

$K_6 = 0.74527926 \times 10^{-8}, \quad K_8 = 0.17266511 \times 10^{-10}.$

The unit of S_m is meters.

The calculating accuracy with this formula is $0.01''$. The accuracy of S_m is 0.1 m (adopting the Krasovskiy ellipsoid).

Example 2: The direct formula for finding the conformal isometric latitude q based on four points

The formula for fitting isometric latitude to the four points $15°$, $30°$, $45°$, and $60°$ takes the form:

$$B = X + K_2 \sin 2X + K_4 \sin 4X + K_6 \sin 6X + K_8 \sin 8X \tag{3.9.7}$$

where:

$$X = 2\arctan(e_1^q) - \frac{\pi}{2}$$

$K_2 = 0.33560699 \times 10^{-2}, \quad K_4 = 0.65699153 \times 10^{-5},$

$K_6 = 0.17796672 \times 10^{-7}, \quad K_8 = 0.16009787 \times 10^{-10}.$

The calculation accuracy with this formula is $0.01''$.

Example 3: The direct formula for finding the equivalent isometric latitude function F based on four points

In the same way, the formula for solving isometric latitude according to the four points $15°$, $30°$, $45°$, and $60°$ takes the form:

$$B = X + K_2 \sin 2X + K_4 \sin 4X + K_6 \sin 6X + K_8 \sin 8X \tag{3.9.8}$$

where:

$X = \arcsin(F/R^2), \quad R^2 = 40\ 591\ 120 \text{ km}^2$

$K_2 = 0.22388826 \times 10^{-2}, \quad K_4 = 0.28853835 \times 10^{-5},$

$K_6 = 0.32000909 \times 10^{-8}, \quad K_8 = 0.75343656 \times 10^{-9}.$

The calculating accuracy with this formula is 0.02.

3.9.2 Approximate method using a power series

From (3.9.1) we know that B is an odd function of X, that is,

$$B = \sum_{i=1}^{n} a_{2i-1} X^{2i-1} \tag{3.9.9}$$

where a_{2i-1} are undetermined coefficients, and the units of B and X are radians.

The inverse formula for finding the footpoint latitude takes the form:

$$X = S_m \cdot 10^{-6} \tag{3.9.10}$$

The inverse formula for finding the conformal isometric latitude takes the form:

$$X = 2\arctan(e_1^q) - \frac{\pi}{2} \tag{3.9.11}$$

The inverse formula for finding the equivalent isometric latitude function takes the form:

$$X = \arcsin(F/R^2) \tag{3.9.12}$$

Just as above, coefficients a_{2i-1} can be determined by finding the values of S_i, q_i, or F_i of some given nodes, developing a system of linear equations, and finally solving the system by elimination about a pivotal element. It is illustrated as follows.

Example

A power series formula for finding the footpoint latitude using the five points 15°, 30°, 40°, 50°, and 55° takes the form:

$$B = a_1 X + a_3 X^3 + a_5 X^5 + a_7 X^7 + a_9 X^9 \tag{3.9.13}$$

where:

$a_1 = 0.15783942, \quad a_3 = -0.13160009 \times 10^{-4},$

$a_5 = 0.67179598 \times 10^{-7}, \quad a_7 = -0.17749357 \times 10^{-9},$

$a_9 = 0.30098146 \times 10^{-12}, \quad X = S_m \cdot 10^{-6}.$

and the unit of S_m is meters. In the range 0–57°, the calculating accuracy with the above formula is 0.0001, and the accuracy of S_m is 0.001 m.

3.10   TRANSFORMATION FORMULAE AMONG SIX KINDS OF LATITUDES

In surveying and mapping, determining the position of a point on the ground and the projection transformation between the ellipsoidal surface and the spherical surface can involve the application of six latitudes (Yang 1995). With the application of space and computer technology to surveying and mapping, it becomes more and more important and valuable in practice to study the six latitudes as well as their transformations. For this reason, in this section we delve into the transformation relationships among the six latitudes, presenting practical algorithms and some application examples.

In Section 3.4 we defined reduced latitude u, geographical latitude B, and geocentric latitude ϕ. In Section 3.5 we also defined conformal isometric latitude φ, equidistant latitude ψ and equivalent latitude τ.

3.10.1   Transformation relationships among the six latitudes

All these latitudes relate to geographical latitude B. In practical application we must often inversely solve the formula for latitude B. Thus it is our objective here to derive trigonometric series for the differences between latitude B and the other latitudes.

For the first derivation, we derive the expansion formula for $(B - \phi)$ in terms of B. From a standard trigonometric equation we have:

$$\tan(B - \phi) = \frac{\tan B - \tan\phi}{1 + \tan B \tan\phi}$$

(3.10.1)

According to (3.4.8) we have:

$$\tan\phi = \sqrt{1 - e^2} \, \tan u$$

Since (3.4.16) takes the form:

$$\tan u = \sqrt{1 - e^2} \, \tan B$$

we obtain:

$$\tan\phi = (1 - e^2) \tan B$$

(3.10.2)

Substituting (3.10.2) into (3.10.1) yields:

$$\tan(B - \phi) = \frac{e^2 \tan B}{1 + (1 - e^2) \tan^2 B} = \frac{e^2 \sin 2B}{2 - e^2 + e^2 \cos 2B}$$

Let:

$$m = \frac{e^2}{2 - e^2}$$

then:

$$\tan(B - \phi) = \frac{m \sin 2B}{1 + m \cos 2B}$$

(3.10.3)

$$\tan(B - \phi) = m \sin 2B (1 + m \cos 2B)^{-1}$$
$$= m \sin 2B - m^2 \sin 2B \cos 2B + m^3 \sin 2B \cos^2 2B - \ldots$$

(3.10.4)

Taking into account the series:

$$B - \phi = \tan(B - \phi) - \frac{1}{3} \tan^3(B - \phi) + \ldots$$

and replacing it with (3.10.4),

$$B - \phi = m \sin 2B - \frac{m^2}{2} \sin 4B + \frac{m^3}{3}(3 \sin 2B - 4 \sin^3 2B) - \ldots$$

finally we obtain:

$$B - \phi = m \sin 2B - \frac{m^2}{2} \sin 4B + \frac{m^3}{3} \sin 6B - \ldots$$

(3.10.5)

For the second instance, we derive the expansion formula for $(B - \phi)$ in terms of ϕ. From (3.10.2) we have:

$$\tan B = \frac{1}{1 - e^2} \tan\phi$$

Substituting into (3.10.1) gives:

$$\tan(B - \phi) = \frac{e^2 \tan\phi}{1 - e^2 + \tan^2\phi} = \frac{e^2 \sin2\phi}{2 - e^2 - e^2 \cos2\phi}$$

Thus we get:

$$\tan(B - \phi) = \frac{m \sin2\phi}{1 - m \cos2\phi} \qquad (3.10.6)$$

By using the above method we can obtain:

$$B - \phi = m \sin2\phi + \frac{m^2}{2}\sin4\phi + \frac{m^3}{3}\sin6\phi + \ldots \qquad (3.10.7)$$

For the third choice, we derive the expansion formula for $(B - u)$ in terms of B.

$$\tan(B - u) = \frac{\tan B - \tan u}{1 + \tan B \tan u} \qquad (3.10.8)$$

From (3.4.16),

$$\tan u = \sqrt{1 - e^2} \, \tan B$$

Substituting into (3.10.8) yields:

$$\tan(B - u) = \frac{(1 - \sqrt{1 - e^2}) \tan B}{1 + \sqrt{1 - e^2} \, \tan^2 B}$$

$$= \frac{(1 - \sqrt{1 - e^2}) \sin2B}{2\cos^2 B + 2\sqrt{1 - e^2} \, \sin^2 B}$$

$$= \frac{(1 - \sqrt{1 - e^2}) \sin2B}{1 + \sqrt{1 - e^2} + (1 - \sqrt{1 - e^2}) \cos^2 B - (1 - \sqrt{1 - e^2}) \sin^2 B}$$

$$= \frac{(1 - \sqrt{1 - e^2}) \sin2B}{1 + \sqrt{1 - e^2} + (1 - \sqrt{1 - e^2}) \cos2B}$$

Let:

$$n = \frac{1 - \sqrt{1 - e^2}}{1 + \sqrt{1 - e^2}}$$

then,

$$\tan(B - u) = \frac{n \sin2B}{1 + n \cos2B} \qquad (3.10.9)$$

Following (3.10.3) we obtain:

$$B - u = n \sin2B - \frac{n^2}{2}\sin4B + \frac{n^3}{3}\sin6B + \ldots \qquad (3.10.10)$$

For the fourth method, we derive the expansion formula for $(B - u)$ in terms of u.
Substituting (3.4.16) into (3.10.8), after transformation we get:

$$\tan(B - u) = \frac{n \sin2u}{1 - n \cos2u} \qquad (3.10.11)$$

Following (3.10.6) we obtain:

$$B - u = n\sin 2u + \frac{n^2}{2}\sin 4u + \frac{n^3}{3}\sin 6u + \ldots \qquad (3.10.12)$$

For the fifth derivation, the expansion formula of $(u - \phi)$ is expressed in terms of u. From (3.4.8),

$$\tan\phi = \sqrt{1 - e^2}\,\tan u$$

Following (3.10.8) we obtain:

$$u - \phi = n\sin 2u - \frac{n^2}{2}\sin 4u + \frac{n^3}{3}\sin 6u + \ldots \qquad (3.10.13)$$

For the sixth case, the expansion formula for $(u - \phi)$ is expressed in terms of ϕ. From (3.4.8) we have:

$$\tan u = \frac{1}{\sqrt{1 - e^2}}\,\tan\phi$$

Following (3.10.11) we get:

$$u - \phi = n\sin 2\phi + \frac{n^2}{2}\sin 4\phi + \frac{n^3}{3}\sin 6\phi + \ldots \qquad (3.10.14)$$

Up to now, we have derived the transformation relationships among reduced latitude u, geographical latitude B, and geocentric latitude φ. In the following we continue to derive the transformation relationships between geographical latitude B and conformal latitude φ, equidistant latitude ψ, and equivalent isometric latitude τ.

For the seventh derivation, the expansion formula for $(B - \varphi)$ is expressed in terms of B:

$$B - \varphi = A_2\sin 2B + A_4\sin 4B + A_6\sin 6B + A_8\sin 8B \qquad (3.10.15)$$

where the coefficients A_2, A_4, A_6, A_8 are as in formula (3.5.9).

$$B - \psi = A_2\sin 2B + A_4\sin 4B + A_6\sin 6B + A_8\sin 8B \qquad (3.10.16)$$

where the coefficients A_2, A_4, A_6, A_8 are as in formulae (3.5.23) and (3.5.20).

$$B - \tau = A_2\sin 2B + A_4\sin 4B + A_6\sin 6B + A_8\sin 8B \qquad (3.10.17)$$

where the coefficients A_2, A_4, A_6, A_8 are as in formula (3.5.34).

The expansion formulae for $(B - \varphi)$, $(B - \psi)$, and $(B - \tau)$, respectively, are expressed in terms of φ, ψ, and τ and computing examples, see Section 3.7.

3.10.2 Practical algorithms and examples of their application

Practical equation applicable to the Krasovskiy ellipsoid

Given elements of the Krasovskiy ellipsoid:

$$a = 6\ 378\ 245\ \text{m}, \quad e^2 = 0.006693421623$$

the following algorithms have two forms, one of which is expressed in radians and the other in seconds of arc:

$$B - \phi = m_2 \sin 2B - m_4 \sin 4B + m_6 \sin 6B \left. \right\} \tag{3.10.18}$$
$$B - \phi = m_2 \sin 2\phi + m_4 \sin 4\phi + m_6 \sin 6\phi$$

where:

$$m_2 = 3.3579489 \times 10^{-3} \text{ m}, \quad m_4 = 5.63791 \times 10^{-6} \text{ m}, \quad m_6 = 1.26 \times 10^{-8} \text{ m}$$

or:

$$m_2 = 692''.6267, \quad m_4 = 1''.1629, \quad m_6 = 0''.0026$$

$$B - u = n_2 \sin 2B - n_4 \sin 4B + n_6 \sin 6B \left. \right\} \tag{3.10.19}$$
$$B - u = n_2 \sin 2u + n_4 \sin 4u + n_6 \sin 6u$$

$$u - \phi = n_2 \sin 2u - n_4 \sin 4u + n_6 \sin 6u \left. \right\} \tag{3.10.20}$$
$$u - \phi = n_2 \sin 2\phi + n_4 \sin 4\phi + n_6 \sin 6\phi$$

where:

$$n_2 = 1.6789792 \times 10^{-3} \text{ m}, \quad n_4 = 1.40949 \times 10^{-6} \text{ m}, \quad n_6 = 1.58 \times 10^{-9} \text{ m}$$
$$n_2 = 346.''3143, \quad n_4 = 0''.2907, \quad n_6 = 0''.0003$$

$$B - \varphi = A_2 \sin 2B + A_4 \sin 4B + A_6 \sin 6B \tag{3.10.21}$$

where:

$$A_2 = 3.3560727 \times 10^{-3} \text{ m}, \quad A_4 = -4.6932 \times 10^{-6} \text{ m}, \quad A_6 = 8.27 \times 10^{-9} \text{ m}$$

or:

$$A_2 = 692''.2397, \quad A_4 = -0''.9680, \quad A_6 = 0''.0017$$

$$B - \varphi = B_2 \sin 2\varphi + B_4 \sin 4\varphi + B_6 \sin 6\varphi \tag{3.10.22}$$

where:

$$B_2 = 3.3560696 \times 10^{-3} \text{ m}, \quad B_4 = 6.5700 \times 10^{-6} \text{ m}, \quad B_6 = 1.8 \times 10^{-8} \text{ m}$$

or:

$$B_2 = 692''.2390, \quad B_4 = 1''.3552, \quad B_6 = 0''.0037$$

$$B - \psi = A_2 \sin 2B + A_4 \sin 4B + A_6 \sin 6B \tag{3.10.23}$$

where:

$$A_2 = 2.5184661 \times 10^{-3} \text{ m}, \quad A_4 = -2.6428 \times 10^{-6} \text{ m}, \quad A_6 = 3.45 \times 10^{-9} \text{ m}$$

or:

$$A_2 = 519''.4790, \quad A_4 = -0''.5451, \quad A_6 = 0''.0007$$

$$B - \psi = B_2 \sin 2\psi + B_4 \sin 4\psi + B_6 \sin 6\psi \tag{3.10.24}$$

where:

$$B_2 = 2.5184648 \times 10^{-3} \text{ m}, \quad B_4 = 3.6999 \times 10^{-6} \text{ m}, \quad B_6 = 7.9 \times 10^{-9} \text{ m}$$

or:

$$B_2 = 519''.4707, \quad B_4 = 0''.7632, \quad B_6 = 0''.0015$$

$$B - \tau = A_2 \sin 2B + A_4 \sin 4B + A_6 \sin 6B \tag{3.10.25}$$

where:

$$A_2 = 2.5388885 \times 10^{-3} \text{ m}, \quad A_4 = -2.1302 \times 10^{-6} \text{ m}, \quad A_6 = 2.55 \times 10^{-9} \text{ m}$$

or:

$$A_2 = 461''.8039, \quad A_4 = -0''.4394, \quad A_6 = 0''.0005$$

$$B - \tau = B_2 \sin 2\tau + B_4 \sin 4\tau + B_6 \sin 6\tau \tag{3.10.26}$$

where:

$$B_2 = 2.2388877 \times 10^{-3} \text{ m}, \quad B_4 = 2.8824 \times 10^{-6} \text{ m}, \quad B_6 = 5.1 \times 10^{-9} \text{ m}$$

or:

$$B_2 = 461''.8037, \quad B_4 = 0''.5945, \quad B_6 = 0''.0011$$

Practical examples

Next we take a practical equation for the Krasovskiy ellipsoid as an example to illustrate the application.

Example 1: The conversion among reduced latitude u, geographical latitude B, and geocentric latitude ϕ

1. Given $u = 30°$, calculate B and ϕ
   From the second expression of (3.10.19) we obtain $B = 30°05'00''.1687$.
   From the first expression of (3.10.21) we obtain $\phi = 29°55'00''.3347$.

2. Given $\phi = 40°$, calculate B and u
   From the second expression of (3.10.18) we obtain $B = 40°11'22''.4996$.
   From the second expression of (3.10.20) we obtain $u = 40°05'41''.1521$.

3. Given $B = 50°$, calculate ϕ and u.
   From the first expression of (3.10.18) we obtain $\phi = 49°48'37''.5003$.
   From the first expression of (3.10.19) we obtain $u = 49°54'18''.8478$.

Example 2: Conversion between geographical latitude B and conformal latitude φ, equidistant latitude ψ, or equivalent latitude τ

1. Conversion between conformal latitude φ and geographical latitude B.
   Given $B = 45°$, from (3.10.21) we get $\varphi = 44°48'27''.762$.
   Given $\varphi = 44°48'27''.762$, from (3.10.22) we get $B = 44°59'59''.9999$.

2. Conversion between equidistant latitude ψ and geographical latitude B.
   Given $B = 45°$, from (3.10.23) we get $\psi = 44°51'20''.5297$.
   Given $\psi = 44°51'20''.5297$, from (3.10.24) we obtain $B = 45°$.

3. Conversion between equivalent latitude τ and geographical latitude B.

Given $B = 45°$, from (3.10.25) we get $\tau = 44°52'18''.1966$.

Given $\tau = 44°52'18''.1966$, from (3.10.26) we obtain $B = 44°59'59''.9998$.

3.11 COMMONLY USED POWER SERIES IN MAP PROJECTION TRANSFORMATION

For functions such as $y = f(x)$, the Taylor series take the form:

$$y = y_0 + \left(\frac{dy}{dx}\right)_0 \Delta x + \frac{1}{2}\left(\frac{d^2 y}{dx^2}\right)_0 \Delta x^2 + \frac{1}{6}\left(\frac{d^3 y}{dx^3}\right)\Delta x^3 + \frac{1}{24}\left(\frac{d^4 y}{dx^4}\right)\Delta x^4$$

$$+ \frac{1}{120}\left(\frac{d^5 y}{dx^5}\right)\Delta x^5 + \frac{1}{720}\left(\frac{d^6 y}{dx^6}\right)\Delta x^6 + \dots$$

(3.11.1)

where Δx can be taken as ΔB, ΔS, or Δq, and y can be taken as $\cos B$, $\dfrac{1}{\cos B}$, t, η^2, N, $\dfrac{1}{N}$, B, S, or q.

Below are listed commonly used derivatives for series expansion:

$$\left.\begin{array}{l}
\dfrac{d \cos B}{dB} = -t \cos B, \quad \dfrac{d}{dB}\dfrac{1}{\cos B} = \dfrac{1}{\cos B}t, \quad \dfrac{dt}{dB} = 1 + t^2 \\[3mm]
\dfrac{d\eta^2}{dB} = -2t\eta^2, \quad \dfrac{dN}{dB} = \dfrac{Nt\eta^2}{1+\eta^2} = Nt(\eta^2 - \eta^4 + \eta^6) \\[3mm]
\dfrac{d}{dB}\dfrac{1}{N} = -\dfrac{t\eta^2}{N(1+\eta^2)} = -\dfrac{1}{N}t(\eta^2 - \eta^4 + \eta^6) \\[3mm]
\dfrac{dS}{dB} = M = \dfrac{N}{1+\eta^2} = N(1 - \eta^2 + \eta^4 - \eta^6) \\[3mm]
\dfrac{dq}{dB} = \dfrac{1}{(1+\eta^2)\cos B} = \dfrac{1}{\cos B}(1 - \eta^2 + \eta^4 - \eta^6)
\end{array}\right\}$$

(3.11.2)

where:

$$t = \tan B, \quad \eta^2 = e'^2 \cos^2 B$$

and:

$$\left.\begin{array}{ll}
\dfrac{d \cos B}{dS} = -\dfrac{1}{N}t \cos B(1 + \eta^2), & \dfrac{d}{dS}\dfrac{1}{\cos B} = \dfrac{1}{N \cos B}t(1 + \eta^2) \\[3mm]
\dfrac{dt}{dS} = \dfrac{1}{N}(1 + t^2 + \eta^2 + t^2\eta^2), & \dfrac{d\eta^2}{dS} = -\dfrac{2}{N}t(\eta^2 + \eta^4) \\[3mm]
\dfrac{dN}{dS} = t\eta^2, & \dfrac{d}{dS}\dfrac{1}{N} = -\dfrac{1}{N^2}t\eta^2 \\[3mm]
\dfrac{dB}{dS} = \dfrac{1}{N}(1 + \eta^2), & \dfrac{dq}{dS} = \dfrac{1}{N \cos B}
\end{array}\right\}$$

(3.11.3)

and:

$$\left.\begin{array}{ll}
\dfrac{d\cos B}{dq} = -t(1+\eta^2)\cos^2 B, & \dfrac{d}{dq}\dfrac{1}{\cos B} = t(1+\eta^2) \\[2mm]
\dfrac{dt}{dq} = (1+t^2+\eta^2+t^2\eta^2)\cos B, & \dfrac{d\eta^2}{dq} = -2t(\eta^2+\eta^4)\cos B \\[2mm]
\dfrac{dN}{dq} = Nt\eta^2\cos B, & \dfrac{d}{dq}\dfrac{1}{N} = -\dfrac{1}{N}t\eta^2\cos B \\[2mm]
\dfrac{dB}{dq} = (1+\eta^2)\cos B, & \dfrac{dS}{dq} = N\cos B
\end{array}\right\} \qquad (3.11.4)$$

By means of the above equations we can obtain commonly used expressions for power series expansions.

Example 1: Power series expansion for ΔB

$$S = S_0 + N_0(1-\eta_0^2+\eta_0^4-\eta_0^6)\Delta B + \frac{3}{2}N_0 t_0(\eta_0^2-2\eta_0^4)\Delta B^2$$

$$+\frac{1}{2}N_0(\eta_0^2-t_0^2\eta_0^2-2\eta_0^4+7t_0^2\eta_0^4)\Delta B^3+\frac{1}{2}N_0 t_0(-\eta_0^2)\Delta B^4 \qquad (3.11.5)$$

$$q = q_0 + \frac{1}{\cos B_0}(1-\eta_0^2+\eta_0^4-\eta_0^6)\Delta B + \frac{1}{2\cos B_0}t_0(1+\eta_0^2-3\eta_0^4)\Delta B^2$$

$$+\frac{1}{6\cos B_0}(1+2t_0^2+\eta_0^2-3\eta_0^4+6t_0^2\eta_0^4)\Delta B^3$$

$$+\frac{1}{24\cos B_0}t_0(5+6t_0^2-\eta_0^2)\Delta B^4+\frac{1}{120\cos B_0}(5+28t_0^2+24t_0^4)\Delta B^5 \qquad (3.11.6)$$

$$+\frac{1}{720\cos B_0}t_0(61+180t_0^2+120t_0^4)\Delta B^6$$

Example 2: Power series expansion for ΔS

$$B = B_0 + \frac{1}{N_0}(1+\eta_0^2)\Delta S + \frac{3}{2N_0^2}t_0(-\eta_0^2-\eta_0^4)\Delta S^2$$

$$+\frac{1}{2N_0^3}(-\eta_0^2+t_0^2\eta_0^2-2\eta_0^4+6t_0^2\eta_0^4)\Delta S^3+\frac{1}{2N_0^4}t_0\eta_0^2\Delta S^4 \qquad (3.11.7)$$

$$q = q_0 + \frac{1}{N_0\cos B_0}\Delta S + \frac{1}{2N_0^2\cos B_0}t_0\Delta S^2 + \frac{1}{6N_0^3\cos B_0}(1+2t_0^2$$

$$+\eta_0^2)\Delta S^3+\frac{1}{24N_0^4\cos B_0}t_0(5+6t_0^2+\eta_0^2)\Delta S^4$$

$$+\frac{1}{120N_0^5\cos B_0}(5+28t_0^2+24t_0^4)\Delta S^5 \qquad (3.11.8)$$

$$+\frac{1}{720N_0^6\cos B_0}t_0(61+180t_0^2+120t_0^4)\Delta S^6$$

Example 3: Power series expansion for Δq

$$B = B_0 + (1 + \eta_0^2)\cos B_0 \Delta q + \frac{1}{2}t_0(-1 - 4\eta_0^2 - 3\eta_0^4)\cos^2 B_0 \Delta q^2$$

$$+ \frac{1}{6}(-1 + t_0^2 - 5\eta_0^2 + 13t_0^2\eta_0^2 - 7\eta_0^4 + 27t_0^2\eta_0^4)\cos^3 B_0 \Delta q^3$$

$$+ \frac{1}{24}t_0(5 - t_0^2 + 56\eta_0^2 - 40t_0^2\eta_0^2)\cos^4 B_0 \Delta q^4 + \frac{1}{120}(5 - 18t_0^2$$

$$+ t_0^4)\cos^5 B_0 \Delta q^5 + \frac{1}{720}t_0(-61 + 58t_0^2 - t_0^4)\cos^6 B_0 \Delta q^6$$

(3.11.9)

$$S = S_0 + N_0 \cos B_0 \Delta q + \frac{1}{2}N_0 t_0(-1)\cos^2 B_0 \Delta q^2 + \frac{1}{6}N_0(-1 + t_0^2$$

$$- \eta_0^2)\cos^3 B_0 \Delta q^3 + \frac{1}{24}N_0 t_0(5 - t_0^2 + 9\eta_0^2)\cos^4 B_0 \Delta q^4$$

(3.11.10)

$$+ \frac{1}{120}N_0(5 - 18t_0^2 + t_0^4)\cos^5 B_0 \Delta q^5 + \frac{1}{720}N_0 t_0(-61$$

$$+ 58t_0^2 - t_0^4)\cos^6 B_0 \Delta q^6$$

CHAPTER FOUR

Analytical Transformation

4.1 AZIMUTHAL PROJECTION

Azimuthal projections are widely applied. They are used mainly for regional and some special-use maps.

We usually adopt an oblique azimuthal projection except for the normal azimuthal projections that are used in polar areas. The transformation for azimuthal projections will be discussed in the following subsections.

4.1.1 Direct transformation between two different azimuthal projections

The direct expression of the relationship which transforms from coordinates (x, y) on an azimuthal projection to coordinates (X, Y) on another azimuthal projection is called direct transformation. Direct transformation is appropriate for a common projection center point. We take the following cases as examples.

Example 1: Find the expression of coordinate relationships for transformation from the azimuthal equidistant projection to the azimuthal conformal projection.

The coordinate formulae for the normal azimuthal equidistant projection take the form:

$$x = R\left(\frac{\pi}{2} - \varphi\right)\sin\lambda, \quad y = R\left(\frac{\pi}{2} - \varphi\right)\cos\lambda \qquad (4.1.1)$$

The coordinate formulae for the normal azimuthal conformal projection have the form:

$$X = 2R\tan\left(\frac{\pi}{4} - \frac{\varphi}{2}\right)\sin\lambda, \quad Y = 2R\tan\left(\frac{\pi}{4} - \frac{\varphi}{2}\right)\cos\lambda \qquad (4.1.2)$$

From (4.1.1), we have:

$$\varphi = \frac{\pi}{2} - \frac{\sqrt{x^2 + y^2}}{R}, \quad \lambda = \arctan\frac{x}{y} \qquad (4.1.3)$$

Substituting (4.1.3) into (4.1.2), the form can be expressed:

$$X = 2R \tan\left(\frac{\sqrt{x^2 + y^2}}{2R}\right) \sin\left(\arctan\frac{x}{y}\right) = 2R \tan\left(\frac{\sqrt{x^2 + y^2}}{2R}\right) \sin\left(\arcsin\frac{x}{\sqrt{x^2 + y^2}}\right)$$

$$= 2R \tan\left(\frac{\sqrt{x^2 + y^2}}{2R}\right) \frac{x}{\sqrt{x^2 + y^2}}$$

$$Y = 2R \tan\left(\frac{\sqrt{x^2 + y^2}}{2R}\right) \cos\left(\arctan\frac{x}{y}\right) = 2R \tan\left(\frac{\sqrt{x^2 + y^2}}{2R}\right) \cos\left(\arccos\frac{y}{\sqrt{x^2 + y^2}}\right)$$

$$= 2R \tan\left(\frac{\sqrt{x^2 + y^2}}{2R}\right) \frac{y}{\sqrt{x^2 + y^2}}$$

Hence we have:

$$X = \frac{2Rx}{\sqrt{x^2 + y^2}} \tan\left(\frac{\sqrt{x^2 + y^2}}{2R}\right), \quad Y = \frac{2Ry}{\sqrt{x^2 + y^2}} \tan\left(\frac{\sqrt{x^2 + y^2}}{2R}\right) \qquad (4.1.4)$$

The circumstance for an oblique projection is as follows.

The coordinate formulae for the azimuthal equidistant projection have the form:

$$x = R\, Z \sin a, \quad y = R\, Z \cos a \qquad (4.1.5)$$

The coordinate formulae for the azimuthal conformal projection take the form:

$$X = 2R \tan\frac{Z}{2} \sin a, \quad Y = 2R \tan\frac{Z}{2} \cos a \qquad (4.1.6)$$

From equation (4.1.5), the form can be written:

$$Z = \frac{\sqrt{x^2 + y^2}}{2R}, \quad a = \arctan\frac{x}{y} \qquad (4.1.7)$$

Substituting equation (4.1.7) into equation (4.1.6), we have:

$$X = \frac{2Rx}{\sqrt{x^2 + y^2}} \tan\left(\frac{\sqrt{x^2 + y^2}}{2R}\right), \quad Y = \frac{2Ry}{\sqrt{x^2 + y^2}} \tan\left(\frac{\sqrt{x^2 + y^2}}{2R}\right) \qquad (4.1.8)$$

It is clear that normal and oblique projection formulae are the same when the projection center points are the same.

Example 2: Find the expression for the coordinate relationships for transformation from the azimuthal conformal projection to the azimuthal equivalent (equal-area) projection.

The coordinate formulae for the azimuthal conformal projection have the form:

$$x = 2R \tan\frac{Z}{2} \sin a, \quad y = 2R \tan\frac{Z}{2} \cos a \qquad (4.1.9)$$

The coordinate formulae for the azimuthal equivalent projection take the form:

$$X = 2R \sin\frac{Z}{2} \sin a, \quad Y = 2R \sin\frac{Z}{2} \cos a \qquad (4.1.10)$$

From equation (4.1.9), we have:

$$\tan a = \frac{x}{y}, \quad \tan\frac{Z}{2} = \frac{\sqrt{x^2 + y^2}}{2R} \tag{4.1.11}$$

After simple transformation, the form can be written as:

$$\cos a = \frac{y}{\sqrt{x^2 + y^2}}, \quad \sin a = \frac{x}{\sqrt{x^2 + y^2}} \tag{4.1.12}$$

$$\sin\frac{Z}{2} = \frac{1}{\sqrt{1 + \cot^2\frac{Z}{2}}} = \frac{\sqrt{x^2 + y^2}}{\sqrt{4R^2 + x^2 + y^2}} \tag{4.1.13}$$

Introducing (4.1.12) and (4.1.13) into (4.1.10), we have:

$$X = \frac{2Rx}{\sqrt{4R^2 + x^2 + y^2}}, \quad Y = \frac{2Ry}{\sqrt{4R^2 + x^2 + y^2}} \tag{4.1.14}$$

Example 3: Find the expression for the coordinate relationship for transformation from the gnomonic projection to the azimuthal conformal projection.

The coordinate formulae for the gnomonic projection have the form:

$$x = R\tan Z \sin a, \quad y = R\tan Z \cos a \tag{4.1.15}$$

The coordinate formulae for the azimuthal conformal projection take the form:

$$X = 2R\tan\frac{Z}{2}\sin a, \quad Y = 2R\tan\frac{Z}{2}\cos a \tag{4.1.16}$$

From equation (4.1.15), we have:

$$\tan Z = \frac{\sqrt{x^2 + y^2}}{R}, \quad \tan a = \frac{x}{y} \tag{4.1.17}$$

After a simple transformation, they can be expressed:

$$\cos a = \frac{y}{\sqrt{x^2 + y^2}}, \quad \sin a = \frac{x}{\sqrt{x^2 + y^2}}$$

$$\tan\frac{Z}{2} = \frac{\tan Z}{1 + \sec Z} = \frac{\tan Z}{1 + \sqrt{1 + \tan^2 Z}} = \frac{\sqrt{x^2 + y^2}}{R + \sqrt{R^2 + x^2 + y^2}}$$

Hence:

$$X = \frac{2Rx}{R + \sqrt{R^2 + x^2 + y^2}}, \quad Y = \frac{2Ry}{R + \sqrt{R^2 + x^2 + y^2}} \tag{4.1.18}$$

According to the above methods, we can establish an expression for the coordinate relationships for transformation among azimuthal projections with different properties, but the processes are complicated. So we introduce a comprehensive formula (2.9.37) for azimuthal projections, and discuss transformation relationships among different azimuthal projections.

$$x = \rho \sin \delta, \quad y = \rho \cos \delta \qquad (4.1.19)$$

where:

$$\rho = \left(\frac{2}{1 + \cos Z}\right)^k R \sin Z, \quad \delta = a$$

$K = 1$ is the azimuthal conformal projection;

$K = \dfrac{3}{4}$ is the azimuthal Breusing projection;

$K = \dfrac{2}{3}$ is the an approximate azimuthal equidistant projection;

$K = \dfrac{1}{2}$ is the azimuthal equivalent projection;

$K = 0$ is the azimuthal orthographic projection.

We rewrite the equation for ρ of equation (4.1.19) in the form:

$$\rho = \frac{2R \sin \dfrac{Z}{2} \cos \dfrac{Z}{2}}{\cos^{2k} \dfrac{Z}{2}}$$

Squaring the two sides, we have:

$$\rho^2 = 4R^2 \frac{1 - \cos^2 \dfrac{Z}{2}}{\cos^{2(2k-1)} \dfrac{Z}{2}} = 4R^2 \frac{1 - C^2}{C^{2L}} \qquad (4.1.20)$$

where:

$$C = \cos \frac{Z}{2}, \quad L = 2K - 1$$

Hence:

$$x = 2R \frac{\sqrt{1 - C^2}}{C^L} \sin a, \quad y = 2R \frac{\sqrt{1 - C^2}}{C^L} \cos a \qquad (4.1.21)$$

But another set of coordinate formulae for azimuthal projections has the form:

$$X = 2R \frac{\sqrt{1 - C^2}}{C^{L'}} \sin a, \quad y = 2R \frac{\sqrt{1 - C^2}}{C^{L'}} \cos a \qquad (4.1.22)$$

Its expression of the coordinate transformation relationships is written in the form:

$$X = C^{L-L'} x, \quad Y = C^{L-L'} y \qquad (4.1.23)$$

Then we find the relation of X and (x, y).

Noting that $\rho^2 = x^2 + y^2$, and substituting into equation (4.1.20), we have:

$$x^2 + y^2 = 4R^2 \frac{1 - C^2}{C^{2L}}$$

Hence:

$$\left(\frac{x^2 + y^2}{4R^2}\right) C^{2L} + C^2 - 1 = 0 \tag{4.1.24}$$

Equations (4.1.23) and (4.1.24) are the common expressions of the relationships for azimuthal projection transformation.

As examples consider the following cases.

The coordinate transformation from the azimuthal conformal projection to other azimuthal projections

When $K = 1$ in an azimuthal conformal projection, then $L = 2K - 1 = 1$, so equation (4.1.24) has the form:

$$\left(\frac{x^2 + y^2}{4R^2}\right) C^2 + C^2 - 1 = 0$$

i.e.,

$$C = \frac{2R}{\sqrt{4R^2 + x^2 + y^2}} \tag{4.1.25}$$

Introducing the above into equation (4.1.23), we have an expression of the coordinate transformation relationship from the azimuthal conformal projection to other azimuthal projections:

$$X = \left[\frac{2R}{\sqrt{4R^2 + x^2 + y^2}}\right]^{1-L'} x, \quad Y = \left[\frac{2R}{\sqrt{4R^2 + x^2 + y^2}}\right]^{1-L'} y \tag{4.1.26}$$

where $L' = 2K' - 1$ is constant for the transformed projection.

For example, the transformation formulae from the azimuthal conformal projection to the azimuthal equivalent projection ($L' = 2K' - 1 = 2 \times \frac{1}{2} - 1 = 0$) are expressed as:

$$X = \frac{2Rx}{\sqrt{4R^2 + x^2 + y^2}}, \quad Y = \frac{2Ry}{\sqrt{4R^2 + x^2 + y^2}} \tag{4.1.27}$$

The transformation formula from the azimuthal conformal projection to the azimuthal Breusing projection ($L' = 2K' - 1 = 2 \times \frac{3}{4} - 1 = \frac{1}{2}$) is written as:

$$\left.\begin{array}{l} X = \left[\dfrac{2R}{\sqrt{4R^2 + x^2 + y^2}}\right]^{1/2} x = \sqrt{x_v}\,\sqrt{x_c} \\[4mm] Y = \left[\dfrac{2R}{\sqrt{4R^2 + x^2 + y^2}}\right]^{1/2} y = \sqrt{y_v}\,\sqrt{y_c} \end{array}\right\} \tag{4.1.28}$$

where subscript v stands for equivalent and subscript c for conformal.

The coordinate transformation from the azimuthal equivalent projection to other azimuthal projections

In the azimuthal equivalent projection, when $K = \frac{1}{2}$, then $L = 2K - 1 = 0$, hence equation (4.1.24) can be expressed as:

$$\frac{x^2 + y^2}{4R^2} + C^2 - 1 = 0$$

i.e.,

$$C = [4R^2 - (x^2 + y^2)]^{1/2}/2R \tag{4.1.29}$$

Substituting the above into equation (4.1.23), the expression of coordinate transformation relationship from the azimuthal equivalent to other azimuthal projections takes the form:

$$X = \left[\frac{2R}{\sqrt{4R^2 - (x^2 + y^2)}}\right]^{L'} x, \quad Y = \left[\frac{2R}{\sqrt{4R^2 - (x^2 + y^2)}}\right]^{L'} y \tag{4.1.30}$$

where $L' = 2K' - 1$ is constant for the transformed projection.

For example, the transformation formulae from the azimuthal equivalent projection to the azimuthal conformal projection ($L' = 2K' - 1 = 2 \times 1 - 1 = 1$) can be written:

$$X = \frac{2Rx}{\sqrt{4R^2 - (x^2 + y^2)}}, \quad Y = \frac{2Ry}{\sqrt{4R^2 - (x^2 + y^2)}} \tag{4.1.31}$$

For example, the transformation formulae from the azimuthal equivalent projection to the azimuthal Breusing projection ($L' = 2K' - 1 = 2 \times \frac{3}{4} - 1 = \frac{1}{2}$) have the form:

$$\left.\begin{aligned}X &= \left[\frac{2R}{\sqrt{4R^2 - (x^2 + y^2)}}\right]^{1/2} & x &= \sqrt{x_v}\,\sqrt{x_c} \\[2mm] Y &= \left[\frac{2R}{\sqrt{4R^2 - (x^2 + y^2)}}\right]^{1/2} & y &= \sqrt{y_v}\,\sqrt{y_c}\end{aligned}\right\} \tag{4.1.32}$$

where subscript v stands for equivalent and subscript c for conformal.

4.1.2　Inverse transformation for azimuthal projections

Formulae for azimuthal projections have the form:

$$\left.\begin{aligned}y &= \rho\cos\delta, & x &= \rho\sin\delta \\ \rho &= f(Z), & \delta &= a\end{aligned}\right\} \tag{4.1.33}$$

Their inverse transformation takes the form: $(x, y) \rightarrow (Z, a) \rightarrow (\varphi, \lambda)$.

From (4.1.33), we have $\tan a = \dfrac{x}{y}$, thus:

$$a = \arctan\left(\frac{x}{y}\right) \tag{4.1.34}$$

Also from (4.1.33), we have $\rho = \sqrt{x^2 + y^2}$, thus:

$$Z = f^{-1}(\rho) = f^{-1}\left(\sqrt{x^2 + y^2}\right) \qquad (4.1.35)$$

For the stereographic projection, we have $\rho = 2R\tan\dfrac{Z}{2}$, thus:

$$Z = 2\arctan\left(\frac{\sqrt{x^2 + y^2}}{2R}\right) \qquad (4.1.36)$$

For the azimuthal equidistant projection, we have $\rho = RZ$, thus:

$$Z = \left(\frac{\sqrt{x^2 + y^2}}{R}\right) \qquad (4.1.37)$$

For the azimuthal equivalent projection, we have $\rho = 2R\sin\dfrac{Z}{2}$, thus:

$$Z = 2\arcsin\left(\frac{\sqrt{x^2 + y^2}}{2R}\right) \qquad (4.1.38)$$

For the gnomonic projection, we have $\rho = R\tan Z$, thus:

$$Z = \arctan\left(\frac{\sqrt{x^2 + y^2}}{R}\right) \qquad (4.1.39)$$

For the orthographic projection, we have $\rho = R\sin Z$, thus:

$$Z = \arcsin\left(\frac{\sqrt{x^2 + y^2}}{R}\right) \qquad (4.1.40)$$

When a polar coordinate (Z, a) for the stereographic projection is given, according to formula (3.4.36), the geographic coordinate (φ, λ) can be found.

Transformation from an azimuthal projection to other projections can adopt the above inverse transformation method, that is, to find the geographic coordinate (φ, λ) of the point, then introduce this into the transformed projection coordinate formula.

When the projection center points are identical, it is very convenient to adopt the above-mentioned direct transformation. When the projection center points are not the same, the transformation between the same azimuthal projections or different azimuthal projections is complicated. Now we shall discuss this problem, for example, the transformation for the azimuthal conformal projection when the projection center points are not identical.

Suppose the coordinate formulae for the value (φ_0, λ_0) of the projection center point of the azimuthal conformal projection have the form:

$$x = 2R\tan\frac{Z}{2}\sin a, \quad y = 2R\tan\frac{Z}{2}\cos a \qquad (4.1.41)$$

and the coordinate formulae for the value (φ_0', λ_0') of the projection center point of the other azimuthal conformal projection have the form:

$$x' = 2R'\tan\frac{Z'}{2}\sin a', \quad y' = 2R'\tan\frac{Z'}{2}\cos a' \qquad (4.1.42)$$

For transformation from (x, y) to (x', y'), we can process the following steps:

- First step: $(x, y) \rightarrow (Z, a)$
  From equation (4.1.34), (4.1.36), (Z, a) can be found.
- Second step: $(Z, a) \rightarrow (\varphi, \lambda)$
  From equation (3.4.36), (φ, λ) can be found.
- Third step: $(\varphi, \lambda) \rightarrow (Z', a')$
  From equation (3.4.32), (Z', a') can be found.
- Fourth step: $(Z', a') \rightarrow (x', y')$
  From equation (4.1.42), we can calculate (x', y').

Transformation for a different projection center point of another azimuthal projection can simulate the above steps.

4.2   CYLINDRICAL PROJECTION

4.2.1   Direct transformation between two cylindrical projections

We will discuss a common standard parallel, taking the earth as a sphere.

Example 1: Expression for the coordinate relationships for transformation from the cylindrical conformal projection to the cylindrical equivalent (equal-area) projection

The case of a normal projection   The coordinate formulae for the normal cylindrical conformal projection have the form:

$$x = r_0\lambda, \quad y = r_0 \ln\tan\left(\frac{\pi}{4} + \frac{\varphi}{2}\right) \tag{4.2.1}$$

The coordinate formulae for the normal cylindrical equivalent projection take the form:

$$X = r_0\lambda, \quad Y = \frac{R^2}{r_0}\sin\varphi \tag{4.2.2}$$

From equation (4.2.1), we have:

$$\varphi = 2\arctan(e_1^{y/r_0}) - \frac{\pi}{2} \tag{4.2.3}$$

where $e_1 = 2.718281828.\ldots$
Substituting the above into equation (4.2.2), the form can be expressed as:

$$Y = \frac{R^2}{r_0}\sin\left(2\arctan e_1^{y/r_0} - \frac{\pi}{2}\right) = -\frac{R^2}{r_0}\cos(2\arctan e_1^{y/r_0})$$

$$= -\frac{R^2}{r_0}[\cos^2(\arctan e_1^{y/r_0}) - \sin^2(\arctan e_1^{y/r_0})]$$

$$= -\frac{R^2}{r_0}\left[\frac{1}{1 + e_1^{2y/r_0}} - \frac{e_1^{2y/r_0}}{1 + e_1^{2y/r_0}}\right] = \frac{R^2}{r_0} \cdot \frac{e_1^{2y/r_0} - 1}{e_1^{2y/r_0} + 1}$$

Hence,

$$X = x, \quad Y = \frac{R^2}{r_0} \cdot \frac{e_1^{2y/r_0} - 1}{e_1^{2y/r_0} + 1} \tag{4.2.4}$$

The case of an oblique projection The coordinate formulae for the oblique cylindrical conformal projection have the form:

$$x = r_0(\pi - a), \quad y = r_0 \ln \cot \frac{Z}{2} \tag{4.2.5}$$

The coordinate formulae for the oblique cylindrical equivalent projection take the form:

$$X = r_0(\pi - a), \quad Y = \frac{R^2}{r_0} \cos Z \tag{4.2.6}$$

Introducing equation (4.2.5) into equation (4.2.6), we have:

$$X = x, \quad Y = \frac{R^2}{r_0} \cdot \frac{e_1^{2y/r_0} - 1}{e_1^{2y/r_0} + 1} \tag{4.2.7}$$

Example 2: Expression for the coordinate relationships for transformation from the cylindrical equivalent projection to the cylindrical conformal projection

The coordinate formulae for the cylindrical equivalent projection have the form:

$$x = r_0\lambda, \quad y = \frac{R^2}{r_0} \sin\varphi$$

The coordinate formulae for the cylindrical conformal projection take the form:

$$X = r_0\lambda, \quad Y = r_0 \ln \tan\left(\frac{\pi}{4} + \frac{\varphi}{2}\right)$$

$$\therefore \quad Y = r_0 \ln \tan\left(\frac{\pi}{4} + \frac{\varphi}{2}\right) = \frac{r_0}{2} \ln \frac{1 + \sin\varphi}{1 - \sin\varphi}$$

Hence,

$$X = x, \quad Y = \frac{r_0}{2} \ln \frac{R^2 + r_0 y}{R^2 - r_0 y} \tag{4.2.8}$$

If the standard parallel is not the same, the transformation is still simple. Suppose the radius of the secant parallel of the original projection is r_0, and after transformation the radius of the secant parallel of the projection is r_0'. Then the expression of the coordinate relationship for transformation from the cylindrical conformal projection to the cylindrical equivalent projection takes the form:

$$X = \frac{r_0'}{r_0} x, \quad Y = \frac{r_0}{r_0'} \cdot \frac{R^2}{r_0} \cdot \frac{e_1^{2y/r_0} - 1}{e_1^{2y/r_0} + 1} \tag{4.2.9}$$

and the expression for the coordinate transformation relationship from the cylindrical equivalent projection to the cylindrical conformal projection has the form:

$$X = \frac{r_0'}{r_0}x, \quad Y = \frac{r_0'}{r_0} \cdot \frac{r_0}{2} \ln \frac{R^2 + r_0 y}{R^2 - r_0 y}$$

(4.2.10)

4.2.2 Inverse transformation for a cylindrical projection

For the ellipsoid, the inverse the transformation method is suitable for transformation among cylindrical projections of different properties. Now we discuss this problem, for example, find the inverse transformation of the normal cylindrical conformal projection.

The coordinate formulae for the cylindrical conformal projection have the form:

$$x = r_0 l, \quad y = r_0 \ln U$$

(4.2.11)

where:

$$U = \tan\left(\frac{\pi}{4} + \frac{B}{2}\right)\left(\frac{1 - e\sin B}{1 + e\sin B}\right)^{e/2}$$

Using equation (3.7.18), inversely find latitude B. The following equations hold:

$$\varphi = 2\arctan\left(e_1^{y/r_0}\right) - \frac{\pi}{2}$$

(4.2.12)

and

$$l = \frac{x}{r_0}$$

(4.2.13)

As another example, we look at the inverse transformation of the normal cylindrical equidistant projection. The coordinate formulae for the normal cylindrical equidistant projection have the form:

$$x = r_0 l, \quad y = S_m$$

(4.2.14)

Using equation (3.7.20), inversely find latitude B, using the following equation:

$$\varphi = y \times 0.1570460641219 \times 10^{-6}$$

(4.2.15)

where the unit of x is meters, and

$$l = \frac{x}{r_0}$$

As another example, we consider the inverse transformation of the normal cylindrical equivalent projection. The coordinate formulae for the normal cylindrical equivalent projection have the form:

$$x = r_0 l, \quad y = \frac{1}{r_0}F$$

(4.2.16)

where $F = \int_0^B M r \, dB$ is the area on the ellipsoid for a longitude difference of 1 radian from the equator to latitude B.

Using equation (3.7.22), inversely find latitude B, using the equation:

$$\varphi = \arcsin(F/R^2) \tag{4.2.17}$$

where the units of $F = r_0 y$ and of $R^2 = 40\ 591\ 120.14123$ are km^2.

However:

$$l = \frac{x}{r_0}$$

4.3 CONIC PROJECTION

4.3.1 Transformations from a secant parallel to a tangent parallel and for different zones

Transformation from a secant parallel to a tangent parallel

Conic conformal projection from a secant parallel to a tangent parallel In the conic conformal projection, secant and tangent parallels have the relationship of a similarity transformation, i.e.,

$$x_{sec} = n_0 x_{tan}, \quad y_{sec} = n_0 y_{tan} \tag{4.3.1}$$

where the subscript 'sec' stands for the secant parallel and subscript 'tan' for the tangent parallel.

So we have:

$$x_{tan} = \frac{1}{n_0} x_{sec}, \quad y_{tan} = \frac{1}{n_0} y_{sec} \tag{4.3.2}$$

where n_0 is the minimum proportion of length of the secant conic projection.

Conic equidistant projection from a secant parallel to a tangent parallel For the conic equidistant projection, the relationship of the secant parallel and the tangent parallel has the form:

$$\rho_{sec} = \rho_{tan}, \quad \alpha_{sec} = n_0 \alpha_{tan} \tag{4.3.3}$$

If the coordinates of the secant projection are (x, y), we have:

$$x = \rho \sin(\alpha_{sec} l), \quad y = \rho_0 - \rho \cos(\alpha_{sec} l) \tag{4.3.4}$$

Suppose the coordinates of the projection with the tangent parallel are (x', y'), then the form can be written:

$$x' = \rho \sin(\alpha_{tan} l), \quad y' = \rho_0 - \rho \cos(\alpha_{tan} l) \tag{4.3.5}$$

From equation (4.3.4), we have:

$$\rho = \sqrt{(\rho_0 - y)^2 + x^2}, \quad \alpha_{sec} l = \arctan \frac{x}{\rho_0 - y} \tag{4.3.6}$$

Hence:

$$x' = \rho \sin\left(\frac{1}{n_0} \arctan \frac{x}{\rho_0 - y}\right)$$

$$y' = \rho_0 - \rho \cos\left(\frac{1}{n_0} \arctan \frac{x}{\rho_0 - y}\right)$$

(4.3.7)

where:

$$\rho = \sqrt{(\rho_0 - y)^2 + x^2}$$

Conic equivalent projection from a secant parallel to a tangent parallel In the conic equivalent projection, the relationship of the secant parallel and the tangent parallel has the form:

$$\rho_{sec} = m_0 \rho_{tan}, \quad \alpha_{sec} = n_0^2 \alpha_{tan}$$

(4.3.8)

If the coordinates of a secant projection are (x, y), we have:

$$x = \rho_{sec} \sin(\alpha_{sec} l), \quad y = \rho_0 - \rho_{sec} \cos(\alpha_{sec} l)$$

(4.3.9)

And if the coordinates of a projection with a tangent parallel are (x', y'), we have:

$$x' = \rho_{tan} \sin(\alpha_{tan} l), \quad y' = \rho_{0'} - \rho_{tan} \cos(\alpha_{tan} l)$$

(4.3.10)

From equation (4.3.9), we have:

$$\rho_{sec} = \sqrt{(\rho_0 - y)^2 + x^2}, \quad \alpha_{sec} l = \arctan \frac{x}{\rho_0 - y}$$

(4.3.11)

Hence:

$$x' = \rho_{0'} - n_0 \sqrt{(\rho_0 - y)^2 + x^2} \cos\left(\frac{1}{n_0^2} \arctan \frac{x}{\rho_0 - y}\right)$$

$$y' = n_0 \sqrt{(\rho_0 - y)^2 + x^2} \sin\left(\frac{1}{n_0^2} \arctan \frac{x}{\rho_0 - y}\right)$$

(4.3.12)

Transformation between different zones

Transformation with different zones for a tangent conic conformal projection Suppose the coordinates of point P are (x, y) in zone B_0 and the coordinates are (x', y') in zone B_0'.
The coordinate formulae in zone B_0 have the form:

$$x = \rho \sin(\alpha l), \quad y = \rho_0 - \rho \cos(\alpha l)$$

(4.3.13)

The coordinate formulae in zone B_0' have the form:

$$x' = \rho' \sin(a'l), \quad y' = \rho_0' - \rho' \cos(a'l)$$

(4.3.14)

In equation (4.3.13), we have:

$$\rho = \rho_0 \left(\frac{U_0}{U}\right)^\alpha$$

(4.3.15)

Hence:

$$\ln U = \frac{1}{\alpha}\ln\rho_0 - \frac{1}{\alpha}\ln\rho + \ln U_0$$

$$= \frac{1}{\alpha}\ln\rho_0 + \ln U_0 - \frac{1}{2\alpha}\ln[(\rho_0 - y)^2 + x^2]$$

and:

$$\ln\rho' = \ln\rho_0' + \alpha'\ln U_0' - \alpha'\ln U$$

$$= \ln\rho_0' + \alpha'\ln U_0' - \frac{\alpha'}{\alpha}\ln\rho_0 - \alpha'\ln U_0 + \frac{\alpha'}{2\alpha}\ln[(\rho_0 - y)^2 + x^2]$$

i.e.,

$$\ln\rho' = K + \frac{L}{2}\ln(x^2 + \Delta y^2) \qquad (4.3.16)$$

where:

$$K = \ln\rho_0' + 2\ln U_0' - \frac{\alpha'}{\alpha}\ln\rho_0 - \alpha'\ln U_0, \quad L = \frac{\alpha'}{\alpha}, \quad \Delta y = \rho_0 - y$$

Also,

$$\tan(\alpha l) = \frac{x}{\Delta y}$$

or

$$l = \frac{1}{\alpha}\arctan\frac{x}{\Delta y} \qquad (4.3.17)$$

$$\therefore \quad \cos(\alpha'l) = \cos\left(L\arctan\frac{x}{\Delta y}\right), \quad \sin(\alpha'l) = \sin\left(L\arctan\frac{x}{\Delta y}\right) \qquad (4.3.18)$$

Hence:

$$x' = e_1^C\sin\left(L\arctan\frac{x}{\Delta y}\right), \quad y' = \rho_0' - e_1^C\cos\left(L\arctan\frac{x}{\Delta y}\right) \qquad (4.3.19)$$

where:

$$C = K + \frac{L}{2}\ln(x^2 + \Delta y^2), \quad L = \frac{\alpha'}{\alpha}, \quad K = \ln\rho_0' - L\ln\rho_0 + a'(\ln U_0' - \ln U),$$

$$a' = \sin B_0', \quad \alpha = \sin B_0$$

Transformation between different zones for a tangent equidistant conic projection In zone B_0, we have:

$$\rho = \sqrt{x^2 + \Delta y^2}, \quad S = C - \sqrt{x^2 + \Delta y^2} \qquad (4.3.20)$$

where:

$$C = \rho_0 + S_0$$

and in zone B'_0 we have:

$$\rho' = C' - S = C' - C + \sqrt{x^2 + \Delta y^2}$$

or:

$$\rho' = \Delta C + \sqrt{x^2 + \Delta y^2} \tag{4.3.21}$$

Noting equation (4.3.18), we have:

$$x' = (\Delta C + \sqrt{x^2 + \Delta y^2}\,)\sin\left(L\arctan\frac{x}{\Delta y}\right)$$

$$y' = \rho'_0 - (\Delta C + \sqrt{x^2 + \Delta y^2}\,)\cos\left(L\arctan\frac{x}{\Delta y}\right) \tag{4.3.22}$$

where:

$$\Delta C = C' - C, \quad L = \frac{a'}{\alpha}, \quad C' = S'_0 + N'_0\cot B'_0, \quad C = S_0 + N_0\cot B_0,$$

$$a' = \sin B'_0, \quad \alpha = \sin B_0$$

Transformation between different zones for a tangent conic equivalent projection   In zone B_0, we have:

$$\rho^2 = \frac{2}{a}(C - F), \quad C = F_0 + \frac{1}{2}r_0\,N_0\cot B_0 \tag{4.3.23}$$

$$\therefore \quad F = C - \frac{\alpha\rho^2}{2} = C - \frac{\alpha}{2}(x^2 + \Delta y^2) \tag{4.3.24}$$

But for standard parallel B'_0,

$$\rho'^2 = \frac{2}{a'}(C' - F) = \frac{2}{a'}[C' - C + \frac{\alpha}{2}(x^2 + \Delta y^2)] \tag{4.3.25}$$

Noting equation (4.1.87), we have:

$$x'^2 = \frac{2}{\alpha'}\left[\Delta C + \frac{\alpha}{2}(x^2 + \Delta y^2)\right]\sin^2\left(L\arctan\frac{x}{\Delta y}\right)$$

$$\Delta y'^2 = \frac{2}{\alpha}\left[\Delta C + \frac{\alpha}{2}(x^2 + \Delta y^2)\right]\cos^2\left(L\arctan\frac{x}{\Delta y}\right) \tag{4.3.26}$$

where:

$$\Delta C = C' - C, \quad L = \frac{\alpha'}{\alpha}, \quad C' = F'_0 + \frac{1}{2}r'_0 N'_0\cot B'_0, \quad C = F_0 + \frac{1}{2}r_0\,N_0\cot B_0,$$

$$a' = \sin B'_0, \quad \alpha = \sin B_0$$

4.3.2   Inverse transformation for conic projections

In the conic projection, we have:

$$x = \rho\sin\delta, \quad y = \rho_s - \rho\cos\delta$$

where $\delta = \alpha l$.

From the above we have:

$$\tan\delta = \frac{x}{\rho_S - y} = \frac{x}{\Delta y}$$

thus:

$$l = \frac{1}{\alpha}\arctan\frac{x}{\Delta y} \tag{4.3.27}$$

In the conic conformal projection, we have:

$$\ln\rho = \ln C - \alpha\ln U \tag{4.3.28}$$

i.e.,

$$q = \frac{1}{\alpha}(\ln C - \ln\rho) \tag{4.3.29}$$

where:

$$\rho = \sqrt{x^2 + \Delta y^2}$$

Using equation (3.7.18), inversely find latitude B, using the following equation:

$$\varphi = 2\arctan(e_1^{\frac{1}{\alpha}(\ln C - \ln\rho)}) - \frac{\pi}{2} \tag{4.3.30}$$

In a conical equidistant projection, the form can be expressed as:

$$\rho = C - S_m$$

$$\therefore \quad S_m = C - \rho \tag{4.3.31}$$

where:

$$\rho = \sqrt{x^2 + \Delta y^2}$$

Using equation (3.7.20), inversely find latitude B, for which there is the following equation:

$$\varphi = (C - \rho) \times 0.1570460641219 \times 10^{-6} \tag{4.3.32}$$

where the unit of $(C - \rho)$ is meters.

The conic equivalent projection can be written:

$$\rho^2 = \frac{2}{\alpha}(C - F)$$

$$\therefore \quad F = C - \frac{\alpha\rho^2}{2} \tag{4.3.33}$$

where:

$$\rho^2 = x^2 + \Delta y^2$$

Using equation (3.7.22), inversely find latitude B, for which there is the following equation:

$$\varphi = \arcsin(F/R^2) \tag{4.3.34}$$

where the units of $F = C - \dfrac{\alpha\rho^2}{2}$ and $R^2 = 40\ 591\ 120.141234\ \text{km}^2$.

4.3.3 Direct transformation between two different conic projections

Suppose the coordinates (x, y) of point P of some conic projection are known, it is required to transform to the coordinates (x', y') of another conic projection, and the tangent parallels on two projections are B_0.

So we have:

$$x = \rho \sin \delta, \quad \Delta y = \rho \cos \delta \tag{4.3.35}$$

and:

$$x' = \rho' \sin \delta, \quad \Delta y' = \rho' \cos \delta \tag{4.3.36}$$

From equations (4.3.35) and (4.3.36), we have:

$$\frac{\Delta y}{x} = \frac{\Delta y'}{x'} \tag{4.3.37}$$

From equation (4.3.36), we have:

$$x'^2 \left[1 + \left(\frac{\Delta y'}{x'} \right)^2 \right] = \rho'^2, \quad \Delta y'^2 \left[1 + \left(\frac{x'}{\Delta y'} \right)^2 \right] = \rho'^2$$

Taking into account equation (4.3.37), we have:

$$x' = \Delta y' \cdot \frac{x}{\Delta y}, \quad \Delta y' = \rho' \left[1 + \left(\frac{x}{\Delta y} \right)^2 \right]^{-1/2} \tag{4.3.38}$$

or:

$$x' = \rho' \left[1 + \left(\frac{\Delta y}{x} \right)^2 \right]^{-1/2}, \quad \Delta y' = x' \cdot \frac{\Delta y}{x} \tag{4.3.39}$$

Equations (4.3.38) and (4.3.39) are common expressions of the relationships for transforming conic projections with different properties but with the same tangent parallels B_0.

When $\frac{x}{\Delta y} < 1$, it is proper to adopt equation (4.3.38). When $\frac{x}{\Delta y} > 1$, equation (4.3.39) should be adopted. The key for using the above equations is to find the expression of the relationship of ρ' and $(x, \Delta y)$. Now we discuss two types of methods. For example, we look at the transformation from a tangent conic conformal projection to a tangent conic equidistant projection. From equations (3.7.18) and (4.3.30), find B; then we have:

$$\rho' = C - S_m$$

Using equation (4.3.38) or (4.3.39), $(x', \Delta y')$ can be found.

If we do not have a tangent conic projection and B_0 is not the same, l and B can be found according to inverse transformation. First using the above equations to find latitude B, we then find l from formula (4.3.27), finally introduce (B, l) into the transformed conic projection formulae to find $(x', \Delta y')$.

4.4 TRANSFORMATION AMONG AZIMUTHAL, CYLINDRICAL AND CONIC PROJECTIONS

4.4.1 Direct transformation between azimuthal and cylindrical projections

The transformation is suitable for projections with the same property.

Example 1

Find an expression for the transformation relationship from an azimuthal conformal projection to a cylindrical conformal projection.

The coordinate formulae for the azimuthal conformal projection have the form:

$$x = \rho \sin \delta, \quad y = \rho \cos \delta \qquad (4.4.1)$$

where:

$$\rho = 2R \tan \frac{Z}{2}, \quad \delta = a$$

The coordinate formulae for a cylindrical conformal projection take the form:

$$X = r_0 a, \quad Y = r_0 \ln \cot \frac{Z}{2} \qquad (4.4.2)$$

From equation (4.4.1), we have:

$$\tan a = \frac{x}{y}, \quad \rho = \sqrt{x^2 + y^2} \qquad (4.4.3)$$

and:

$$a = \arctan \frac{x}{y}, \quad \cot \frac{Z}{2} = \frac{2R}{\sqrt{x^2 + y^2}} \qquad (4.4.4)$$

Introducing the above into equation (4.1.2), we have:

$$\left. \begin{aligned} X &= r_0 \arctan \frac{x}{y} \\[2mm] Y &= r_0 \ln 2R - \frac{r_0}{2} \ln (x^2 + y^2) \end{aligned} \right\} \qquad (4.4.5)$$

Example 2

Find the expression for the coordinate relationship for transformation from the cylindrical conformal projection to the azimuthal conformal projection.

The coordinate formulae for the cylindrical conformal projection have the form:

$$x = r_0 a, \quad y = r_0 \ln \cot \frac{Z}{2}$$

The coordinate formulae for the azimuthal conformal projection take the form:

$$X = 2R \tan \frac{Z}{2} \sin a, \quad Y = 2R \tan \frac{Z}{2} \cos a$$

From equation (4.4.2), we have:

$$\tan\frac{Z}{2} = e_1^{-y/r_0}, \quad a = \frac{x}{r_0} \tag{4.4.6}$$

Introducing the above equation into equation (4.4.1), we have

$$X = 2Re_1^{-y/r_0}\sin\frac{x}{r_0}, \quad Y = 2Re_1^{-y/r_0}\cos\frac{x}{r_0} \tag{4.4.7}$$

Example 3

Transformation from the azimuthal equidistant projection to the cylindrical equidistant projection.

The coordinate formulae for the azimuthal equidistant projection have the form:

$$x = R\,Z\sin a, \quad y = R\,Z\cos a \tag{4.4.8}$$

The coordinate formulae for the cylindrical equidistant projection take the form

$$x' = r_0 a, \quad y' = RZ \tag{4.4.9}$$

So we have:

$$x' = r_0\arctan\frac{x}{y}, \quad y' = \sqrt{x^2 + y^2} \tag{4.4.10}$$

$$x = x'\sin\frac{x'}{r_0}, \quad y = x'\cos\frac{x'}{r_0} \tag{4.4.11}$$

Example 4

Transformation from the azimuthal equivalent projection to the cylindrical equivalent projection.

The coordinate formulae for the azimuthal equivalent projection have the form:

$$x = 2R\sin\frac{Z}{2}\sin a, \quad y = 2R\sin\frac{Z}{2}\cos a \tag{4.4.12}$$

The coordinate formulae for the cylindrical equivalent projection take the form:

$$x' = r_0 a, \quad y' = \frac{R^2}{r_0}\cos Z \tag{4.4.13}$$

So we have:

$$\left.\begin{array}{l} x' = r_0\arctan\dfrac{x}{y} \\[2mm] y' = \dfrac{R^2}{r_0}\left(1 - \dfrac{x^2 + y^2}{2R^2}\right) = \dfrac{1}{2r_0}[2R^2 - (x^2 + y^2)] \end{array}\right\} \tag{4.4.14}$$

$$x = 2\sqrt{\frac{R^2 - r_0 y'}{2}}\sin\frac{x'}{r_0}, \quad y = 2\sqrt{\frac{R^2 - r_0 y'}{2}}\cos\frac{x'}{r_0} \tag{4.4.15}$$

4.4.2   Direct transformation between cylindrical and conical projections

Example 5

Transformation from the conic conformal projection to the cylindric conformal projection.
The coordinate formulae for the conic conformal projection have the form:

$$x = \rho \sin \delta, \quad y = \rho_s - \rho \cos \delta \qquad\qquad (4.4.16)$$

where:

$$\rho = \frac{C}{U^a}, \quad \delta = \alpha l$$

The coordinate formulae for the cylindrical conformal projection take the form:

$$x' = r_0 l, \quad y' = r_0 \ln U \qquad\qquad (4.4.17)$$

From equation (4.4.16), we have:

$$\ln \rho = \ln C - \alpha \ln U$$

i.e.,

$$\ln U = \frac{1}{\alpha}(\ln C - \ln \rho) \qquad\qquad (4.4.18)$$

where:

$$\rho = \sqrt{(\rho_s - y)^2 + x^2}$$

and:

$$l = \frac{1}{\alpha} \arctan \frac{x}{\rho_s - y} \qquad\qquad (4.4.19)$$

Thus we obtain an expression of the coordinate relationship for transformation from the conic conformal projection to the cylindrical conformal projection:

$$x' = \frac{r_0}{\alpha} \arctan \frac{x}{\rho_s - y}, \quad y' = \frac{r_0}{\alpha}(\ln C - \ln \rho) \qquad\qquad (4.4.20)$$

where:

$$\rho = \sqrt{(\rho_s - y)^2 + x^2}$$

From equation (4.4.17), we also have:

$$\ln U = \frac{y'}{r_0}$$

i.e.,

$$U = e_1^{y'/r_0} \quad \text{and} \quad l = \frac{x'}{r_0}$$

Introducing the above formula into equation (4.4.16), we obtain an expression for the coordinate relationship for the transformation from the cylindrical conformal projection to the conic conformal projection:

$$x = \frac{C}{\frac{\alpha}{r_0}y'} \sin\left(\frac{\alpha}{r_0}x'\right), \quad y = \rho_s - \frac{C}{\frac{\alpha}{r_0}y'} \cos\left(\frac{\alpha}{r_0}x'\right) \tag{4.4.21}$$

Example 6

Transformation from the conic equidistant projection to the cylindrical equidistant projection.

The coordinate formulae for the conic equidistant projection have the form:

$$x = (C - S_m)\sin(\alpha l), \quad y = \rho_s - (C - S_m)\cos(\alpha l) \tag{4.4.22}$$

The coordinate formulae for the cylindrical equidistant projection take the form:

$$x' = r_0 l, \quad y' = S_m \tag{4.4.23}$$

so we obtain:

$$x' = \frac{r_0}{\alpha} \arctan\frac{x}{\rho_s - y}, \quad y' = C - \sqrt{(\rho_s - y)^2 + x^2} \tag{4.4.24}$$

and:

$$x = (C - y')\sin\frac{\alpha}{r_0}x', \quad y = \rho_s - (C - y')\cos\frac{\alpha}{r_0}x' \tag{4.4.25}$$

Example 7

Transformation from the conic equivalent projection to the cylindrical equivalent projection.

The coordinate formulae for the conic equivalent projection have the form:

$$x = \rho\sin\delta, \quad y = \rho_s - \rho\cos\delta \tag{4.4.26}$$

where:

$$\rho^2 = \frac{2}{\alpha}(C - F), \quad \delta = \alpha l, \quad F = \int_0^B MrdB$$

The coordinate formulae for the cylindrical equivalent projection take the form:

$$x' = r_0 l, \quad y' = \frac{1}{r_0}F \tag{4.4.27}$$

So we obtain:

$$x' = \frac{r_0}{\alpha} \arctan\frac{x}{\rho_s - y}, \quad y' = \frac{1}{r_0}\left\{C - \frac{\alpha}{2}[(\rho_s - y)^2 + x^2]\right\} \tag{4.4.28}$$

and:

$$x = \sqrt{\frac{2}{a}(C - r_0 y')} \sin\frac{a}{r_0}x', \quad y = \rho_s - \sqrt{\frac{2}{a}(C - r_0 y')} \cos\frac{a}{r_0}x' \tag{4.4.29}$$

4.4.3 Direct transformation between azimuthal and conical projections

Example 8

Transformation from the azimuthal conformal projection to the conic conformal projection.

The coordinate formulae for the azimuthal conformal projection have the form:

$$x = \frac{C}{U} \sin l, \quad y = \frac{C}{U} \cos l \tag{4.4.30}$$

The coordinate formulae for the conic conformal projection take the form:

$$x' = \frac{C'}{U^\alpha} \sin(\alpha l), \quad y' = \rho_s - \frac{C'}{U^\alpha} \cos(\alpha l) \tag{4.4.31}$$

From equation (4.4.30), we have:

$$\frac{C}{U} = \sqrt{x^2 + y^2}$$

or:

$$U = \frac{C}{\sqrt{x^2 + y^2}} \quad \text{and} \quad l = \arctan\frac{x}{y}$$

Introducing the above equation into equation (4.4.31), we obtain an expression for the coordinate relationship for transformation from the azimuthal conformal projection to the conic conformal projection:

$$\left.\begin{array}{l} x' = \dfrac{C'}{C^\alpha}(x^2 + y^2)^{\alpha/2} \sin\left(\alpha \cdot \arctan\dfrac{x}{y}\right) \\[3mm] y' = \rho_s - \dfrac{C'}{C^\alpha}(x^2 + y^2)^{\alpha/2} \cos\left(\alpha \cdot \arctan\dfrac{x}{y}\right) \end{array}\right\} \tag{4.4.32}$$

From equation (4.4.31), we have:

$$\ln U = \frac{1}{\alpha}\left\{\ln C' - \frac{1}{2}[(\rho_s - y')^2 + x'^2]\right\}, \quad l = \frac{1}{\alpha}\arctan\frac{x'}{\rho_s - y'}$$

Introducing the above equation into equation (4.4.30), we obtain an expression for the coordinate relationship for transformation from the conic conformal projection to the azimuthal conformal projection:

$$\left.\begin{array}{l} x = \dfrac{C}{\dfrac{1}{e_1^{\frac{1}{\alpha}}\left\{\ln C' - \frac{1}{2}[(\rho_s - y')^2 + x'^2]\right\}}} \sin\left(\dfrac{1}{\alpha}\arctan\dfrac{x'}{\rho_s - y'}\right) \\[6mm] y = \dfrac{C}{\dfrac{1}{e_1^{\frac{1}{\alpha}}\left\{\ln C' - \frac{1}{2}[(\rho_s - y')^2 + x'^2]\right\}}} \cos\left(\dfrac{1}{\alpha}\arctan\dfrac{x'}{\rho_s - y'}\right) \end{array}\right\} \tag{4.4.33}$$

4.4.4 Transformation for a conical projection with different zones, using a cylindrical projection as an intermediate

Above we have discussed transformation methods and formulae for calculating conic projections between different zones. Now we introduce a cylindrical projection as an intermediate to achieve a transformation among conic projections in different zones.

Transformation for conical conformal projections with different zones

The projection constants a, C, ρ_s in zone (B_0, L_0) of the conic conformal projection and the projection constants a', C', ρ_s' in zone (B_0', L_0') of the conic conformal projection are assumed known. For the transformation from coordinates (x, y) in zone (B_0, L_0) to coordinates (X, Y) in zone (B_0', L_0') take steps as follows.

First: Transformation from the conic conformal projection to the cylindrical conformal projection.

Using equation (4.4.20), we have:

$$x_1 = \frac{r_0}{\alpha} \arctan \frac{x}{\rho_s - y}, \quad y_1 = \frac{r_0}{\alpha}(\ln C - \ln \rho) \tag{4.4.34}$$

Second: Coordinate transformation for the datum point:

$$x_2 = r_0'\left(\frac{x_1}{r_0} + L_0 - L_0'\right), \quad y_2 = \frac{r_0'}{r_0} y_1 \tag{4.4.35}$$

Third: Transformation from the cylindrical conformal projection to the conic conformal projection.

From equation (4.4.21), we have:

$$X = \frac{C}{e_1^{\frac{\alpha'}{r_0'} y_2}} \sin\left(\frac{\alpha'}{r_0'} x_2\right), \quad Y = \rho_s' - \frac{C'}{e_1^{\frac{\alpha'}{r_0'} y_2}} \cos\left(\frac{\alpha'}{r_0'} x_2\right) \tag{4.4.36}$$

Introducing equations (4.4.34) and (4.4.35) into the above equation, rearranging to obtain coordinate transformation formulae from coordinates (x, y) in zone (B_0, L_0) to coordinates (X, Y) in zone (B_0', L_0'):

$$\begin{aligned}
X &= \frac{C'}{e_1^{\frac{\alpha'}{\alpha}(\ln C - \ln \rho)}} \sin\left(\frac{\alpha'}{\alpha} \arctan \frac{x}{\rho_s - y} + \alpha'(L_0 - L_0')\right) \\[2ex]
Y &= \rho_s' - \frac{C'}{e_1^{\frac{\alpha'}{\alpha}(\ln C - \ln \rho)}} \cos\left(\frac{\alpha'}{\alpha} \arctan \frac{x}{\rho_s - y} + \alpha'(L_0 - L_0')\right)
\end{aligned} \tag{4.4.37}$$

Transformation for different zones of other conic projections can emulate the above methods.

Examples

Now we present a practical example about coordinate transformation for a conic conformal projection with different projection constants.

Projection constants: $\alpha = 0.4719381$, $C = 3\,009.818$ cm, $\rho_s = 2\,613.462$ cm of the conic conformal projection are known and coordinates are shown in Table 4.1.

Table 4.1 Coordinates for a conic conformal projection

$L \rightarrow$		$0°$	$2°$	$4°$	$6°$
$B\downarrow$	Coord.				
$22°$	x	0	41.221	82.431	123.619
	y	111.118	111.457	112.476	114.173

The above coordinates originate in table 12 of Yang (1979b).

Now we transform the above coordinates to coordinates for a conic conformal projection with projection constants $\alpha' = 0.4463426$, $C' = 3\,163.6409$ cm, $\rho'_s = 2\,750$ cm. Using equation (4.4.37), the results of coordinate transformation are as shown in Table 4.2 where (X, Y) are the transformed coordinates and (X', Y') are theoretical values from tables. The Krasovskiy ellipsoid is used. The map scale is 1:1 million.

Table 4.2 Coordinate transformation to a second conic conformal projection

$L \rightarrow$		$0°$	$2°$	$4°$	$6°$
$B\downarrow$	Coord.				
$22°$	X	0	41.3904	82.7710	124.1319
	Y	93.296	93.6175	94.5853	96.1969
	X'	0	41.391	82.771	124.131
	Y'	93.296	93.618	94.585	96.197

4.5 PSEUDOAZIMUTHAL, PSEUDOCYLINDRICAL AND PSEUDOCONIC PROJECTIONS

4.5.1 Inverse transformation for pseudoazimuthal projections

Pseudoazimuthal projection for maps of the Pacific

From equation (2.5.1), the coordinate formulae for pseudoazimuthal projections have the form:

$$\left. \begin{array}{l} x = \rho\sin\delta, \quad y = \rho\cos\delta \\ \rho = 3R\sin\dfrac{Z}{3}, \quad \delta = a - C \cdot \dfrac{Z}{Z_n}\sin 2a \end{array} \right\} \qquad (4.5.1)$$

In view of the above:

$$\rho = \sqrt{x^2 + y^2}, \quad \tan\delta = \frac{x}{y} \qquad (4.5.2)$$

and:

$$Z = 3\arcsin\frac{\rho}{3R}, \quad \delta = \arctan\frac{x}{y} \qquad (4.5.3)$$

After (Z, δ) are given, using Newton's iteration method we can find a.

Accordingly, letting

$$f(a) = a - C \cdot \frac{Z}{Z_n} \sin 2a - \delta$$

Hence:

$$f'(a) = 1 - \frac{ZC}{Z_n} \cdot Z \cos 2a$$

Thus using Newton's iteration method, the formula for computing a can be written:

$$\left. \begin{aligned} a_{i+1} &= a_i - \frac{f(a_i)}{f'(a_i)} \quad (i = 0, 1, 2, \ldots) \\ a_0 &= \delta, \quad |a_{i+1} - a_i| < \varepsilon \end{aligned} \right\}$$

(4.5.4)

where ε is the convergence limit to obtain the desired accuracy and:

$$f(a_i) = a_i - \frac{C}{Z_n} \cdot Z \sin 2a_i - \delta$$

$$f'(a_i) = 1 - \frac{2C}{Z_n} \cdot Z \cos 2a_i$$

After (Z, a) are given, using equations (3.4.36), we can find (φ, λ).

Example: Pseudoazimuthal projection constant coefficients for a map of the Pacific are given as $C = 0.1$, $Z_n = 120°$, $R = 6\,364\,472$ m.

The coordinates of specific points at a map scale of 1:10 million are based on $\varphi_0 = 25°$, $\lambda_0 = -30°$. In Table 4.3 the formulae for computing coordinate y are $y = \rho \cos \delta + y_0$, $y_0 = 120.000$ cm.

Table 4.3 Coordinates for points on a pseudoazimuthal map of the Pacific

	λ	$-30°$	$-20°$	$-10°$	$0°$	$10°$
φ	$\Delta\lambda$	0	10	20	30	40
$30°$	x	0	9.571	19.072	28.371	37.351
	y	125.554	126.001	127.250	129.326	132.286

Corresponding inverse coordinates are obtained on the basis of equations (4.5.2), (4.5.3), (4.5.4), and (3.4.36); see Table 4.4.

Table 4.4 Corresponding inverse latitude/longitude from (x, y) in Table 4.3

x	0	9.571	19.072	28.371	37.351
y	125.554	126.001	127.250	129.326	132.286
φ	29°59'59".13	29°59'55".99	29°59'58".23	29°59'56".23	29°59'57".81
λ	-30°00'00".00	-20°00'00".22	-10°00'00".88	0°00'00".74	9°59'58".30

When using iteration equations (4.5.4), the determination of the initial value a_0 should take note of the following.

Letting $\delta = \arctan \dfrac{x}{y - y_0}$, when $x > 0$ and $\delta > 0$, $a_0 = \delta$; when $x > 0$ and $\delta < 0$, $a_0 = \pi + \delta$; when $x < 0$ and $\delta > 0$, $a_0 = \pi + \delta$; and when $x < 0$ and $\delta < 0$, $a_0 = 2\pi + \delta$.

Pseudoazimuthal projection for maps of China

From equation (2.5.2), the coordinate formulae for pseudoazimuthal projections for maps of China take the form:

$$\left. \begin{array}{l} x = \rho\sin\delta, \quad y = \rho\cos\delta \\ \rho = RZ, \quad \delta = a + 0.011697143 \cdot Z\sin3(15° + a) \end{array} \right\} \tag{4.5.5}$$

Hence:

$$Z = \frac{\rho}{R}, \quad \delta = \arctan\frac{x}{y} \tag{4.5.6}$$

and a can be found with Newton's iteration formula (4.2.4), where:

$$\left. \begin{array}{l} f(a_i) = a_i + 0.011697143 \cdot Z\sin3(15° + a_i) - \delta \\ f'(a_i) = 1 + 0.035091429 \cdot Z\cos3(15° + a_i) \end{array} \right\} \tag{4.5.7}$$

The inverse coordinate transformation can be carried out with equations (4.5.2), (4.5.6), (4.5.4), (4.5.7), and (3.4.36).

4.5.2 Inverse transformation for pseudocylindrical projections

Taking the Eckert VI equivalent sinusoidal projection as an example to explain the inverse transformation method, from equation (2.6.3) we have:

$$\left. \begin{array}{l} x = \dfrac{2R\lambda}{\sqrt{\pi + 2}}\cos^2\dfrac{\alpha}{2}, \quad y = \dfrac{2R}{\sqrt{\pi + 2}}\alpha \\[2ex] \sin\alpha + \alpha = \dfrac{\pi + 2}{2}\sin\varphi \end{array} \right\} \tag{4.5.8}$$

From the above we have:

$$\alpha = \frac{y\sqrt{\pi + 2}}{2R} \tag{4.5.9}$$

We can obtain the inverse coordinate transformation formulae for the Eckert VI sinusoidal projection:

$$\varphi = \arcsin\left[\frac{2(\sin\alpha + \alpha)}{\pi + 2}\right], \quad \lambda = \frac{x\sqrt{\pi + 2}}{2R\cos^2\dfrac{\alpha}{2}} \tag{4.5.10}$$

where:

$$\alpha = \frac{y\sqrt{\pi + 2}}{2R}$$

Example 1: Computation of coordinates for the Eckert VI sinusoidal projection

Using Newton's iteration method to find a, it can be written:

$$\left.\begin{array}{l} \alpha_{i+1} = \alpha_i - \dfrac{f(a_i)}{f'(a_i)}, \quad |\alpha_{i+1} - \alpha_i| < 10^{-8} \\[4mm] \alpha_0 = \dfrac{\pi + 2}{2} \sin\varphi, \quad (i = 0, 1, 2, \ldots) \end{array}\right\}$$ (4.5.11)

where:

$$f(\alpha_i) = \sin\alpha_i + a_i - \frac{\pi + 2}{2}\sin\varphi, \quad f'(\alpha_i) = 1 + \cos\alpha_i$$

The results of solving for α using the above are expressed in Table 4.5.

Table 4.5 Inverse coordinate conversion for latitude points on any meridian of the pseudocylindrical Eckert VI sinusoidal projection

φ	0°	15°	30°	45°	60°	75°	90°
α	0°	19°14′28″.7	38°12′31″.6	56°25′11″.8	72°49′19″.1	85°11′00″.1	90°

Given a map scale $\mu_0 = 10^{-7}$, $100\mu_0 R = 63.71116$ cm, we can compute coordinates according to equations (4.5.8) as shown in Table 4.6.

Table 4.6 Inverse coordinate conversion for points of longitude at latitude 15° on the pseudocylindrical Eckert VI sinusoidal projection

φ	λ	0°	15°	30°	45°
15°	x	0	14.30088	28.60175	42.90263
	y	18.87159	18.87159	18.87159	18.87159

Example 2: Inverse coordinate transformation for the Eckert VI sinusoidal projection

Coordinates of Table 4.6 are inversely transformed with equations (4.5.10) as shown in Table 4.7.

Table 4.7 Inverse coordinate conversion for (x, y) in Table 4.6

x	0	14.30088	28.60175	42.90263
y	18.87159	18.87159	18.87159	18.87159
φ	15°00′00″.03	15°00′00″.03	15°00′00″.03	15°00′00″.03
λ	0°	15°00′00″.02	29°59′59″.99	45°00′00″.01

4.5.3 Inverse transformation for pseudoconic projections

Taking the Bonne projection as an example to illustrate its inverse transformation, from equations (2.7.1), we have:

$$x = \rho \sin \delta, \quad y = q - \rho \cos \delta$$
$$\rho = C - S, \quad C = N_0 \cot B_0 + S_0$$
$$\delta = \frac{r}{\rho} \cdot l \qquad\qquad\qquad\qquad\qquad\qquad (4.5.12)$$

where S is the arc length on the meridian; N is the radius of the prime vertical; q is constant.

From the above we have:

$$\rho = \sqrt{x^2 + (q - y)^2}, \quad \tan \delta = \frac{x}{q - y} \qquad\qquad (4.5.13)$$

and:

$$S = C - \rho \qquad\qquad\qquad\qquad\qquad\qquad\qquad (4.5.14)$$

Hence we obtain:

$$l = \frac{\rho \delta}{r} = \frac{\sqrt{x^2 + (q - y)^2}}{r} \arctan \frac{x}{q - y} \qquad\qquad (4.5.15)$$

If $S = C - \rho$ is given, we can inversely find latitude B by equation (3.7.20).

Then using equations (4.5.13), (4.5.14), (3.7.20), and (4.5.15), latitude B and longitude difference l can be inversely found.

Example: Given projection constants $C = 22\,630\,127$ m, $q = 20\,738\,918$ m for the Bonne projection, we can compute coordinates with equations (4.5.12) using the (x, y) detailed in Table 4.8. Results are expressed using inverse transformation formulae as shown in Table 4.9.

Table 4.8 Sample rectangular coordinates for the Bonne projection

x	−300 000	−262 500	−300 000	−262 500
y	600 000	600 000	650 000	650 000

Table 4.9 Inverse latitude/longitude values for the rectangular coordinates of Table 4.8

B	22°29′52″.29	22°30′09″.32	22°56′57″.51	22°57′14″.58
l	−2°54′56″.06	−2°33′04″.28	−2°55′30″.56	−2°33′34″.48

4.6   POLYCONIC PROJECTION

4.6.1   Inverse transformation for the ordinary polyconic projection

From equations (2.8.1), we have:

$$x = \rho \sin \delta, \quad y = S + \rho(1 - \cos \delta)$$
$$\rho = N \cot B, \quad \delta = l \sin B \qquad\qquad\qquad (4.6.1)$$

where S is the arc length of the meridian.

From the above, we have:

$$\rho^2 = x^2 + (S + \rho - y)^2 \qquad\qquad\qquad\qquad (4.6.2)$$

and:

$$\tan\delta = \frac{x}{S + \rho - y}$$ (4.6.3)

Letting:

$$f(B) = x^2 + (S + \rho - y)^2 - \rho^2$$

Note that:

$$\rho = N\cot B = \frac{r}{\sin B}$$

Hence:

$$\frac{d\rho}{dB} = \frac{r'\sin B - r\cos B}{\sin^2 B} = \frac{-M\sin^2 B - N\cos^2 B}{\sin^2 B} = -M - N\cot^2 B$$

and:

$$
\begin{aligned}
f'(B) &= 2(S + \rho - y)(S' + \rho') - 2\rho\rho' = 2(S + \rho - y)S' + 2(S - y)\rho' \\
&= 2(S + \rho - y)\cdot M + 2(S - y)(-M - \rho\cot B) \\
&= 2\rho(M - S\cot B + y\cot B)
\end{aligned}
$$

Thus the formula for finding latitude B by Newton's iterative method is obtained:

$$
\left.
\begin{aligned}
B_{i+1} &= B_i - \frac{f(B_i)}{f'(B_i)} \quad (i = 0, 1, 2, \ldots) \\
B_0 &= \frac{x}{a}, |B_{i+1} - B_i| < \varepsilon
\end{aligned}
\right\}
$$ (4.6.4)

where:

$$f(B_i) = x^2 + (S_i + \rho_i - y)^2 - \rho_i^2,$$
$$f'(B_i) = 2\rho_i(M_i - S_i\cot B_i + y\cot B_i),$$
$$\rho_i = N_i\cot B_i, M \text{ is the radius of curvature of the meridian.}$$

From equations (4.6.1) and (4.6.3), the formula for computing the longitude difference l takes the form:

$$l = \frac{1}{\sin B}\arctan\frac{x}{S + \rho - y}$$ (4.6.5)

Examples: Given coordinates of the ordinary polyconic projection as shown in Table 4.10. The results of the inverse coordinate transformation of the values in Table 4.10 using equations (4.6.4) and (4.6.5) are as shown in Table 4.11.

Table 4.10   Sample coordinates for the ordinary polyconic projection

B	20°	20°	20°	20°
l	0°	1°	2°	3°
x	0	104 648.2	209 292.7	313 929.8
y	2 212 405.7	2 212 718.0	2 213 655.1	2 215 216.7

Table 4.11 Inverse transformation of (x, y) values in Table 4.10

B	19°59′59″.99	19°59′59″.99	20°00′00″.00	19°59′59″.99
l	0°	0°59′59″.999	1°59′59″.998	3°00′00″.000

4.6.2 Inverse transformation for the orthogonal (rectangular) polyconic projection

From equation (2.8.3), we have:

$$x = \rho\sin\delta, \quad y = S + \rho(1 - \cos\delta),$$
$$\rho = N\cot B, \quad \tan\frac{\delta}{2} = \frac{1}{2}l\sin B \qquad (4.6.6)$$

Latitude B of the projection can be calculated using equations (4.6.4).
The formula for computing inversely the longitude difference l can be written:

$$l = 2\tan\frac{\dfrac{\delta}{2}}{\sin B} \qquad (4.6.7)$$

where:

$$\tan\frac{\delta}{2} = \frac{\tan\delta}{1 + \sqrt{1 + \tan^2\delta}}, \quad \tan\delta = \frac{x}{S + \rho - y}$$

Example: Given the coordinates in Table 4.12 for the orthogonal polyconic projection.

Table 4.12 Sample coordinates for the orthogonal polyconic projection

B	l	0°	10°	20°	30°
10°	x	0	1 096 160.5	2 190 812.1	3 282 452.7
	y	1 105 874.6	1 122 485.4	1 172 272.3	1 255 098.1

The results of the inverse coordinate transformation of the (x, y) coordinates in Table 4.12 using equations (4.6.4) and (4.6.7) are given in Table 4.13.

Table 4.13 Inverse coordinate transformation of coordinates in Table 4.12

B	9°59′59″.999	9°59′59″.998	10°00′00″.00	9°59′59″.998
l	0°	10°00′00″.00	20°00′00″.00	29°59′59″.99

4.6.3 Inverse transformation for the family of polyconic projections

Now we discuss inverse transformation for polyconic projections that are appropriate for the transverse Mercator projection (often called the Gauss–Krüger projection). From equations (2.8.10) and (2.8.12), we have:

$$x = \rho\sin\delta, \quad y = S + \rho(1 - \cos\delta)$$
$$\rho = N\cot B, \quad \delta = l\sin B + \frac{l^3}{6}\sin B\cos^2 B(1 + \eta^2) \qquad (4.6.8)$$

Latitude B for this projection can also be computed using equations (4.6.4). Now we discuss the method for computing longitude difference l.

Letting:

$$f(l) = l \sin B + \frac{l^3}{6} \sin B \cos^2 B (1 + \eta^2) - \delta$$

Hence:

$$f'(l) = \sin B + \frac{l^2}{2} \sin B \cos^2 B (1 + \eta^2)$$

Then the formulae for finding the longitude difference l using Newton's iterative method are obtained as follows:

$$\left. \begin{aligned} &l_{i+1} = l_i - \frac{f(l_i)}{f'(l_i)} \quad (i = 0, 1, 2, \ldots) \\[2mm] &l_0 = \frac{\delta}{\sin B}, \quad |l_{i+1} - l_i| < \varepsilon \end{aligned} \right\} \tag{4.6.9}$$

where:

$$f(l_i) = l_i \sin B + \frac{l_i^3}{6} \sin B \cos^2 B (1 + \eta^2) - \delta$$

$$f'(l_i) = \sin B + \frac{l_i^2}{2} \sin B \cos^2 B (1 + \eta^2)$$

$$\delta = \arctan \frac{x}{S + \rho - y}$$

4.6.4 Inverse transformation for the modified polyconic projection (Yang 1989a)

From equations (2.8.5), the direct coordinate formulae for the modified polyconic projection as used for the International Map of the World (IMW) have the form:

$$\left. \begin{aligned} x &= x_S + \frac{x_N - x_S}{4}(B - B_S) \\[2mm] y &= y_S + \frac{y_N - y_S}{4}(B - B_S) \end{aligned} \right\} \tag{4.6.10}$$

where:

$$x_S = \rho_S \sin \delta_S, \quad y_S = \rho_S - \rho_S \cos \delta_S$$

$$x_N = \rho_N \sin \delta_N, \quad y_N = y_0 + \rho_N (1 - \cos \delta_N)$$

$$\rho_j = N_j \cot B_j, \quad \delta_j = l \sin B_j, \quad j = s, N$$

$$x_0 = \Delta S - 0.0271 \cos^2 \frac{B_S + B_N}{2} \text{(cm)}$$

$N = a/\sqrt{1 - e^2 \sin^2 B}$ is the radius of the prime vertical, $\Delta S = S_N - S_S$ is the difference along the arc length of the meridian, its unit is cm (translated into the length at a scale of 1:1 million). The units for B, B_S are degrees.

From the above, we have:

$$y = y_S + \frac{y_N - y_S}{x_N - x_S}(x - x_S)$$

This is the equation for the meridians of this projection.

Letting:

$$f(l) = y_S + \frac{y_N - y_S}{x_N - x_S}(x - x_S) - y$$

Hence:

$$f'(l) = y_S'$$
$$+ \frac{[(y_N' - y_S')(x - x_S) - x_S'(y_N - y_S)](x_N - x_S) - (y_N - y_S)(x - x_S)(x_N' - x_S')}{(x_N - x_S)^2}$$

$$= y_S' + \frac{1}{x_N - x_S}[(y_N' - y_S')(x - x_S) - x_S'(y_N - y_S)] - \frac{y - y_S}{x_N - x_S}(x_N' - x_S')$$

$$= y_S' + \frac{x - x_S}{x_N - x_S}(y_N' - y_S') - \frac{y_N - y_S}{x_N - x_S}x_S' - \frac{y - y_S}{x_N - x_S}(x_N' - x_S')$$

$$= y_S' + \frac{x - x_S}{x_N - x_S}(y_N' - y_S') - \frac{y - y_S}{x_N - x_S}x_N' + \frac{y - y_N}{x_N - x_S}x_S'$$

Then using Newton's iterative method, the formula for computing the longitude difference l for the modified polyconic projection can be expressed as:

$$\left. \begin{array}{l} l_{i+1} = l_i - \dfrac{f(l_i)}{f'(l_i)} \quad (i = 0, 1, 2, \ldots) \\[3mm] l_0 = \dfrac{1}{\sin B_S}\arcsin\dfrac{x}{\rho_S}, \quad |l_{i+1} - l_i| < \varepsilon \end{array} \right\} \tag{4.6.11}$$

where:

$$f(l_i) = y_S + \frac{y_N - y_S}{x_N - x_S}(x - x_S) - y$$

$$f'(l_i) = y_S' + \frac{x - x_S}{x_N - x_S}(y_N' - y_S') - \frac{y - y_S}{x_N - x_S}x_N' + \frac{y - y_N}{x_N - x_S}x_S'$$

$$x_S' = \rho_S \cos\delta_S \sin B_S, \quad x_N' = \rho_N \cos\delta_N \sin B_N$$

$$y_S' = \rho_S \sin\delta_S \sin B_S, \quad y_N' = \rho_N \sin\delta_N \sin B_N$$

$$\rho_j = N_j \cot B_j, \quad j = s, N$$

After the longitude difference l is given, the formulae for the inverse computation of the latitude B for the modified polyconic projection can be written as:

$$B = B_S + \frac{4(x - x_S)}{x_N - x_S} \tag{4.6.12}$$

or:

$$B = B_S + \frac{4(y - y_S)}{y_N - y_S} \tag{4.6.13}$$

Examples

On the J-50 sheet of the IMW, Table 4.14 lists sample coordinates for computation of the modified polyconic projection. Inverse computation of coordinates for the modified polyconic projection are given in Table 4.15.

Table 4.14 Sample coordinates for the modified polyconic projection (IMW)

B	l	0°	1°	2°	3°
38°	x	0	8.777855638	17.55469945	26.32951972
	y	22.19128918	22.2383641	22.3795834	22.61493075

Table 4.15 Inverse computation of coordinates in Table 4.14

B	38°00′00″.0035	38°00′00″.0114	38°00′00″.0053	37°59′59″.9902
l	0°	0°59′59″.99884	2°00′00″.00037	2°59′59″.98574

4.6.5 Inverse transformation for the Hammer–Aitoff projection

From equation (2.8.13), the coordinate formulae for Hammer–Aitoff projection have the form:

$$x = \frac{2\sqrt{2}\,R\cos\varphi\sin\dfrac{\lambda}{2}}{\sqrt{1 + \cos\varphi\cos\dfrac{\lambda}{2}}}, \quad y = \frac{\sqrt{2}\,R\sin\varphi}{\sqrt{1 + \cos\varphi\cos\dfrac{\lambda}{2}}} \qquad (4.6.14)$$

From the above, we have:

$$y^2(1 + \cos\varphi\cos\frac{\lambda}{2}) = 2R^2\sin^2\varphi$$

or:

$$\cos\frac{\lambda}{2} = \frac{1}{y^2\cos\varphi}(2R^2\sin^2\varphi - y^2) \qquad (4.6.15)$$

$$\sin\frac{\lambda}{2} = \frac{\sqrt{y^4\cos^2\varphi - (2R^2\sin^2\varphi - y^2)^2}}{y^2\cos\varphi} \qquad (4.6.16)$$

Introducing equation (4.6.16) into the two expressions of equation (4.6.14), we have:

$$x = \frac{2\sqrt{2}\,Ry\cos\varphi}{\sqrt{2}\,R\sin\varphi} \cdot \frac{\sqrt{y^4\cos^2\varphi - (2R^2\sin^2\varphi - y^2)^2}}{y^2\cos\varphi}$$

$$= \frac{2}{y\sin\varphi}\sqrt{y^4\cos^2\varphi - (2R^2\sin\varphi - y^2)^2}$$

The equation for the coordinates of parallels of latitude is obtained as follows:

$$x = \frac{2}{y}\sqrt{4R^4\cos^2\varphi - (y^2 - 2R^2)^2}$$
(4.6.17)

From the above we have:

$$\frac{x^2y^2}{4} = 4R^4\cos^2\varphi - (y^2 - 2R^2)^2$$

or:

$$x^2y^2 = 16R^4(1 - \sin^2\varphi) - 4(y^4 - 4R^2y^2 + 4R^4)$$
$$x^2y^2 = 16R^2y^2 - 4y^4 - 16R^4\sin^2\varphi$$

Thus the formula for computing inversely latitude φ is obtained:

$$\varphi = \arcsin\left(\frac{y\sqrt{16R^2 - 4y^2 - x^2}}{4R^2}\right)$$
(4.6.18)

From equations (4.6.14), we have:

$$\frac{y}{x} = \frac{\sin\varphi}{2\cos\varphi\sin\dfrac{\lambda}{2}}$$

i.e.,

$$\sin\frac{\lambda}{2} = \frac{x}{2y}\tan\varphi$$
(4.6.19)

The formula for computing inversely the longitude difference λ may then be obtained:

$$\lambda = 2\arcsin\left(\frac{x}{2y}\tan\varphi\right)$$
(4.6.20)

Along the equator, at $y = 0$, on the basis of the two expressions of equations (4.6.14), we have:

$$x = 4R\sin\frac{\lambda}{4}$$

The formula for finding the longitude difference λ along the equator is obtained:

$$\lambda = 4\arcsin\left(\frac{x}{4R}\right)$$
(4.6.21)

When the longitude difference $\lambda > 180°$, in view of equations (4.6.15) and (4.6.16), the formula for computation of longitude difference λ has the form:

$$\tan\frac{\lambda}{2} = \frac{\sqrt{y^4\cos^2\varphi - (2R^2\sin^2\varphi - y^2)^2}}{2R^2\sin^2\varphi - y^2}$$
(4.6.22)

Letting $\tan\dfrac{\lambda}{2} = D$, when $D < 0$, then:

$$\lambda = 2\pi + 2\arctan D$$
(4.6.23)

Examples

Computation of coordinates for the Hammer–Aitoff projection is shown in Table 4.16, where for coordinate computation, the scale is taken as $\mu_0 = 1{:}10$ million, and $100\,\mu_0 R = 63.7116$ cm. Inverse coordinate computation is shown in Table 4.17.

Table 4.16 Sample coordinates for the Hammer–Aitoff projection

B	l	0°	60°	120°	180°
38°	x	0	37.6335766	69.79207899	90.10118659
	y	63.71116001	65.18326672	67.79207897	78.02991646

Table 4.17 Inverse computation of latitude/longitude from coordinates of Table 4.16

B	59°59′59″.9999	59°59′59″.9999	59°59′59″.9999	59°59′59″.9999
L	0	59°59′59″.9999	119°59′59″.999	179°59′59″.999

4.7 DOUBLE PROJECTION

The method of transformation from the ellipsoid to the sphere and then transformation from the sphere to the plane is called double projection. In the case of oblique projections, if the requirement is to transform from the ellipsoid to the plane, the sphere is often taken as an intermediate surface, i.e. using the double projection method to achieve an oblique projection from the ellipsoid to the plane. Now we discuss inverse transformation from the ellipsoid to the sphere and then the inverse transformation from the sphere to the plane.

4.7.1 Inverse transformation for conformal projection from the ellipsoid to the sphere

From equations (2.10.1), the coordinate formulae for conformal projection from the ellipsoid to the sphere take the form:

$$\tan\left(45° + \frac{\varphi}{2}\right) = \beta U^{\alpha}, \quad \Delta\lambda = \alpha l \tag{4.7.1}$$

where:

$$\tan\varphi_0 = \frac{R}{N_0}\tan B_0, \quad \alpha = \frac{\sin B_0}{\sin\varphi_0} \ (\text{or } \alpha = \sqrt{1 + e'^2 \cos^4 B_0}\,)$$

$$\beta = \tan\left(45° + \frac{\varphi_0}{2}\right)U_0^{-\alpha}, \quad R = \sqrt{M_0 N_0}$$

From the above, we have:

$$l = \frac{1}{\alpha}\Delta\lambda \tag{4.7.2}$$

and:

$$\ln \tan \left(45° + \frac{\varphi}{2} \right) = \ln \beta + \alpha \ln U$$

i.e.,

$$q = \frac{1}{\alpha} \left[\ln \tan \left(45° + \frac{\varphi}{2} \right) - \ln \beta \right] \qquad (4.7.3)$$

If the isometric latitude q is given, latitude B can be found using equation (3.7.18). Equations (4.7.2), (4.7.3), and (3.7.18) are formulae for computing the inverse coordinates for a conformal projection from the ellipsoid to the sphere.

4.7.2 Inverse transformation for equivalent projection from the ellipsoid to the sphere

From equation (2.10.4), the coordinate formulae for the equivalent (equal-area) projection from the ellipsoid to the sphere take the form:

$$\Delta \lambda = l, \quad \varphi = \arcsin \left(\frac{F}{R^2} \right) \qquad (4.7.4)$$

where:

$$F = \int_0^B M r dB, \quad R^2 = \int_0^{90°} M r dB$$

From the above, we have:

$$l = \Delta \lambda \qquad (4.7.5)$$

and:

$$F = R^2 \sin \varphi \qquad (4.7.6)$$

If the equivalent isometric latitude F is given, latitude B can be found using equation (3.7.22).

Equations (4.7.5), (4.7.6), and (3.7.22) are formulae for computing inverse coordinates for the equivalent projection from the ellipsoid to the sphere.

4.7.3 Inverse transformation for equidistant projection from the ellipsoid to the sphere

From equations (2.10.6), the coordinate formulae for the equidistant projection from the ellipsoid to the sphere take the form:

$$\Delta \lambda = l, \quad \varphi = \frac{S}{R} \qquad (4.7.7)$$

where:

$$S = \int_0^B M dB, \quad R = \frac{2}{\pi} \int_0^{90°} M dB$$

From the above, we have:

$$l = \Delta\lambda \tag{4.7.8}$$

and:

$$S = R\varphi \tag{4.7.9}$$

If the equidistant isometric latitude S is given, latitude B can be found using equation (3.7.20). Equations (4.7.8), (4.7.9), and (3.7.20) are formulae for computing inverse coordinates for the equidistant projection from the ellipsoid to the sphere.

4.7.4   Inverse transformation for the projection from the sphere for the datum surface of the earth (or between two spherical surfaces)

The double projection method consists of ellipsoid → sphere → plane. Its inverse transformation consists of plane → sphere → ellipsoid.

In the above we discussed the inverse transformation of a projection from the ellipsoid to the sphere. Now we take conic projections as examples to apply inverse transformation methods.

Inverse transformation for conformal conic projection

The coordinate formulae for conformal conic projection from the sphere for the datum surface of the earth have the form:

$$\left. \begin{array}{l} x = \rho\sin\delta, \quad y = \rho_s - \rho\cos\delta \\[2mm] \rho = C \cdot \cot^{\alpha}\left(45° + \dfrac{\varphi}{2}\right), \quad \delta = \alpha \cdot \Delta\lambda \end{array} \right\} \tag{4.7.10}$$

From the above equations, we have:

$$\rho = \sqrt{x^2 + (\rho_s - y)^2}, \quad \tan\delta = \frac{x}{\rho_s - y} \tag{4.7.11}$$

Its inverse transformation equations are then obtained as follows:

$$\Delta\lambda = \frac{1}{\alpha}\arctan\left(\frac{x}{\rho_s - y}\right) \tag{4.7.12}$$

and:

$$\varphi = 2\arctan(e_1^{\frac{1}{\alpha}(\ln C - \ln\rho)}) - \frac{\pi}{2} \tag{4.7.13}$$

where:

$$\ln\rho = \frac{1}{2}\ln[x^2 + (\rho_s - y)^2]$$

and:

$$e_1 = 2.718281828\ldots$$

Inverse transformation for equivalent conic projection

The coordinate formulae for equivalent conic projection from the sphere for the datum surface of the earth have the form:

$$x = \rho\sin\delta, \quad y = \rho_s - \rho\cos\delta$$
$$\rho^2 = \frac{2}{\alpha}(C - R^2\sin\varphi), \quad \delta = \alpha \cdot \Delta\lambda$$

(4.7.14)

From the above and noting equations (4.7.11), the inverse transformation formulae are obtained:

$$\Delta\lambda = \frac{1}{\alpha}\arctan\left(\frac{x}{\rho_s - y}\right)$$
$$\varphi = \arcsin\left[\frac{1}{R^2}\left(C - \frac{\alpha\rho^2}{2}\right)\right]$$

(4.7.15)

where:

$$\rho^2 = x^2 + (\rho_s - y)^2$$

Inverse transformation for equidistant conic projection

The coordinate formulae for equidistant conic projection from the sphere for the datum surface of the earth have the form:

$$x = \rho\sin\delta, \quad y = \rho_s - \rho\cos\delta$$
$$\rho = C - R\varphi, \quad \delta = \alpha \cdot \Delta\lambda$$

(4.7.16)

From equations (4.7.11) and (4.7.16), the inverse transformation formulae are obtained:

$$\Delta\lambda = \frac{1}{\alpha}\arctan\left(\frac{x}{\rho_s - y}\right)$$
$$\varphi = \frac{1}{R}(C - \rho)$$

(4.7.17)

where:

$$\rho = \sqrt{x^2 + (\rho_s - y)^2}$$

CHAPTER FIVE

Analytical Transformation for Conformal Projection

5.1 GENERAL MODEL FOR CONFORMAL PROJECTION TRANSFORMATION

5.1.1 Differential equations for conformal projection transformation

On the basis of map projection theory, we know that the general coordinate formula for conformal projections can be expressed as:

$$y + ix = f(q + il) \tag{5.1.1}$$

Suppose the coordinate formula for some conformal projection has the form:

$$y + ix = f_1(q + il) \tag{5.1.2}$$

and the coordinate formula for another conformal projection has the form:

$$Y + iX = f_2(q + il) \tag{5.1.3}$$

Eliminating (q, l) from formulae (5.1.1) and (5.1.2), we then obtain a coordinate transformation formula between two conformal projections:

$$Y + iX = f(y + ix) \tag{5.1.4}$$

From the theory of analytic functions of a complex variable, we know that equation (5.1.4) satisfies the Cauchy–Riemann criterion, i.e.,

$$\frac{\partial Y}{\partial y} = \frac{\partial X}{\partial x}, \quad \frac{\partial X}{\partial y} = -\frac{\partial Y}{\partial x} \tag{5.1.5}$$

Obtaining the partial derivative of y relative to the first expression of (5.1.5), and the partial derivative of x relative to the second expression, and adding these results, we have:

$$\frac{\partial^2 Y}{\partial^2 y} + \frac{\partial^2 Y}{\partial^2 x} = 0 \tag{5.1.6}$$

Obtaining the partial derivative of x relative to the first expression of (5.1.5), and the partial derivative of y relative to the second expression, subtracting their results, we then have:

$$\frac{\partial^2 X}{\partial^2 y} + \frac{\partial^2 X}{\partial^2 x} = 0 \tag{5.1.7}$$

in which (5.1.6) and (5.1.7) are the differential equations for conformal projection trans-formation, and are called the Laplace equations.

If the boundary values of the coordinates (x, y) of the original conformal map pro-jection and the coordinates (X, Y) of the map on the new conformal projection being compiled in the transformation range are given, then the Laplace equations for the first kind of boundary value problem have the following form:

$$\frac{\partial^2 X}{\partial^2 x} + \frac{\partial^2 X}{\partial^2 y} = 0 \quad (x, y) \in \Omega$$

$$X|_\Gamma = D'(x, y) \tag{5.1.8}$$

$$\frac{\partial^2 Y}{\partial^2 x} + \frac{\partial^2 Y}{\partial^2 y} = 0 \quad (x, y) \in \Omega$$

$$Y|_\Gamma = D(x, y) \tag{5.1.9}$$

where Ω is inside the transformation range; Γ is the boundary of the transformation range; D and D' are the known values of the boundary.

From the theory of differential equations, the solution of the first kind of boundary value problem of the Laplace equation is called a Dirichlet problem. So it is clear that their boundary conditions are different when transformation is among different conformal projections, and the Dirichlet problems are different. So the problem which transforms from a conformal projection to another conformal projection can be concluded to be a problem of the solution of the Dirichlet problem.

If the plane polar coordinates (ρ, δ) of an existing conformal map projection are given, the relationship of plane rectangular coordinates and plane polar coordinates can be expressed:

$$x = \rho \sin \delta, \quad y = \rho_0 - \rho \cos \delta \tag{5.1.10}$$

and:

$$\rho = \sqrt{x^2 + (\rho_0 - y)^2}, \quad \delta = \arctan\left(\frac{x}{\rho_0 - y}\right) \tag{5.1.11}$$

Formulae (5.1.8) and (5.1.9) are easily transformed, but we can obtain the differen-tial equations for a conformal projection transformation that take polar coordinates as parameters:

$$\frac{\partial^2 Y}{\partial \rho^2} + \frac{1}{\rho}\frac{\partial Y}{\partial \rho} + \frac{1}{\rho^2}\frac{\partial^2 Y}{\partial \delta^2} = 0 \quad (\rho, \delta) \in \Omega$$

$$Y|_\Gamma = H(\rho, \delta) \tag{5.1.12}$$

$$\frac{\partial^2 X}{\partial \rho^2} + \frac{1}{\rho}\frac{\partial X}{\partial \rho} + \frac{1}{\rho^2}\frac{\partial^2 X}{\partial \delta^2} = 0 \quad (\rho, \delta) \in \Omega$$

$$X|_\Gamma = H'(\rho, \delta) \tag{5.1.13}$$

where H and H' are the known functions along the boundary.

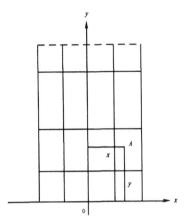

Figure 5.1 Coordinate system for Mercator projection

For transformation between conformal projections having simple boundaries (for example, an arc or a line), we can use the above differential equations to find their analytic transformation relationships. For example, the transformation from conformal conic projection to Mercator projection.

Suppose a conformal conic projection is shown in polar coordinates (ρ, δ), and the Mercator projection is shown in rectangular coordinates (X, Y), we solve it according to (5.1.12).

As shown in Figures 5.1 and 3.1, we can find the above transformation. The Y coordinates of the Mercator projection are relative only to function ρ, not δ. So (5.1.12) degenerates to an ordinary differential equations as follows:

$$\frac{\partial^2 Y}{\partial \rho^2} + \frac{1}{\rho}\frac{\partial Y}{\partial \rho} = 0 \tag{5.1.14}$$

$$Y|_{\rho=\rho_0} = 0$$

and:

$$\lim_{\rho \to 0} Y = \infty \tag{5.1.15}$$

where (5.1.14) and (5.1.15) consist of a Dirichlet problem of transformation from a conformal conic projection to the Mercator projection. The solution is written:

$$Y = C \cdot \ln\left(\frac{\rho_0}{\rho}\right) \tag{5.1.16}$$

where C is constant.

In accordance with the character of conformal projections, X and Y are a pair of conjugate harmonic functions. On the basis of the method for solution of conjugate harmonic functions in the theory of analytic functions, we can find the function X.

Noting (5.1.10), from the first expression of (5.1.5), we have:

$$\frac{\partial Y}{\partial \rho} = -\frac{1}{\rho}\frac{\partial X}{\partial \delta} \tag{5.1.17}$$

From (5.1.17) and (5.1.16), we can find:

$$X = C \cdot \delta \tag{5.1.18}$$

If we note that $X = r_0\, l$ for a cylindrical projection, and $\delta = \alpha\, l$ for a conic projection, then from (5.1.18), we have:

$$C = \frac{r_0}{\alpha} \tag{5.1.19}$$

Substituting (5.1.11) into (5.1.16) and (5.1.18), and noting (5.1.19), analytic expressions for transformation from a conformal conic projection to the Mercator projection are obtained, as

$$X = \frac{r_0}{\alpha} \arctan \frac{x}{\rho_0 - y}, \quad Y = \frac{r_0}{\alpha} \ln \frac{\rho_0}{\sqrt{x^2 + (\rho_0 - y)^2}} \tag{5.1.20}$$

and the transformation from the Mercator projection to the conformal conic projection can emulate this method. Usually, however, the solution of the Dirichlet problem when it is a transformation between two conformal projections is very difficult.

5.1.2 General formula for conformal projection transformation

From map projection theory, we know that the general formula for conformal projections can be written:

$$y + ix = f(q + il) \tag{5.1.21}$$

where (q, l) are the isometric coordinates. They can also be called Mercator coordinates.

In view of equation (5.1.21), any conformal projection can be given with Mercator coordinates. So (5.1.21) can also be seen as the general transformation formula between any conformal projection and the Mercator projection.

In order to study the transformation relationship between any two conformal projections, we introduce complex variables as follows:

$$\omega = q + il, \quad z = y + ix, \quad Z = Y + iX \tag{5.1.22}$$

Because (q, l), (x, y) and (X, Y) are all isometric coordinates, so are complex variables ω, z and Z and all are analytic functions.

Suppose the coordinate formulae for some conformal projection have the form:

$$y + ix = f_1(q + il) \tag{5.1.23}$$

which can be rewritten as:

$$z = f_1(\omega) \tag{5.1.24}$$

and coordinate formulae for another conformal projection take the form:

$$Y + iX = f_2(q + il) \tag{5.1.25}$$

which can be written as:

$$Z = f_2(\omega) \tag{5.1.26}$$

Direct transformation between two conformal projections

Eliminating (q, l) from equations (5.1.23) and (5.1.25), we obtain the coordinate transformation formulae between two conformal projections:

$$Y + iX = f(y + ix) \tag{5.1.27}$$

or:

$$Z = f(z) \tag{5.1.28}$$

An analytic function of a single value $f(z)$ can only be expanded in a unique power series $(z - z_0)$. Suppose z_0 is an arbitrary point in the transformation range, with the expansion point 0 locating the initial arbitrary position, and R is the distance from point z_0 to the boundary. Then the power series is uniformly convergent in the circle $C:|z - z_0| < R$, and can be expressed as follows inside the circle:

$$Z = Z_0 + c_1(z - z_0) + c_2(z - z_0)^2 + \ldots + c_n(z - z_0)^n + \ldots \tag{5.1.29}$$

where:

$$c_K = \frac{1}{K!}\left(\frac{d^K Z}{dz^K}\right)_0 \quad (K = 1, 2, \ldots) \tag{5.1.30}$$

Noting from equation (5.1.29) that:

$$Z_0 = Y_0 + iX_0, \quad z_0 = y_0 + ix_0, \quad c_K = a_K + ib_K \tag{5.1.31}$$

Introducing (5.1.22) and (5.1.31) into (5.1.29), equating the imaginary and real parts of the equality, we have:

$$\left.\begin{aligned}
Y =\ & Y_0 + a_1\Delta y - b_1\Delta x + a_2(\Delta y^2 - \Delta x^2) - b_2 \cdot 2\Delta x\Delta y + a_3(\Delta y^3 - 3\Delta y\Delta x^2) \\
& - b_3(3\Delta y^2\Delta x - \Delta x^3) + a_4(\Delta y^4 - 6\Delta y^2\Delta x^2 + \Delta x^4) - b_4(4\Delta y^3\Delta x - 4\Delta y\Delta x^3) \\
& + a_5(\Delta y^5 - 10\Delta y^3\Delta x^2 + 5\Delta y\Delta x^4) - b_5(5\Delta y^4\Delta x - 10\Delta y^2\Delta x^3 + \Delta x^5) + \ldots \\
X =\ & X_0 + b_1\Delta y + a_1\Delta x + b_2(\Delta y^2 - \Delta x^2) + a_2 \cdot 2\Delta x\Delta y + b_3(\Delta y^3 - 3\Delta y\Delta x^2) \\
& + a_3(3\Delta y^2\Delta x - \Delta x^3) + b_4(\Delta y^4 - 6\Delta x^2\Delta y^2 + \Delta x^4) + a_4(4\Delta y^3\Delta x - 4\Delta y\Delta x^3) \\
& + b_5(\Delta y^5 - 10\Delta y^3\Delta x^2 + 5\Delta y\Delta x^4) + a_5(5\Delta y^4\Delta x - 10\Delta y^2\Delta x^3 + \Delta x^5) + \ldots
\end{aligned}\right\} \tag{5.1.32}$$

where

$$\Delta x = x - x_0, \quad \Delta y = y - y_0$$

If c_K and $(z - z_0)$ are used to introduce trigonometric functions as follows,

$$\left.\begin{aligned}
c_K &= \sigma_K(\cos\omega_K + i\sin\omega_K) \\
a_K &= \sigma_K\cos\omega_K, \quad b_K = \sigma_K\sin\omega_K \\
z - z_0 &= \rho(\cos\theta + i\sin\theta) \\
\Delta x &= \rho\sin\theta, \quad \Delta y = \rho\cos\theta
\end{aligned}\right\} \tag{5.1.33}$$

then introducing (5.1.33) into (5.1.29), we have:

$$\left.\begin{aligned}
Y =\ & Y_0 + \sigma_1\rho\cos(\theta + \omega_1) + \sigma_2\rho^2\cos(2\theta + \omega_2) \\
& + \sigma_3\rho^3\cos(3\theta + \omega_3) + \sigma_4\rho^4\cos(4\theta + \omega_4) + \ldots \\
X =\ & X_0 + \sigma_1\rho\sin(\theta + \omega_1) + \sigma_2\rho^2\sin(2\theta + \omega_2) \\
& + \sigma_3\rho^3\sin(3\theta + \omega_3) + \sigma_4\rho^4\sin(4\theta + \omega_4) + \ldots
\end{aligned}\right\} \tag{5.1.34}$$

In (5.1.32), letting:

$$\left.\begin{aligned}
P_1 &= \Delta y, & Q_1 &= \Delta x \\
P_2 &= \Delta y^2 - \Delta x^2, & Q_2 &= 2\Delta x\Delta y \\
P_3 &= \Delta y^3 - 3\Delta y\Delta x^2, & Q_3 &= 3\Delta y^2\Delta x - \Delta x^3 \\
P_4 &= \Delta y^4 - 6\Delta x^2\Delta y^2 + \Delta x^4, & Q_4 &= 4\Delta y^3\Delta x - 4\Delta y\Delta x^3
\end{aligned}\right\} \tag{5.1.35}$$

$\ldots$

then (5.1.32) can be rewritten as follows:

$$\left. \begin{array}{l} Y = Y_0 + \sum_{K=1}^{n} (a_K P_K - b_K Q_K) \\[3mm] X = X_0 + \sum_{K=1}^{n} (b_K P_K + a_K Q_K) \end{array} \right\} \qquad (5.1.36)$$

where:

$$P_{K+1} = \Delta y P_K - \Delta x Q_K, \quad Q_{K+1} = \Delta x P_K + \Delta y Q_K, \quad p_1 = \Delta y, \quad Q_1 = \Delta x$$

In fact, letting:

$$c_K = a_k + i b_k = \frac{1}{K!} \left(\frac{d^K Z}{dz^K} \right)_0$$

$$(z - z_0)^K = P_K + i Q_K, \quad \text{then,}$$

$$(z - z_0)^{K+1} = P_{K+1} + i Q_{K+1} = (P_K + i Q_K)(z - z_0)$$

$$= (P_K + i Q_K)(\Delta y + i \Delta x) = \Delta y P_K - \Delta x Q_K + i(\Delta x P_K + \Delta y Q_K)$$

So we have:

$$P_{K+1} = \Delta y P_K - \Delta x Q_K, \quad Q_{K+1} = \Delta x P_K + \Delta y Q_K$$

The above (5.1.36) is general transformation formulae between two conformal projections, and can also be called conformal type transformation polynomials.

From (5.1.36), we can see that the key to transformation between two conformal projections is to find

$$c_K = \frac{1}{K!} \left(\frac{d^K Z}{dz^K} \right)_0$$

i.e., to find certain constant coefficients a_K, b_K. So (5.1.32) (or (5.1.36)) can also be called the general formulae of a constant coefficient transformation between two conformal projections for which the expansion point 0 is in an arbitrary position.

In particular, if the expansion point 0 of the power series is located along the central meridian of the projection, then $x_0 = X_0 = 0$ and

$$c_K = \frac{1}{K!} \left(\frac{d^K Y}{dy^K} \right)_0$$

i.e., $b_K = 0$, so (5.1.36) can also be written as follows:

$$Y = Y_0 + \sum_{K=1}^{n} a_K P_K, \quad X = \sum_{K=1}^{n} a_K Q_K \qquad (5.1.37)$$

where:

$$P_{K+1} = \Delta y P_K - x Q_K, \quad Q_{K+1} = x P_K + \Delta y Q_K, \quad P_1 = \Delta y, \quad Q_1 = x, \quad \Delta y = y - y_0$$

$$a_K = \frac{1}{K!} \left(\frac{d^K Y}{dy^K} \right)_0$$

According to (5.1.37), we can find constant coefficients a_K.

Formula (5.1.37) can also be written as follows:

$$\left.\begin{aligned}
Y &= Y_0 + a_1\Delta y + a_2(\Delta y^2 - x^2) + a_3(\Delta y^3 - 3\Delta yx^2) \\
&\quad + a_4(\Delta y^4 - 6\Delta x^2y^2 + x^4) + a_5(\Delta y^5 - 10\Delta y^3x^2 + 5\Delta yx^4) + \cdots \\
X &= a_1x + a_2 \cdot 2\Delta yx + a_3(3\Delta y^2x - x^3) + a_4(4\Delta y^3x - 4\Delta yx^3) \\
&\quad + a_5(5\Delta y^4x - 10\Delta y^2x^3 + x^5) + \cdots
\end{aligned}\right\}$$

(5.1.38)

As an analytic function, however small the line element is given, it is completely stable. Hence the coordinate relationships along the central meridian ($l = 0$) for two conformal projections have the form:

$$\Delta Y = a_1\Delta y + a_2\Delta y^2 + a_3\Delta y^3 + a_4\Delta y^4 + \cdots$$

(5.1.39)

It is clear that, having the circumstance that the expansion point is at 0 along the central meridian, we only know the coordinate expression along the central meridian of two conformal projections. Then coefficients $a_1, a_2, a_3, a_4, \ldots$ are known, and the transformation formulae for the two conformal projections are determined. So we can find the constant coefficients a_k from formula (5.1.39).

Direct transformation for conformal projection

From (4.3.21), we know that the direct coordinate transformation formulae for conformal projection have the form:

$$y + ix = f(q + il)$$

(5.1.40)

or rewriting:

$$z = f(\omega)$$

(5.1.41)

Emulating the above methods to find constant coefficients by the formula for direct transformation for conformal projections when the expansion point 0 is located at an arbitrary position can be expressed as:

$$\left.\begin{aligned}
y &= y_0 + \sum_{K=1}^{n}(a_KP_K - b_KQ_K) \\
x &= x_0 + \sum_{K=1}^{n}(b_KP_K + a_KQ_K)
\end{aligned}\right\}$$

(5.1.42)

where:

$$P_{K+1} = \Delta qP_K - \Delta lQ_K, \quad Q_{K+1} = \Delta lP_K + \Delta qQ_K,$$

$$P_1 = \Delta q = q - q_0, \quad Q_1 = \Delta l = l - l_0, \quad c_k = a_k + ib_k = \frac{1}{k!}\left(\frac{d^k z}{d\omega^k}\right)_0$$

If expansion point 0 is located along the central meridian, then the constant coefficient formulae for direct coordinate transformation for conformal projection are written in the form:

$$y = y_0 + \sum_{K=1}^{n}a_KP_K, \quad x = \sum_{K=1}^{n}a_KQ_K$$

(5.1.43)

where:

$$P_{K+1} = \Delta q P_K - l Q_K, \quad Q_{K+1} = l P_K + \Delta q Q_K, \quad P_1 = \Delta q, \quad Q_1 = l,$$

$$a_k = \frac{1}{k!}\left(\frac{d^k y}{dq^k}\right)_0$$

Inverse transformation for conformal projection

For equation (4.3.44), its inverse coordinate transformation formula has the form:

$$q + il = F(y + ix) \tag{5.1.44}$$

or rewriting:

$$\omega = F(z) \tag{5.1.45}$$

Emulating the above methods, formulae for constant coefficients of an inverse transformation for a conformal projection, when the expansion point is located at an arbitrary position, can be expressed as:

$$\left.\begin{array}{l} q = q_0 + \sum\limits_{K=1}^{n}(a_K P_K - b_K Q_K) \\[4mm] l = l_0 + \sum\limits_{K=1}^{n}(b_K P_K + a_K Q_K) \end{array}\right\} \tag{5.1.46}$$

where:

$$P_{K+1} = \Delta y P_K - \Delta x Q_K, \quad Q_{K+1} = \Delta x P_K + \Delta y Q_K, \quad P_1 = \Delta y, \quad Q_1 = \Delta x$$

$$c_k = a_k + ib_k = \frac{1}{k!}\left(\frac{d^k \omega}{dz^k}\right)_0$$

Isometric latitude q may be calculated, using (3.7.18); thus we can find latitude B.

If the expansion point 0 is located along the central meridian, then the formula for calculating the constant coefficients for the inverse coordinate transformation for a conformal projection can be written as:

$$q = q_0 + \sum_{K=1}^{n} a_K P_K, \quad l = \sum_{K=1}^{n} a_K Q_K \tag{5.1.47}$$

where:

$$P_{K+1} = \Delta y P_K - x Q_K, \quad Q_{K+1} = x P_K + \Delta y Q_K, \quad P_1 = \Delta y, \quad Q_1 = x$$

$$a_k = \frac{1}{K!}\left(\frac{d^k q}{dy^k}\right)_0$$

As above, the expansion point 0 is located along the central meridian, and the direct coordinate transformation formula for conformal projection along the central meridian has the form:

$$\Delta y = a_1 \Delta q + a_2 \Delta q^2 + a_3 \Delta q^3 + a_4 \Delta q^4 + \dots \tag{5.1.48}$$

The expression for the relationship between coordinates for an inverse transformation of a conformal projection along the central meridian takes the form:

$$\Delta q = a_1 \Delta y + a_2 \Delta y^2 + a_3 \Delta y^3 + a_4 \Delta y^4 + \dots \tag{5.1.49}$$

So we can deduce all kinds of power series expansion formulae for constant coefficients of the coordinate transformation for conformal projections.

Variable coefficient transformation for conformal projections

In (5.1.32), letting $\Delta y = 0$, i.e. $y = y_0$, hence:

$$\left.\begin{array}{l} Y = Y_0 - b_1\Delta x - a_2\Delta x^2 + b_3\Delta x^3 + a_4\Delta x^4 + \cdots \\ X = X_0 + a_1\Delta x - b_2\Delta x^2 - a_3\Delta x^3 + b_4\Delta x^4 + \cdots \end{array}\right\} \tag{5.1.50}$$

This formula's expansion point 0 changes along y, i.e. X_0, Y_0 and a_n, b_n change along y, so we call (5.1.50) a general formula for a variable coefficient transformation between two conformal projections.

If the expansion point 0 is located along the central meridian, then the formulae for a variable coefficient transformation for two conformal projections have the form:

$$\left.\begin{array}{l} Y = Y_0 - a_2 x^2 + a_4 x^4 + \cdots \\ X = a_1 x - a_3 x^3 + a_5 x^5 + \cdots \end{array}\right\} \tag{5.1.51}$$

As above, the general formulae for the variable coefficients of a direct transformation for conformal projections take the form:

$$\left.\begin{array}{l} x = x_0 - b_1\Delta l - a_2\Delta l^2 + b_3\Delta l^3 + a_4\Delta l^4 + \cdots \\ y = y_0 + a_1\Delta l - b_2\Delta l^2 - a_3\Delta l^3 + b_4\Delta l^4 + \cdots \end{array}\right\} \tag{5.1.52}$$

When the expansion point 0 is located along the central meridian, we have:

$$\left.\begin{array}{l} y = y_0 - a_2 l^2 + a_4 l^4 + \cdots \\ x = a_1 l - a_3 l^3 + a_5 l^5 + \cdots \end{array}\right\} \tag{5.1.53}$$

The general formulae of a variable coefficient inverse transformation for conformal projections take the form:

$$\left.\begin{array}{l} q = q_0 - b_1\Delta x - a_2\Delta x^2 + b_3\Delta x^3 + a_4\Delta x^4 + \cdots \\ l = l_0 + a_1\Delta x - b_2\Delta x^2 - a_3\Delta x^3 + b_4\Delta x^4 + \cdots \end{array}\right\} \tag{5.1.54}$$

When the expansion point 0 is located along the central meridian, we have:

$$\left.\begin{array}{l} q = q_0 - a_2 x^2 + a_4 x^4 + \cdots \\ l = a_1 x - a_3 x^3 + a_5 x^5 + \cdots \end{array}\right\} \tag{5.1.55}$$

To the above coefficient transformations, based on $(y - y_0) = 0$, latitude B_0 can be found inversely. Then the calculation is conducted with formula (5.1.50) or (5.1.51). This method which integrates an inverse transformation with a direct transformation is also called a synthetic transformation method.

The foregoing discusses general formulae for the coordinate transformation of conformal projections, in all sorts of circumstances. Now we will specifically study coordinate transformation formulae and applications among different conformal projections.

5.2   DIRECT TRANSFORMATION BETWEEN TWO CONFORMAL PROJECTIONS

From (5.1.37), we know that constant coefficient direct transformation formulae between two conformal projections, while the expansion point 0 is located along the central meridian, can be expressed as:

$$Y = Y_0 + \sum_{K=1}^{6} a_K P_K, \quad X = \sum_{K=1}^{6} a_K Q_K \tag{5.2.1}$$

where:

$$P_{K+1} = \Delta y P_K - x Q_K, \quad Q_{K+1} = x P_K + \Delta y Q_K, \quad P_1 = \Delta y, \quad Q_1 = x$$

5.2.1 Transverse Mercator projection and Mercator projection

Transverse Mercator projection (Gauss–Krüger) → Mercator projection

From (5.1.39), given the coordinate relationship as a formula between two conformal projections with a common central meridian, we determine the above constant coefficients a_K.

The initial coordinates along the central meridian of the Gauss–Krüger projection are given as follows:

$$\Delta y = S - S_0 = \Delta S \tag{5.2.2}$$

and the initial coordinates along the central meridian of the Mercator projection are given as follows:

$$\Delta Y = r_0 q - r_0 q_0 = r_0 \Delta q \tag{5.2.3}$$

In order to establish the expression of the relationship of the coordinates for the Gauss–Krüger and the Mercator projection along the central meridian, we now expand Δq as a power series of ΔS:

$$\Delta q = \left(\frac{dq}{dS}\right)_0 \Delta S + \frac{1}{2}\left(\frac{d^2 q}{dS^2}\right)_0 \Delta S^2 + \frac{1}{6}\left(\frac{d^3 q}{dS^3}\right)_0 \Delta S^3 + \frac{1}{24}\left(\frac{d^4 q}{dS^4}\right)_0 \Delta S^4$$
$$+ \frac{1}{120}\left(\frac{d^5 q}{dS^5}\right)_0 \Delta S^5 + \frac{1}{720}\left(\frac{d^6 q}{dS^6}\right)_0 \Delta S^6 \tag{5.2.4}$$

From (3.11.8), we have:

$$\left.\begin{aligned}
\left(\frac{dq}{dS}\right)_0 &= \frac{1}{N_0 \cos B_0} \\[2mm]
\left(\frac{d^2 q}{dS^2}\right)_0 &= \frac{1}{N_0^2 \cos B_0} t_0 \\[2mm]
\left(\frac{d^3 q}{dS^3}\right)_0 &= \frac{1}{N_0^3 \cos B_0}(1 + 2t_0^2 + \eta_0^2) \\[2mm]
\left(\frac{d^4 q}{dS^4}\right)_0 &= \frac{1}{N_0^4 \cos B_0} t_0(5 + 6t_0^2 + \eta_0^2 - 4\eta_0^4) \\[2mm]
\left(\frac{d^5 q}{dS^5}\right)_0 &= \frac{1}{N_0^5 \cos B_0}(5 + 28t_0^2 + 24t_0^4 + 6\eta_0^2 + 8t_0^2\eta_0^4) \\[2mm]
\left(\frac{d^6 q}{dS^6}\right)_0 &= \frac{1}{N_0^6 \cos B_0} t_0(61 + 180t_0^2 + 120t_0^4 + 46\eta_0^2 + 48t_0^2\eta_0^2)
\end{aligned}\right\} \tag{5.2.5}$$

where: $t_0 = \tan B_0$, $\eta_0^2 = e'^2 \cos^2 B_0$.

Considering (5.2.2) and (5.2.3), the coordinate expression showing the relation between the two projections along the central meridian is obtained as:

$$\Delta Y = N_0 \cos B_0 \Delta q = \Delta y + \frac{1}{2N_0} t_0 \Delta y^2 + \frac{1}{6N_0^2}(1 + 2t_0^2 + \eta_0^2)\Delta y^3$$

$$+ \frac{1}{24N_0^3} t_0(5 + 6t_0^2 + \eta_0^2 - 4\eta_0^4)\Delta y^4 + \frac{1}{120N_0^4}(5 + 28t_0^2 + 24t_0^4 + 6\eta_0^2 \qquad (5.2.6)$$

$$+ 8t_0^2\eta_0^4)\Delta y^5 + \frac{1}{720N_0^5} t_0(61 + 180t_0^2 + 120t_0^4 + 46\eta_0^2 + 48t_0^2\eta_0^2)\Delta y^6$$

So we have formulae for calculating the constant coefficients of the coordinate transformation (5.2.1) from the Gauss–Krüger projection to the Mercator projection:

$$
\left.
\begin{aligned}
a_1 &= 1 \\[2mm]
a_2 &= \frac{1}{2N_0} t_0 \\[2mm]
a_3 &= \frac{1}{6N_0^2}(1 + 2t_0^2 + \eta_0^2) \\[2mm]
a_4 &= \frac{1}{24N_0^3} t_0(5 + 6t_0^2 + \eta_0^2 - 4\eta_0^4) \\[2mm]
a_5 &= \frac{1}{120N_0^4}(5 + 28t_0^2 + 24t_0^4 + 6\eta_0^2 + 8t_0^2\eta_0^4) \\[2mm]
a_6 &= \frac{1}{720N_0^5} t_0(61 + 180t_0^2 + 120t_0^4 + 46\eta_0^2 + 48t_0^2\eta_0^2)
\end{aligned}
\right\}
\qquad (5.2.7)
$$

First we use (5.2.1) and (5.2.7) for transformation with the central meridian of the Mercator projection at L_0 and the latitude of the secant parallel at B_0.

If we specify that after transformation, the central meridian of the Mercator projection is L_K, and the latitude of the secant parallel is B_K, then we need continually to adopt the zone coordinate transformation of the Mercator projection; its transformation formula is written as follows:

$$y_M = \frac{r_K}{r_0} Y, \quad x_M = \frac{r_K}{r_0} X + r_K (L_0 - L_K) \qquad (5.2.8)$$

where r_0, r_K are the radii of the parallels for latitudes B_0, B_K, respectively; and L_0, L_K are the longitudes of the central meridians of the Gauss–Krüger and the Mercator projections, respectively.

Mercator projection → transverse Mercator projection (Gauss–Krüger)

So we expand Δq as a power series of ΔS as follows:

$$\Delta S = \left(\frac{dS}{dq}\right)_0 \Delta q + \frac{1}{2}\left(\frac{d^2S}{dq^2}\right)_0 \Delta q^2 + \frac{1}{6}\left(\frac{d^3S}{dq^3}\right)_0 \Delta q^3 + \frac{1}{24}\left(\frac{d^4S}{dq^4}\right)_0 \Delta q^4$$

$$+ \frac{1}{120}\left(\frac{d^5S}{dq^5}\right)_0 \Delta q^5 + \frac{1}{720}\left(\frac{d^6S}{dq^6}\right)_0 \Delta q^6$$

$$(5.2.9)$$

From (3.11.10), we have:

$$
\begin{aligned}
\left(\frac{dS}{dq}\right)_0 &= N_0 \cos B_0 \\[2mm]
\left(\frac{d^2S}{dq^2}\right)_0 &= -N_0 t_0 \cos^2 B_0 \\[2mm]
\left(\frac{d^3S}{dq^3}\right)_0 &= N_0(-1 + t_0^2 - \eta_0^2) \cos^3 B_0 \\[2mm]
\left(\frac{d^4S}{dq^4}\right)_0 &= N_0 t_0 (5 - t_0^2 + 9\eta_0^2 + 4\eta_0^4) \cos^4 B_0 \\[2mm]
\left(\frac{d^5S}{dq^5}\right)_0 &= N_0(5 - 18t_0^2 + t_0^4 + 14\eta_0^2 - 58t_0^2\eta_0^2) \cos^5 B_0 \\[2mm]
\left(\frac{d^6S}{dq^6}\right)_0 &= N_0 t_0 (-61 + 58t_0^2 - t_0^4 - 270\eta_0^2 + 330t_0^2\eta_0^2) \cos^6 B_0
\end{aligned}
\qquad (5.2.10)
$$

From (5.2.2) and (5.2.3), we have:

$$
\Delta S = \Delta Y, \quad \Delta q = \frac{\Delta y}{N_0 \cos B_0}
\qquad (5.2.11)
$$

Noting (5.2.9), and introducing (5.2.10) into (5.2.8), we have:

$$
\begin{aligned}
\Delta Y = \Delta y &- \frac{t_0}{2N_0}\Delta y^2 + \frac{1}{6N_0^2}(-1 + t_0^2 - \eta_0^2)\Delta y^3 \\[2mm]
&+ \frac{t_0}{24N_0^3}(5 - t_0^2 + 9\eta_0^2 - 4\eta_0^4)\Delta y^4 + \frac{1}{120N_0^4}(5 - 18t_0^2 + t_0^4 + 14\eta_0^2 \\[2mm]
&- 58t_0^2\eta_0^2)\Delta y^5 + \frac{t_0}{720N_0^5}(-61 + 58t_0^2 - t_0^4 - 270\eta_0^2 + 330t_0^2\eta_0^2)\Delta y^6
\end{aligned}
\qquad (5.2.12)
$$

The constant coefficients for transforming coordinates in (5.2.1) from the Mercator projection to the Gauss–Krüger projection are thus obtained as:

$$
\begin{aligned}
a_1 &= 1, \quad a_2 = -\frac{1}{2N_0}t_0 \\[2mm]
a_3 &= \frac{1}{6N_0^2}(-1 + t_0^2 - \eta_0^2) \\[2mm]
a_4 &= \frac{1}{24N_0^3}t_0(5 - t_0^2 + 9\eta_0^2 + 4\eta_0^4) \\[2mm]
a_5 &= \frac{1}{120N_0^4}(5 - 18t_0^2 + t_0^4 + 14\eta_0^2 - 58t_0^2\eta_0^2) \\[2mm]
a_6 &= \frac{1}{720N_0^5}t_0(-61 + 58t_0^2 - t_0^4 - 270\eta_0^2 + 330t_0^2\eta_0^2)
\end{aligned}
\qquad (5.2.13)
$$

Using (5.2.1) and (5.2.13) to calculate, if latitudes B_K and B_0 of the secant parallel of the Mercator projection are not the same, and longitude L_K of the central meridian of the Mercator projection and longitude L_0 of the central meridian of the Gauss–Krüger projection are different, then the forward zone coordinate transformations for the Mercator projection are needed. Those transformation formulae have the form:

$$\Delta y = \frac{r_0}{r_K} y_M - r_0 q_0, \quad x = \frac{r_0}{r_K} x_M + r_0 (L_K - L_0) \tag{5.2.14}$$

Then we use (5.2.1) and (5.2.13) to carry out the coordinate transformation.

Applications

Example 1 Using the formula for constant coefficients to obtain coordinate transformations from the Mercator to the Gauss–Krüger projection.

Given: $B_K = 4°$, $L_K = 117°$, Mercator coordinates $x = 388\,682.0384$ m, $y = 0$ for the point $B = 0°$, $L = 120°30'$.

Find: Gauss–Krüger coordinates (x_G, y_G) for the point at $L_0 = 117°$, in a Gauss 6° zone.

Computing the result: $x_G = 389\,868.9976$ m, $y_G = -0.0015$ m.

Observe that (B_K, L_K) are the secant parallels and the central meridian of the Mercator coordinates. If $B_0 = 0°$, formula (5.2.13) may be used to compute the constant coefficients.

Example 2 Using the formulae for constant coefficients, transform coordinates from the Gauss–Krüger to the Mercator projection.

Given: $L_0 = 117°$, Gauss–Krüger coordinates $x = 389\,868.997$ m, $y = 0$ for a point at $B = 0°$, $L = 120°30'$ in a 6° zone.

Find: Mercator coordinates x_M, y_M of the point at $B_K = 10°$, $L_K = 120°$.

Computing the result: $x_M = 54\,820.609$ cm, $y_M = 0.001675$ cm.

Observe that if $B_0 = 2°$, the constant coefficients may be computed from formula (5.2.7).

5.2.2 Mercator projection and a conformal conic projection

Mercator projection → conformal conic projection

The coordinate formulae for the conformal conic projection have the form:

$$\left. \begin{array}{ll} X = \rho \sin \delta, & Y = \rho_s - \rho \cos \delta \\ \rho = \dfrac{C}{U^\alpha}, & \delta = \alpha l \end{array} \right\} \tag{5.2.15}$$

where:

$$C = \rho_0 U_0^\alpha, \quad \alpha = \sin B_0, \quad \rho_0 = n_0 N_0 \cot B_0$$

From the above equations, we have:

$$\rho = \rho_0 U_0^\alpha U^{-\alpha}$$

Taking the logarithms of each side, (5.2.15) can be expressed as:

$$\ln\rho = \ln\rho_0 - \alpha(\ln U - \ln U_0) = \ln\rho_0 - \alpha \cdot \Delta q \tag{5.2.16}$$

When $l = 0°$, from (5.2.15), we have $Y = \rho_s - \rho$ and noting that when $B = B_0$, we have $Y_0 = \rho_s - \rho_0$.

Hence:

$$\rho = \rho_0 - \Delta Y \tag{5.2.17}$$

Considering (5.2.3), we have:

$$\Delta q = \frac{\Delta y}{r_0} \tag{5.2.18}$$

Introducing (5.2.17) and (5.2.18) into (5.2.16), we have:

$$\ln(\rho_0 - \Delta Y) = \ln\rho_0 - \alpha \cdot \frac{\Delta y}{r_0}$$

In a conformal conic projection, when $n_0 \doteq 1$, then the above formula can be rewritten as:

$$\ln\left(1 - \frac{\Delta Y}{\rho_0}\right) = -\frac{\Delta y}{\rho_0} \tag{5.2.19}$$

Rewriting the above equation, we have:

$$1 - \frac{\Delta Y}{\rho_0} = e_1^{-\frac{\Delta y}{\rho_0}}$$

$$\Delta Y = \rho_0\left(1 - e_1^{-\frac{\Delta y}{\rho_0}}\right) \tag{5.2.20}$$

where e_1 is the base of natural logarithms 2.718281828, and $\Delta Y = Y - Y_0$, $\Delta y = y - y_0$. The point of origin of the coordinates (X, Y) of the conformal conic projection is the intersection point of parallel B_s and the central meridian; and the point of origin of the (x, y) coordinates on the Mercator projection is the point of intersection of the equator and the central meridian of the map.

Expanding

$$e_1^{-\frac{\Delta y}{\rho_0}}$$

into a series, we have:

$$\Delta Y = \rho_0\left[1 - \left(1 - \frac{\Delta y}{\rho_0} + \frac{\Delta y^2}{2\rho_0^2} - \frac{\Delta y^3}{6\rho_0^3} + \frac{\Delta y^4}{24\rho_0^4} - \frac{\Delta y^5}{120\rho_0^5} + \cdots\right)\right]$$

So we obtain the coordinate relationship of two projections at expansion point B_0 along the central meridian:

$$\Delta Y = \Delta y - \frac{\Delta y^2}{2\rho_0} + \frac{\Delta y^3}{6\rho_0^2} - \frac{\Delta y^4}{24\rho_0^3} + \frac{\Delta y^5}{120\rho_0^4} - \frac{\Delta y^6}{720\rho_0^5} \tag{5.2.21}$$

where:

$$\rho_0 = N_0 \cot B_0 = \frac{N_0}{t_0}, \quad t_0 = \tan B_0$$

Noting that the relationship of a conformal tangent conic projection and a conformal secant conic projection has the form:

$$\Delta Y_{sec} = n_0 \Delta Y_{tan} \qquad (5.2.22)$$

we obtain the constant coefficients for the coordinate transformation formula (5.2.1) from the Mercator projection to a conformal secant conic projection:

$$\left.\begin{array}{ll} a_1 = n_0, & a_2 = -\dfrac{n_0}{2N_0}t_0 \\[3mm] a_3 = \dfrac{n_0}{6N_0^2}t_0^2, & a_4 = -\dfrac{n_0}{24N_0^3}t_0^3 \\[3mm] a_5 = \dfrac{n_0}{120N_0^4}t_0^4, & a_6 = -\dfrac{n_0}{720N_0^5}t_0^5 \end{array}\right\} \qquad (5.2.23)$$

where n_0 is the minimum linear scale factor along the parallel of latitude B_0 of the conformal conic projection. If constants α and C of the conformal conic projection are given, B_0 and n_0 can be calculated as follows:

$$B_0 = \arcsin \alpha, \quad n_0 = \frac{\alpha \rho_0}{r_0} = \frac{\alpha C}{r_0 U_0^\alpha}$$

If the standard parallel and central meridian of the Mercator projection have coordinates (B_K, L_K), respectively, we need to find $(\Delta y, x)$ from formula (5.2.14), then from (5.2.1) and (5.2.23) to carry out the coordinate transformation.

Conformal conic projection → Mercator projection

Noting that,

$$\Delta q = \frac{\Delta y}{r_0}, \quad \rho = \rho_0 - \Delta y$$

from (5.2.19), we have:

$$-\frac{\Delta Y}{\rho_0} = \ln\left(1 - \frac{\Delta y}{\rho_0}\right) \qquad (5.2.24)$$

Expanding the above into series, it can be written:

$$\Delta Y = \Delta y + \frac{\Delta y^2}{2\rho_0} + \frac{\Delta y^3}{3\rho_0^2} + \frac{\Delta y^4}{4\rho_0^3} + \frac{\Delta y^5}{5\rho_0^4} + \frac{\Delta y^6}{6\rho_0^5} \qquad (5.2.25)$$

Noting the conformal conic projection, we have:

$$\Delta y_{tan} = \frac{1}{n_0}\Delta y_{sec} \qquad (5.2.26)$$

Hence we obtain the constant coefficients of the formula for coordinate transformation (5.2.1) from a conformal secant conic projection to the Mercator projection:

$$\left.\begin{array}{ll} a_1 = \dfrac{1}{n_0}, & a_2 = \dfrac{1}{2n_0^2 N_0} t_0 \\[3mm] a_3 = \dfrac{1}{3n_0^3 N_0^2} t_0^2, & a_4 = \dfrac{1}{4n_0^4 N_0^3} t_0^3 \\[3mm] a_5 = \dfrac{1}{5n_0^5 N_0^4} t_0^4, & a_6 = \dfrac{1}{6n_0^6 N_0^5} t_0^5 \end{array}\right\} \qquad (5.2.27)$$

Using (5.2.1) and (5.2.27), transformation from coordinates (x, y) (B_0 and n_0 as parameters) of the conformal conic projection to obtain coordinates (X, Y) and the latitude B_0 (as the standard parallel) for the Mercator projection is achieved.

Applications

Example 1   Using the constant coefficient formula, calculate the coordinate transformation from a Mercator projection to a 1:1 million conformal conic projection.

    Given a map scale of 1:1 million, $B_K = 10°$, $L_K = 120°$, and Mercator coordinates $x = 5.482$ cm, $y = 0$ for the point $B = 0°$, $L = 120°30'$.

    Find: coordinates (x_c, y_c) of the point on the conformal conic projection at 1:1 million at $L_0 = 117°$.

    Computing the result: $x_c = 38.97418209$ cm, $y_c = 5.572957$ cm.

    Observe that the constant coefficients are found at $B_0 = 2°.000420472$, using formula (5.2.23). The coordinates for a Chinese 1:1 million conformal conic projection can be seen in Yang (1987a).

Example 2   Using the constant coefficient formula, transform coordinates from the conformal conic projection for a 1:1 million map to the Mercator projection.

    Given: Coordinates for a conformal conic projection $x = 18.718$ cm, $y = 50.481$ cm, of the point $B = 56°$, $L = 3°$ for 1:1 million maps.

    Find: Mercator coordinates x_M, y_M of the point (scale is 1:1 million, B_K, L_K and B_0, L_0 are the same).

    Computing the result: $x_M = 19.66493672$, $y_M = 442.9967275$.

    The constant coefficients are found from $B_0 = 54°.01618964$ and formula (5.2.27).

5.2.3   Transverse Mercator projection and conformal conic projection

Transverse Mercator projection (Gauss–Krüger) → *conformal conic projection*

Introducing (5.2.6) into (5.2.21), we have:

$$\Delta Y = \Delta y + \frac{1}{6N_0^2}(1 + \eta_0^2)\Delta y^3 + \frac{1}{24 N_0^3} t_0(1 - 3\eta_0^2 - 4\eta_0^4)\Delta y^4$$

$$+ \frac{1}{120 N_0^4}(5 + 3t_0^2 + 6\eta_0^2 + 3t_0^2\eta_0^2 + 20t_0^2\eta_0^4)\Delta y^5 \qquad (5.2.28)$$

$$+ \frac{1}{720 N_0^5} t_0(21 - 28t_0^2 - 80t_0^4 - 10\eta_0^2 - 40t_0^2\eta_0^2 - 10\eta_0^4)\Delta y^6$$

Noting (5.2.22), constant coefficients for formula (5.2.1) to obtain coordinate transformation from a Gauss–Krüger projection to a conformal secant conic projection, are achieved:

$$a_1 = n_0, \qquad\qquad a_2 = 0$$

$$a_3 = \frac{n_0}{6N_0^2}(1 + \eta_0^2), \quad a_4 = \frac{n_0}{24N_0^3}t_0(1 - 3\eta_0^2 - 4\eta_0^4)$$

$$a_5 = \frac{n_0}{120N_0^4}(5 + 3t_0^2 + 6\eta_0^2 + 3t_0^2\eta_0^2 + 20t_0^2\eta_0^4)$$

$$a_6 = \frac{n_0}{720N_0^5}t_0(21 - 28t_0^2 - 80t_0^4 - 10\eta_0^2 - 40t_0^2\eta_0^2 - 10\eta_0^4)$$

$$(5.2.29)$$

If the central meridians of the Gauss–Krüger projection and the conformal conic projection are not identical, suppose that the central meridian of the Gauss–Krüger projection is L_0, and the central meridian of the conformal conic projection is L_K.

Letting $l' = L_K - L_0$, then we have:

$$y_c = \rho_s - [(\rho_s - Y)\cos\alpha l' + X\sin\alpha l']$$
$$x_c = (\rho_s - Y)\sin\alpha l' - X\cos\alpha l'$$

$$(5.2.30)$$

where:

$$\alpha = \sin B_0, \quad \rho_s = Y_0 + \rho_0, \quad \rho_0 = n_0 N_0 \cot B_0$$

Conformal conic projection → Gauss–Krüger projection

Introducing (5.2.25) into (5.2.12), we have:

$$\Delta Y = \Delta y - \frac{1}{6N_0^2}(1 + \eta_0^2)\Delta y^3 - \frac{1}{24N_0^3}t_0(1 - 3\eta_0^2 - 4\eta_0^4)\Delta y^4$$

$$+ \frac{1}{120N_0^4}(5 - 3t_0^2 + 14\eta_0^2 - 3t_0^2\eta_0^2 + 40t_0^2\eta_0^4)\Delta y^5$$

$$+ \frac{1}{720N_0^5}t_0(14 - 108t_0^2 - 120t_0^4 - 60\eta_0^2 + 120t_0^2\eta_0^2 + 340t_0^2\eta_0^4)\Delta y^6$$

$$(5.2.31)$$

Noting (5.2.26), constant coefficients of the formula for coordinate transformation (5.2.1) from a secant conformal conic projection to Gauss–Krüger projection are obtained:

$$a_1 = \frac{1}{n_0}, \qquad\qquad a_2 = 0$$

$$a_3 = -\frac{1}{6n_0^3N_0^2}(1 + \eta_0^2), \quad a_4 = -\frac{1}{24n_0^4N_0^3}t_0(1 - 3\eta_0^2 - 4\eta_0^4)$$

$$a_5 = \frac{1}{120n_0^5N_0^4}(5 - 3t_0^2 + 14\eta_0^2 - 3t_0^2\eta_0^2 + 40t_0^2\eta_0^4)$$

$$a_6 = \frac{1}{720n_0^6N_0^5}t_0(14 - 108t_0^2 - 120t_0^4 - 60\eta_0^2 + 120t_0^5\eta_0^2 + 340t_0^2\eta_0^4)$$

$$(5.2.32)$$

If the central meridian of the Gauss–Krüger projection is L_0, and the central meridian of conformal conic projection is L_K, we also need to transform in advance the conformal conic projection coordinates (x_c, y_c) as follows,

$$y = \rho_s - [(\rho_s - y_c)\cos(\alpha l') - x_c\sin(\alpha l')]$$
$$x_c = (\rho_s - y_c)\sin(\alpha l') + x_c\cos(\alpha l')$$

$$(5.2.33)$$

where:

$$l' = L_K - L_0$$

Applications

Example 1 Using the constant coefficient formula, transform coordinates from the Gauss–Krüger projection to the conformal conic projection for a 1:1 million map.

Given: Gauss–Krüger coordinates $x = 29.517$ cm, $y = 310.213$ cm (map scale 1:1 million) for the point $B = 28°$, $l = 3°$.

Find: the conformal conic projection coordinates (x_c, y_c) of the point on a 1:1 million map.

Computing the result: $x_c = 29.51499695$, $y_c = 5.93019227$.

Note: Constant coefficients are found using $B_0 = 30°.00685786$ and formula (5.2.29).

Example 2 Using the constant coefficient formula, transform coordinates from a conformal conic projection for 1:1 million maps to the Gauss–Krüger projection.

Given: Conformal conic projection coordinates $x = 29.515$ cm, $y = 5.930$ cm of the point $B = 28°$, $l = 3°$ for 1:1 million maps.

Find: Gauss–Krüger coordinates (x_G, y_G) of the point (scale is 1:1 million).

Computing the result: $x_G = 29.51704048$, $y_G = 310.212817$ cm.

Note: Constant coefficients are found for $B_0 = 30°.00685786$ using formula (5.2.32).

5.2.4 Gauss–Krüger projection and the Roussilhe projection

Gauss–Krüger projection → Roussilhe projection

Suppose the centerpoint of the Roussilhe projection is at (B_0, L_0). From (2.2.15) we know the coordinate relationship for the Roussilhe projection at the central meridian:

$$Y = 2R_0 \tan \frac{\Delta S}{2R_0} \qquad (5.2.34)$$

We can thus obtain the coordinate relationship from the Roussilhe projection to the Gauss–Krüger projection at the central meridian (L_0),

$$Y = 2R_0 \tan \frac{\Delta y}{2R_0} \qquad (5.2.35)$$

Expanding the above to a series as follows,

$$Y = \Delta y + \frac{1}{12R_0^2} \Delta y^3 + \frac{1}{120R_0^4} \Delta y^5 + \dots \qquad (5.2.36)$$

we obtain constant coefficients for the coordinate transformation formula (5.2.1) from the Gauss–Krüger projection to the Roussilhe projection:

$$\left.\begin{array}{l} a_1 = 1, \quad a_2 = 0, \quad a_3 = \dfrac{1}{12R_0^2} \\[2ex] a_4 = 0, \quad a_5 = \dfrac{1}{120R_0^4}, \quad a_6 = 0 \end{array}\right\} \qquad (5.2.37)$$

where:

$$R_0 = \sqrt{M_0 N_0}$$

Roussilhe projection → Gauss–Krüger projection

Noting that $\Delta Y = \Delta S$, from (5.2.34), we have:

$$\Delta Y = 2R_0 \arctan \frac{y}{2R_0} \qquad (5.2.38)$$

Expanding the above in series as follows,

$$\Delta Y = y - \frac{1}{12R_0^2} \Delta y^3 + \frac{1}{80R_0^4} \Delta y^5 - \cdots \qquad (5.2.39)$$

We can obtain constant coefficients for the coordinate transformation formula (5.2.1) from the Roussilhe projection to the Gauss–Krüger projection:

$$\left.\begin{array}{l} a_1 = 1, \quad a_2 = 0, \quad a_3 = -\dfrac{1}{12R_0^2} \\[2mm] a_4 = 0, \quad a_5 = \dfrac{1}{80R_0^4}, \quad a_6 = 0 \end{array}\right\} \qquad (5.2.40)$$

and the coordinates of the expansion point 0 are:

$$\left.\begin{array}{l} x_0 = 0, \quad y_0 = 0 \\ X_0 = 0, \quad Y_0 = S_0 \end{array}\right\} \qquad (5.2.41)$$

where S_0 is the meridian arc length from the equator to latitude B_0.

Applications

Example 1 The coordinate computation of Roussilhe's projection.
Using (2.2.25), (2.2.27) to compute a projection coordinate (scale 1:1 million). When $B_0 = 30°$, we have:

$$a_1 = 552.83495, \quad a_2 = -138.20874, \quad a_3 = -11.692018,$$
$$a_4 = 14.658678, \quad a_5 = 2.4474464$$

For the results of the coordinate calculations, see Table 5.1.

Table 5.1 Coordinates on the Roussilhe projection based on Example 1

	$l \to$	$0°$	$1°$	$2°$	$3°$
$B\downarrow$	Coord.				
$38°$	x	0	9.8394	19.6793	29.5200
	y	−22.1698	−22.1282	−22.0032	−21.7950

Example 2 Coordinate transformation from the Gauss–Krüger projection (scale 1:1 million) to the Roussilhe projection.

Using (5.2.37) to compute constant coefficients with $B_0 = 30°$:

$$x_0 = 0, \quad y_0 = 332.017, \quad X_0 = 0, \quad Y_0 = 0$$

$$a_1 = 1, \quad a_2 = 0, \quad a_3 = 0.20553149 \times 10^{-6},$$

$$a_4 = 0, \quad a_5 = 0.50691833 \times 10^{-13}, \quad a_6 = 0$$

Using (5.2.1) to compute coordinate transformation from a Gauss–Krüger projection (scale 1:1 million) to a Roussilhe projection as shown in Table 5.2.

Table 5.2 Coordinate conversion from the Gauss–Krüger to Roussilhe projection

	$l \rightarrow$	$0°$	$1°$	$2°$	$3°$
$B\downarrow$	Coord.				
$28°$	X	0	9.839774	19.679308	29.520363
	Y	−22.169239	−22.127906	−22.002936	−21.794417

Example 3 Coordinate transformation from the Roussilhe projection to the Gauss–Krüger projection.

Using (5.2.40) to compute the constant coefficients with $B_0 = 30°$:

$$x_0 = 0, \quad y_0 = 0, \quad X_0 = 0, \quad Y_0 = 332.017$$

$$a_1 = 1, \quad a_2 = 0, \quad a_3 = -0.20553149 \times 10^{-6},$$

$$a_4 = 0, \quad a_5 = 0.76037750 \times 10^{-13}, \quad a_6 = 0$$

Use (5.2.1) to compute a coordinate transformation from the Roussilhe projection to the Gauss–Krüger projection as shown in Table 5.3.

Table 5.3 Coordinate conversion from the Roussilhe to the Gauss–Krüger projection

	$l \rightarrow$	$0°$	$1°$	$2°$	$3°$
$B\downarrow$	Coord.				
$28°$	X	0	9.83623	19.67469	29.51664
	Y	309.84923	309.88990	310.01093	310.21241

5.2.5 Mercator projection and the Lagrange projection

Mercator projection → Lagrange projection

First we discuss their closed direct transformation equations.

The coordinate formulae for the Mercator projection can be expressed:

$$x_M = r_K l, \quad y_M = r_K q \tag{5.2.42}$$

From (2.8.6), we know that the coordinate formulae for the Lagrange projection take the form:

$$\left. \begin{array}{l} x_L = \dfrac{K \cos \delta \sin \alpha l}{1 + \cos \delta \cos \alpha l}, \quad y_L = \dfrac{K \sin \delta}{1 + \cos \delta \cos \alpha l} \\[3mm] \tan\left(\dfrac{\pi}{4} + \dfrac{\delta}{2}\right) = \beta U^\alpha \end{array} \right\} \tag{5.2.43}$$

where:

$$K = \frac{n_0 r_0}{\alpha}(1 + \sec \delta_0)$$

n_0 is the linear scale factor or the ratio of length at the centerpoint;

$$\alpha = \sqrt{1 + \frac{1 - n^2}{1 + n^2} \cos^2 B_0}$$

B_0 is the latitude of the centerpoint of the projection, $n = \dfrac{b}{a}$, a is the semimajor axis of the line of equal distortion along the direction of the meridian, and b is the semiminor axis along the direction of the vertex.

$$\beta = \tan\left(\frac{\pi}{4} + \frac{\delta_0}{2}\right) \cdot U_0^{-\alpha}, \quad \tan\frac{\delta_0}{2} = \frac{\sin B_0}{\alpha}$$

From (5.2.43), we have:

$$\delta = 2 \arctan(\beta U^\alpha) - \frac{\pi}{2} \tag{5.2.44}$$

and from (5.2.42), we have:

$$U = e_1^{\frac{y_M}{r_k}}, \quad l = \frac{x_M}{r_k} \tag{5.2.45}$$

Introducing (5.2.44) and (5.2.45) into (5.2.43), we have:

$$x_L = \frac{K \cos \delta \sin \dfrac{\alpha x_M}{r_k}}{1 + \cos \delta \cos \dfrac{\alpha x_M}{r_k}}, \quad y_L = \frac{K \sin \delta}{1 + \cos \delta \cos \dfrac{\alpha x_M}{r_k}} \tag{5.2.46}$$

where:

$$\delta = 2 \arctan\left(\beta e_1^{\frac{\alpha y_M}{r_k}}\right) - \frac{\pi}{2}$$

The above formulae are the direct coordinate relationships in transforming from the Mercator projection to the Lagrange projection.

Noting that,

$$\cos \delta = \sin(90° + \delta) = 2 \sin\frac{90° + \delta}{2} \cos\frac{90° + \delta}{2} = \frac{2 \tan\dfrac{90° + \delta}{2}}{1 + \tan^2\dfrac{90° + \delta}{2}}$$

$$\sin \delta = -\cos(90° + \delta) = 2 \sin^2\frac{90° + \delta}{2} - 1 = \frac{\tan^2\dfrac{90° + \delta}{2} - 1}{\tan^2\dfrac{90° + \delta}{2} + 1} \tag{5.2.47}$$

Introducing

$$\tan\left(\frac{90° + \delta}{2}\right) = \beta U^\alpha = \beta e_1^{\frac{\alpha y_M}{r_k}}$$

into the above equation, we have:

$$\cos\delta = \frac{2\beta e_1^{\frac{\alpha y_M}{r_k}}}{1+\beta^2 e_1^{\frac{\alpha y_M}{r_k}}}, \quad \sin\delta = \frac{\beta^2 e_1^{\frac{2\alpha y_M}{r_k}}-1}{1+\beta^2 e_1^{\frac{2\alpha y_M}{r_k}}} \tag{5.2.48}$$

Introducing the above into (5.2.46), the direct coordinate transformation relationship from the Mercator projection to the Lagrange projection is obtained:

$$\left.\begin{array}{l} y_L = \dfrac{K\left(\beta^2 e_1^{\frac{2\alpha y_M}{r_k}}-1\right)}{1+\beta^2 e_1^{\frac{2\alpha y_M}{r_k}}+2\beta e_1^{\frac{\alpha y_M}{r_k}}\cos\dfrac{\alpha x_M}{r_k}} \\[4ex] x_L = \dfrac{2K\beta e_1^{\frac{\alpha y_M}{r_k}}\sin\dfrac{\alpha x_M}{r_k}}{1+\beta^2 e_1^{\frac{2\alpha y_M}{r_k}}+2\beta e_1^{\frac{\alpha y_M}{r_k}}\cos\dfrac{\alpha x_M}{r_k}} \end{array}\right\} \tag{5.2.49}$$

Lagrange projection → Mercator projection

From (5.2.43), after a simple transformation, the meridian equation for the Lagrange projection takes the form:

$$y_L^2 + 2Kx_L\cot\alpha l + x_L^2 = K^2 \tag{5.2.50}$$

Hence:

$$\tan\alpha l = \frac{2Kx_L}{K^2 - x_L^2 - y_L^2} \tag{5.2.51}$$

Noting (5.2.42), we have:

$$x_M = \frac{r_k}{\alpha}\arctan\left(\frac{2Kx_L}{K^2 - x_L^2 - y_L^2}\right) \tag{5.2.52}$$

From (5.2.43), the equation for the parallels of the Lagrange projection is obtained:

$$y_L^2 - 2Ky_L\csc\delta + x_L^2 = -K^2 \tag{5.2.53}$$

Noting (5.2.47), we have:

$$\sin\delta = \frac{2Ky_L}{K^2 + x_L^2 + y_L^2} = \frac{\tan^2\dfrac{90°+\delta}{2}-1}{\tan^2\dfrac{90°+\delta}{2}+1} \tag{5.2.54}$$

Noting that

$$\tan^2\frac{90°+\delta}{2} = \beta^2 e_1^{\frac{2\alpha y_M}{r_k}}$$

the above equation is simply transformed to:

$$y_M = \frac{r_k}{2\alpha} \ln \frac{(y_L + K)^2 + x_L^2}{(K - y_L)^2 + x_L^2} - \frac{r_K}{\alpha} \ln \beta \tag{5.2.55}$$

Equations (5.2.52) and (5.2.55) are direct coordinate transformation relationships in converting from the Lagrange projection to the Mercator projection.

Applications

Example 1 The application of Lagrange projection to Chinese maps.
When the Lagrange projection is used for Chinese maps, its central point can be selected as $B_0 = 37°30'$, and the scale factor or proportion of length of centerpoint $m_0 = 0.982$, the ratio of the major axis and the minor axis for the distortion isogram ellipse is $n = \frac{b}{a} = 1.2273$. Its projection constants are $\alpha = 0.93426942$, $\beta = 2.45823591$, $K = 18\ 509\ 066.19$ m.

Chinese maps (Hainan islands as plate) use the Lagrange projection; its distortion of length is less than ±1.8%.

Now using the computer, the coordinates of the Lagrange projection for Chinese maps may be calculated as shown in Table 5.4.

Table 5.4 Examples of Lagrange rectangular coordinates used for Chinese maps

$B\rightarrow$		15°	20°	25°	30°
$l\downarrow$	Coord.				
10°	x	110.572917	105.6558176	100.5399455	95.21189135
	y	962.1252721	1 018.521165	1 074.034367	1 128.927182

Example 2 Coordinate transformation from the Mercator projection to the Lagrange projection.
From formula (5.2.49), the coordinate transformation is as shown in Table 5.5 ($r_k = 637.8245$ cm).

Table 5.5 Coordinate transformation from the Mercator to the Lagrange projection

$B\rightarrow$		15°	20°	25°	30°
$l\downarrow$	Coord.				
10°	x_M	111.321	111.321	111.321	111.321
	y_M	167.818	225.846	285.774	348.225
	x	110.5725226	105.6554947	100.5396368	95.21157965
	y	962.1254875	1 018.520755	1 074.034011	1 128.927055

Example 3 Coordinate transformation from the Lagrange projection to the Mercator projection.
From formulae (5.2.55) and (5.2.52), the coordinate transformation is as shown in Table 5.6.

Table 5.6 Coordinates transforming from the Lagrange to the Mercator projection

$B\rightarrow$		15°	20°	25°	30°
$\downarrow\downarrow$ 10°	Coord.				
	x	111.3213611	111.3213606	111.3213292	111.3213878
	y	167.817781	225.8464366	285.7743999	348.2251314

5.3   DIRECT TRANSFORMATION OF CONFORMAL PROJECTIONS

The general formula of direct transformation for conformal projections is $y + ix = f(q + il)$; for example, for the coordinate formulae for the conformal conic projection see formulae (2.4.1) and (2.4.5). And, as another example, for the coordinate formulae for the transverse Mercator projection (Gauss–Krüger) see formula (2.3.20). They all belong to variable coefficient formulae. Now we discuss constant coefficient formulae for conformal projections.

From equations (5.2.43), formulae for constant coefficients used in direct transformation for conformal projections, when the expansion point is located along the central meridian, can be written:

$$x = \sum_{k=1}^{6} a_k Q_k, \quad y = y_0 + \sum_{k=1}^{6} a_k P_k \tag{5.3.1}$$

where:

$$P_{k+1} = \Delta q p_k - l Q_k, \quad Q_{k+1} = l P_k + \Delta q Q_k, \quad p_1 = \Delta q,$$

$$Q_1 = l, \quad \Delta q = q - q_0, \quad a_k = \frac{1}{k!}\left(\frac{d^k y}{dq}\right)_0$$

5.3.1   Transverse Mercator (Gauss–Krüger) projection

Coordinate relations for the Gauss–Krüger projection at central meridian take the form:

$$\left.\begin{array}{l} \Delta y = a_1\Delta q + a_2\Delta q^2 + a_3\Delta q^3 + a_4\Delta q^4 + \cdots \\[2mm] a_k = \frac{1}{k!}\left(\frac{d^k s}{dq^k}\right)_0 \quad \Delta y = \Delta s, \quad y_0 = s_0 \end{array}\right\} \tag{5.3.2}$$

Noting (5.2.9) and (5.2.10), we obtain the formula for constant coefficients used in the direct transformation formula (5.3.1) (Yang 1986b):

$$\left.\begin{array}{l} a_1 = N_0\cos B_0, \quad a_2 = -\frac{1}{2}N_0\cos B_0\sin B_0 \\[4mm] a_3 = -\frac{1}{6}N_0\cos^3 B_0(1 - t_0^2 + \eta_0^2) \\[4mm] a_4 = \frac{1}{24}N_0\cos^4 B_0 \cdot t_0(5 - t_0^2 + 9\eta_0^2 + 4\eta_0^4) \\[4mm] a_5 = \frac{1}{120}N_0\cos^5 B_0(5 - 18t_0^2 + t_0^4 + 14\eta_0^2 - 58t_0^2\eta_0^2) \\[4mm] a_6 = -\frac{1}{720}N_0\cos^6 B_0 \cdot t_0(61 - 58t_0^2 + t_0^4 + 270\eta_0^2 - 330t_0^2\eta_0^2) \end{array}\right\} \tag{5.3.3}$$

Example

Constant coefficients for the direct coordinate transformation for a Gauss–Krüger projection.
   Given: Geographic coordinates $B = 0°$, $l = 3°30'$ and $B = 4°$, $l = 3°30'$, find coordinates
(x_G, y_G) for the Gauss–Krüger projection.
   Computing the result:

$$x_G = 389\ 868.9970\ \text{cm}, \quad x_G = -0.0015\ \text{cm},$$

$$x_G = 388\ 923.2554\ \text{m}, \quad y_G = 443\ 141.6456\ \text{m}$$

The values for searching for the transformation point are $x_G = 389\ 868.997$ m, $y_G = 0$, and
$x_G = 388\ 923.255$ m, $y_G = 443\ 141.647$ m.

5.3.2 The Roussilhe projection

For constant coefficients for the direct transformation of the Roussilhe projection, see
formula (2.2.27). For the application of a direct transformation to the Roussilhe projection, see the third part of Section 5.2.4.
   Observe: From the general formulae for conformal projections, we know that any
conformal projection can be expressed using Mercator coordinates. So when constant
coefficients α'_n from the Mercator projection to certain other conformal projections are
known, then the constant coefficients for direct transformation of this conformal projection are $a_n = a\,r_0^n$.

5.4   INVERSE TRANSFORMATION OF CONFORMAL PROJECTIONS

5.4.1   Gauss-Krüger projection

Variable coefficient formula

For the variable coefficient formula for the inverse transformation of the Gauss–Krüger
projection see formula (2.3.23), where the method to find the footpoint latitude B_f is
under Section 3.7; the other is a direct inverse-solution method, for which see formula
(3.7.20).

Constant coefficient formulae

From (5.1.47), the constant coefficient formulae for inverse transformation of the Gauss–
Krüger projection have the form, when the expansion point is located along the central
meridian, of:

$$q = q_0 + \sum_{k=1}^{6} a_k P_k, \quad l = \sum_{k=1}^{6} a_k Q_k \qquad (5.4.1)$$

where:

$$P_{k+1} = \Delta y \cdot P_k - x \cdot Q_k, \quad Q_{k+1} = x \cdot P_k + \Delta y \cdot Q_k$$
$$P_1 = \Delta y, \quad Q_1 = x$$

Coordinate relationships for the Gauss–Krüger projection along the central meridian take the form:

$$\left.\begin{aligned}
\Delta q &= a_1\Delta y + a_2\Delta y^2 + a_3\Delta y^3 + a_4\Delta y^4 + \dots \\
a_k &= \frac{1}{k!}\left(\frac{d^k q}{dS^k}\right)_0, \quad \Delta y = \Delta S
\end{aligned}\right\} \tag{5.4.2}$$

Using (5.2.4) and (5.2.5), formula (5.3.2) for constant coefficients of the inverse transformation for the Gauss–Krüger projection are obtained as:

$$\left.\begin{aligned}
a_1 &= \frac{1}{N_0\cos B_0}, \quad a_2 = \frac{1}{2N_0^2\cos B_0}t_0 \\
a_3 &= \frac{1}{6N_0^3\cos B_0}(1 + 2t_0^2 + \eta_0^2) \\
a_4 &= \frac{1}{24N_0^4\cos B_0}t_0(5 + 6t_0^2 + \eta_0^2 - 4\eta_0^4) \\
a_5 &= \frac{1}{120N_0^5\cos B_0}(5 + 28t_0^2 + 24t_0^4 + 6\eta_0^2 + 8t_0^2\eta_0^2) \\
a_6 &= \frac{1}{720N_0^6\cos B_0}t_0(61 + 180t_0^2 + 120t_0^4 + 46\eta_0^2 + 48t_0^2\eta_0^2)
\end{aligned}\right\} \tag{5.4.3}$$

When isometric latitude q is known, for the inverse method of solving for latitude B see Sections 3.6, 3.7. One of the inverse transformation formulae is Newton's iterative method, for which see formulae (3.6.4), (3.6.5) and (3.6.6). The other formula is the direct inverse-solution transformation, for which see (3.7.18).

Applications

Example 1　Variable coefficients for an inverse coordinate transformation on the Gauss–Krüger projection.

　　Given: Gauss–Krüger coordinates $x = 333\,307.8984$ m, $y = 442\,921.3436$ m, solve for geographic coordinates B, l.

　　Calculating the result: $B = 4°$, $l = 3°$.

Example 2　Constant coefficients for inverse coordinate transformation on the Gauss–Krüger projection.

　　Given: Gauss–Krüger coordinates $x = 389\,868.9970$ m, $y = 0$, solve for the geographic coordinates B, l.

　　Calculating the result: $B = 0°$, $l = 3°29'59''.99998$.

5.4.2　Mercator projection

For inverse transformation for the Mercator projection, see formulae (4.2.11)–(4.2.13) and (3.7.18).

Application: Given the latitude $B_k = 10°$ for the isometric parallel, map scale $\mu_0 =$ 1:500 000, Mercator coordinates $x = 142.535847$ cm, $y = 685.9388037$ cm, find geographic coordinates B, l.

Calculating the result: $B = 30°, l = 6°30'$.

5.4.3 Conic conformal projections

For inverse transformation for conformal conic projections, see formulae (3.7.18), (4.3.30) and (4.3.27).

Application: Given constant coefficients for the conformal conic projection $\alpha = 0.6328054027$, $C = 2\,498.23506$ cm, $\rho_s = 1\,634.588198$ cm, find the geographic coordinates B, l of the point $x = 35.12276847$ cm, $y = 44.79528551$ cm.

Calculating the result: $B = 38°, l = 2°$.

5.4.4 Roussilhe stereographic projection

Equation (5.4.1) is the formula for constant coefficients of the inverse coordinate transformation for the Roussilhe projection, and its constant coefficients are obtained from (2.2.28).

Given isometric latitude q, we can inversely solve for latitude B according to formula (3.7.18).

Application: Inverse transformation of coordinates $x = 9.8392$, $y = 22.1282$ of Table 5.1.

Calculating the result: $B = 27°59'59''.9865$, $l = 59'59''.996978$.

5.4.5 Lagrange projection

From (5.2.42), (5.2.52), and (5.2.55) the inverse coordinate transformation formulae for the Lagrange projection take the form:

$$l = \frac{1}{\alpha}\arctan\left(\frac{2Kx_L}{K^2 - x_L^2 - y_L^2}\right) \tag{5.4.4}$$

$$q = \frac{1}{2\alpha}\ln\frac{(K + y_L)^2 + x_L^2}{(K - y_L)^2 + x_L^2} - \frac{1}{\alpha}\ln\beta \tag{5.4.5}$$

Given the isometric latitude q, we can inversely solve for latitude B according to formula (3.7.18).

Application: When the Lagrange projection is used for Chinese maps, the coordinates $x = 355.5793534$, $y = 1\,255.373986$ (for projection constants see the Applications subsection in Section 5.2.5, and solution for geographic coordinate B, l).

Calculating the result: $B = 34°59'59''.9998$, $l = 40°$.

5.5 CONFORMAL OBLIQUE CYLINDRICAL AND OBLIQUE CONIC PROJECTIONS

This section uses conformal oblique cylindrical and oblique conic projections as examples to explain the analytical transformation method for oblique cylindrical and oblique conic projection and their development by double projection.

5.5.1 Conformal oblique cylindrical projection

Inverse transformation for a conformal oblique cylindrical projection

From equation (2.3.11), we obtain the coordinate formulae for a conformal oblique cylindrical projection:

$$x = 100\mu_0 Rn_0 (\pi - a) + x_0, \quad y = -100\mu_0 Rn_0 \ln\tan\frac{Z}{2} + y_0 \tag{5.5.1}$$

where μ_0 is map scale; n_0 is minimum proportion of length; R is earth's radius.
From the above, the inverse transformation formulae are obtained as:

$$Z = 2\arctan(e_1^{\frac{y_0-y}{100\mu_0 n_0 R}}), \quad a = \frac{x_0 - x}{100\mu_0 n_0 R} + \pi \tag{5.5.2}$$

Given polar spherical coordinates (Z, a), geographic coordinates (φ, λ) can be computed using formula (3.4.36).

Inverse transformation for a double conformal oblique cylindrical projection

Direct transformation for a double conformal oblique cylindrical projection; that is the conformal projection of the ellipsoid onto the sphere, then conformal projection of the sphere to the oblique cylindrical. From formula (2.10.1), its calculating formulae have the form:

$$\tan\left(45° + \frac{\varphi}{2}\right) = \beta U^a, \quad \Delta\lambda = \alpha l \tag{5.5.3}$$

where:

$$\tan\varphi_0 = \frac{R}{N_0}\tan B_0, \quad \alpha = \frac{\sin B_0}{\sin\varphi_0}, \quad \beta = \tan\left(45° + \frac{\varphi_0}{2}\right)\cdot U_0^{-\alpha}, \quad R = \sqrt{M_0 N_0}$$

$$\Delta\lambda = \lambda - \lambda_0, \quad l = L - L_0, \quad \lambda_0 = L_0$$

Using (5.5.3) and (5.5.1), we can conduct a direct coordinate transformation for a double conformal oblique cylindrical projection.

Using (5.5.2) and (3.4.36), (φ, λ) can be found, and then using (4.7.2) and (4.7.3), we can obtain an inverse coordinate transformation for a double conformal oblique cylindrical projection:

$$l = \frac{1}{\alpha}\Delta\lambda, \quad q = \frac{1}{\alpha}\left[\ln\tan\left(45° + \frac{\varphi}{2}\right) - \ln\beta\right] \tag{5.5.4}$$

Using (5.5.4) and (3.7.18), B, l can be obtained.

Applications

Example 1   Calculation of direct transformation for an oblique conformal cylindrical projection.

New polar geographic coordinates $\varphi_0 = 25°30'$, $\lambda_0 = 15°$ on the great circle of the sphere are given, the earth's radius $R = 6\ 367\ 518$ m , minimum proportion of length $n_0 = 0.995$, map scale $\mu_0 = 1{:}1\,000\,000$, suppose the origin of coordinates is $x_0 = -900$ cm, $y_0 = 500$ cm, we need to find the projection coordinates for an oblique conformal cylindrical projection, when the mapping range is $B_S = 0$, $B_N = 10°$, $L_W = 90°$, $L_E = 92°$, and the graticule interval for meridians and parallels is $\Delta L = \Delta B = 2°$.

Using formulae (3.4.32) and (5.5.1), the results of direct coordinate transformation for the oblique conformal cylindrical projection are expressed in Table 5.7.

Table 5.7 Oblique conformal cylindrical projection transformation

$\varphi \rightarrow$		$0°$	$4°$	$6°$	$10°$
$\lambda \downarrow$	Coord.				
$90°$	x	22.4426598	63.54562721	84.3461133	126.4850342
	y	650.7895174	670.6843862	680.430199	699.4592393

Example 2 Calculation of inverse transformation for an oblique conformal cylindrical projection.

Let us conduct an inverse coordinate transformation on the foregoing coordinates $x = 126.4850342$ cm, $y = 699.4592393$ cm.

Computing the result: $\varphi = 10°$, $\lambda = 90°$.

Example 3 Direct coordinate transformation for a double conformal oblique cylindrical projection.

If coordinates $B_0 = 30°$, $L_0 = 15°$ are given, we can solve for the direct conformal transformation from the ellipsoid to the sphere, then, according to the conditions of example 1, carry on a direct coordinate transformation for an oblique conformal cylindrical projection. The computed results are as follows:

$\varphi_0 = 29°56'15''.0154$, $\alpha = 1.001893418$, $\beta = 1.00105631$

$L = 90°$, $\lambda = 90°08'31''.2228$

$B = 0°$, $\varphi = 03'37''.7645407$, $x = 23.77396682$ cm, $y = 649.6459754$ cm

$B = 10°$, $\varphi = 10°00'45''.3639$, $x = 127.2723546$ cm, $y = 698.0323535$ cm

Example 4 Inverse transformation of the coordinates for the above example.

Computing follows the principle:

$x, y \xrightarrow{\text{inverse}} \varphi, \lambda \xrightarrow{\text{inverse}} B, L,$

with the resulting computations: $B = 0°$, $L = 90°$; $B = 10°$, $L = 90°$.

5.5.2 Conformal oblique conic projection

Inverse transformation for a conformal oblique conic projection

From equations (2.4.20) and (2.4.21), we have the coordinate formulae for a conformal oblique conic projection:

$$\left.\begin{array}{l} x = \rho \sin\delta + y_0, \quad y = \rho_s - \rho \cos\delta \\[2mm] \rho = K \cdot \tan^2\dfrac{Z}{2}, \quad \delta = \alpha(\pi - a) \end{array}\right\} \tag{5.5.5}$$

where:

$$K = \rho_0 \cot^\alpha \frac{Z_0}{2}, \quad \rho_0 = 100\mu_0 n_0 R \tan Z_0, \quad \alpha = \cos Z_0,$$

n_0 is the minimum proportion of length, ρ_s is constant, Z_0 is the almucantar of minimum proportion of length.

From the above, we have:

$$\left.
\begin{array}{l}
Z = 2\arctan\left(\dfrac{\rho}{K}\right)^{1/2}, \quad a = \pi - \dfrac{1}{\alpha}\delta \\[3mm]
\rho = [(\rho_s - y)^2 + (x - x_0)^2]^{1/2}, \quad \delta = \arctan\dfrac{x - x_0}{\rho_s - y}
\end{array}
\right\} \tag{5.5.6}$$

Equations (5.5.6) and (3.4.36) are the computing formulae for the inverse coordinate transformation for the conformal oblique conic projections.

Inverse transformation for a double conformal oblique conic projection

Using (5.5.3) and (5.5.5), direct coordinate computation for a double conformal oblique conic projection can be realized.

Using (5.5.6) and (3.4.36), φ, λ can be obtained. Then using (5.5.4) and (3.7.18), B, l can be found. So we can realize inverse transformation for a double conformal oblique conic projection.

Applications

Example 1   The calculation of direct coordinates for a oblique conformal conic projection.

Given polar geographic coordinates $\varphi_0 = 31°$, $\lambda_0 = 111°$ on the small circle of the sphere, an earth radius of $R = 6\ 372\ 311$ m, the almucantar for a minimum proportion of length. $z_0 = 11°$, its minimum proportion of length $n_0 = 1$, and a map scale $\mu_0 = 10^{-6}$. Suppose the coordinates of the origin are $x_0 = 0$, $y_0 = 0$, we need to find the projection coordinates for an oblique conformal conic projection when the mapping range is $B_S = 15°$, $B_N = 40°$, $L_W = 105°$, $L_E = 130°$ and the graticule of meridian and parallel spacings is $\Delta L = \Delta B = 25°$.

Using formulae (5.5.5) and (3.4.32), the results of a direct coordinate transformation for the oblique conformal conic projection are expressed as follows:

$$\alpha = 0.9816271835, \quad K = 1\ 232.247038$$

Table 5.8   Direct coordinate transformation for an oblique conformal conic projection

$\varphi\rightarrow$ $\lambda\downarrow$		15° Coord.	40°	$\varphi\rightarrow$ $\lambda\downarrow$		10° Coord.	40°
105°	x	−64.68932385	−56.97886214	130°	x	205.3767565	167.4393067
	y	−178.4154088	100.2859431		y	−167.9151895	109.5666472

Example 2   Calculation of inverse coordinate transformation for an oblique conformal conic projection.

First we compute the inverse coordinate transformation to the above coordinates:

$$x = -6\,468\,932\,385 \text{ cm}, \quad y = -178.4154088 \text{ cm}.$$

Computing the result: $\varphi = 15°, \quad \lambda = 105°$.

Example 3   Direct coordinate transformation for a double conformal, conformal conic projection.

Given $B_0 = 30°, L_0 = 15°$, we find the direct conformal transformation from the ellipsoid to sphere, then according to the conditions of example 1, use direct coordinate transformation for the oblique conformal conic projection. The computed results are as follows:

$$\varphi_0 = 29°56'15''.0154, \quad \alpha = 1.001893418, \quad \beta = 1.00105631,$$
$$L = 105°, \quad \lambda = 104°59'19''.102.$$
$$B = 15°, \quad \varphi = 14°59'24''.3926, \quad x = -64.8157212 \text{ cm}, \quad y = -178.5215859 \text{ cm}$$
$$B = 40°, \quad \varphi = 39°55'13''.1565, \quad x = -57.08632432 \text{ cm}, \quad y = 99.40090888 \text{ cm}.$$

Example 4   Inverse transformation for the coordinates of the above example.

Computing the result: $B = 15°00'00''.0001, \quad L = 105°; \quad B = 40°, \quad L = 105°$.

CHAPTER SIX

Numerical Transformation

6.1 GENERAL POLYNOMIAL APPROXIMATIONS

6.1.1 General considerations

Polynomial approximations can be used to establish the relation between two projections when their analytic expressions are difficult to obtain or the analytic expression of the original map is undetermined. That is, one can establish the relationship of the two projections from the theory and method of numerical transformation using some discrete points (also called common points or checkpoints). The method is called the numerical transformation of map projections.

The general equation of map projection can be written according to (3.1.1) as:

$$X = F_1(x, y), \quad Y = F_2(x, y) \tag{6.1.1}$$

The general formulation for equation (6.1.1) is to be given a surface or a set of discrete approximated values $F_{ij} = F(x_i, y_j)$ of function $F = F(x, y)$, then to construct a simpler function $f(x, y)$ to approximate the $F(x, y)$ or the discrete values F_{ij}. It is referred to as an interpolation approximation when $f(x_i, y_j) = F_{ij}$. Generally, since there always exist errors of measurement in F_{ij}, there is no need for the equation $f(x_i, y_j) = F_{ij}$ to be always satisfied. An approximate equality between $f(x_i, y_j)$ and F_{ij} would be sufficient. This method of surface approximation through some given points is called surface fitting.

In the numerical transformation of map projections, there are a number of questions necessary to research and develop. These questions include: the construction of approximating functions, the stability of polynomial approximation, and the precision of the transformation. Therefore, this study has a particular emphasis on the theory and method of numerical transformation of map projections, and certain progress has been made in recent years.

Next we introduce the general methods of polynomial and related transformation commonly used in the numerical transformation of map projection.

A polynomial of degree n is generally written as:

$$X = \sum_{i=0}^{n} \sum_{j=0}^{n} a_{ij} x^i y^j, \quad Y = \sum_{i=0}^{n} \sum_{j=0}^{n} b_{ij} x^i y^j \tag{6.1.2}$$

where $i + j \leq n$.

The (x, y) in the above equation indicates the coordinates of the old map projection; (X, Y) indicates those of the new one.

Equation (6.1.2) is the direct transformation equation from the old projection coordinates (x, y) to those of the new projection (X, Y).

For a bivariate cubic polynomial such as (3.1.3), the coefficients a_{ij}, b_{ij} can be obtained from 10 known common points and by solving two systems of linear algebraic equations with 10 unknown numbers.

The computational formula for N, number of common points of the bivariate polynomial of degree n, takes the form:

$$N = \frac{1}{2}(2 + n)(1 + n)$$

(6.1.3)

For example the number of common points of a bivariate polynomial of degree 4 is $N = 15$.

If the number of common points $m > N$, then the formula for the condition of least-squares in (3.1.4) can be composed, and the systems of linear equations for computing coefficients a_{ij}, b_{ij} can be respectively developed according to the least-squares method. This kind of transformation method belongs to the direct transformation method that determines polynomials by least-squares approximation.

Given a group of rectangular interpolation points on $[a, b] \times [c, d]$ as:

$$(x_s, y_t), \quad s = 1, 2, \ldots, N; \quad t = 1, 2, \ldots, M$$

according to the desired interpolation conditions, a_{ij} satisfies

$$\sum_{i=1}^{N} \sum_{j=1}^{M} a_{ij} \varphi_i(x_s) \psi_j(y_t) = F(x_s, y_t)$$

(6.1.4)

If the interpolation gives a unique solution, then the equation

$$f = \sum_{i=1}^{N} \sum_{j=1}^{M} a_{ij} \varphi_i(x) \psi_j(y)$$

(6.1.5)

is called the product interpolation surface for $F = F(x, y)$. According to the different numbers of the given common points, we can select the product interpolation and the product least-squares method.

The computational formula for N, which is the number of common points of the polynomial of double degree in n, takes the form:

$$N = (n + 1)^2$$

For example, the number of common points of the polynomial of double degree 2 is $N = 9$.

Suppose the general equation for inverse transformation of map projection is as follows:

$$\varphi = F_1(x, y), \quad \lambda = F_2(x, y)$$

(6.1.6)

Then the approximate polynomial discussed above can be suitable for (6.1.6) as well. In this case, the approximate polynomials are called the formulae of the inverse transformation.

6.1.2 Conventional method for determining the coefficients of approximate polynomials

Direct and inverse transformation methods for bivariate cubic polynomials

In the following we take bivariate cubic polynomials as an example to illustrate the direct and inverse numerical transformation methods of determining the coefficients in the approximate polynomials.

From (6.1.2) we know:

$$f = \sum_{i=0}^{3} \sum_{j=0}^{3} a_{ij} x^i y^j, \quad i + j \le 3 \tag{6.1.7}$$

where f indicates (X, Y) or (ϕ, λ) respectively.

Formula (6.1.7) is just the expression of the relation of a direct and an inverse transformation of the bivariate cubic polynomial.

From (6.1.7) we can give the following expressions of the direct and inverse transformation:

$$\left. \begin{aligned} X &= a_{00} + a_{10}x + a_{01}y + a_{20}x^2 + a_{11}xy + a_{02}y^2 \\ &\quad + a_{30}x^3 + a_{21}x^2y + a_{12}xy^2 + a_{03}y^3 \\ Y &= b_{00} + b_{10}x + b_{01}y + b_{20}x^2 + b_{11}xy + b_{02}y^2 \\ &\quad + b_{30}x^3 + b_{21}x^2y + b_{12}xy^2 + b_{03}y^3 \end{aligned} \right\} \tag{6.1.8}$$

and:

$$\left. \begin{aligned} \varphi &= a_{00} + a_{10}x + a_{01}y + a_{20}x^2 + a_{11}xy + a_{02}y^2 \\ &\quad + a_{30}x^3 + a_{21}x^2y + a_{12}xy^2 + a_{03}y^3 \\ \lambda &= b_{00} + b_{10}x + b_{01}y + b_{20}x^2 + b_{11}xy + b_{02}y^2 \\ &\quad + b_{30}x^3 + b_{21}x^2y + b_{12}xy^2 + b_{03}y^3 \end{aligned} \right\} \tag{6.1.9}$$

It can be seen from (6.1.8) and (6.1.9) that the key to a transformation by polynomials between two types of projections is to determine the values of the coefficients a_{ij}, b_{ij}. In order to determine the coefficients a_{ij}, b_{ij} of a bivariate cubic approximating polynomial, the coordinates (x_s, y_s) and (X_s, Y_s) or (x_s, y_s) and (φ_s, λ_s) $(s = 1, 2, \ldots, 10)$ of 10 discrete points should be selected, fixed, and a system of linear equations of order 10 be formed. Then the values of the coefficients a_{ij}, b_{ij} can be obtained by the elimination method for principal elements or other algorithms. Thus the direct and inverse numerical transformation of two kinds of projections will be realized.

For calculating convenience, we introduce the following symbols:

$$A_{i0} = 1, \quad A_{i1} = x_i, \quad A_{i2} = y_i, \quad A_{i3} = x_i^2, \quad A_{i4} = x_i y_i,$$
$$A_{i5} = y_i^2, \quad A_{i6} = x_i^3, \quad A_{i7} = x_i^2 y_i, \quad A_{i8} = x_i y_i^2, \quad A_{i9} = y_i^3.$$
$$c_i = X_i, \quad d_i = Y_i, \quad e_i = \varphi_i, \quad g_i = \lambda_i$$

Using these symbols, (6.1.8) and (6.1.9) can be rewritten as:

$$\sum_{j=0}^{9} A_{ij} a_j = c_i, \quad \sum_{j=0}^{9} A_{ij} b_j = d_i \quad (i = 0, 1, \ldots, 9) \tag{6.1.10}$$

$$\sum_{j=0}^{9} A_{ij} a_j = e_i, \quad \sum_{j=0}^{9} A_{ij} b_j = e_i \quad (i = 0, 1, \ldots, 9) \tag{6.1.11}$$

Next, we use a matrix to illustrate them further. Denote:

$$
A = \begin{pmatrix}
A_{00} & A_{01} & A_{02} & \cdots & A_{09} \\
A_{10} & A_{11} & A_{12} & \cdots & A_{19} \\
\cdots & \cdots & \cdots & \cdots & \cdots \\
A_{90} & A_{91} & A_{92} & \cdots & A_{99}
\end{pmatrix}
$$

$$
\begin{aligned}
& a = (a_0, a_1, a_2, \ldots, a_9)^T, \qquad b = (b_0, b_1, b_2, \ldots, b_9)^T, \\
& c = (X_0, X_1, X_2, \ldots, X_9)^T, \quad d = (Y_0, Y_1, Y_2, \ldots, Y_9)^T, \\
& e = (\varphi_0, \varphi_1, \varphi_2, \ldots, \varphi_9)^T, \qquad g = (\lambda_0, \lambda_1, \lambda_2, \ldots, \lambda_9)^T,
\end{aligned}
$$

(6.1.12)

Hence (6.1.10) and (6.1.11) can be rewritten as the following forms respectively:

$$
Aa = c, \quad Ab = d \tag{6.1.13}
$$

$$
Aa = e, \quad Ab = g \tag{6.1.14}
$$

The flow diagram for programming of bivariate cubic polynomials used for direct numerical transformation is shown in Figure 6.1.

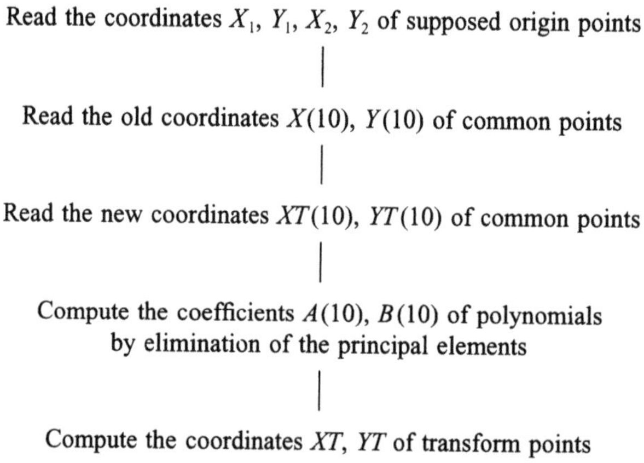

Read the coordinates X_1, Y_1, X_2, Y_2 of supposed origin points

|

Read the old coordinates $X(10)$, $Y(10)$ of common points

|

Read the new coordinates $XT(10)$, $YT(10)$ of common points

|

Compute the coefficients $A(10)$, $B(10)$ of polynomials
by elimination of the principal elements

|

Compute the coordinates XT, YT of transform points

Figure 6.1   Flow diagram for programming of bivariate cubic polynomials

Direct and inverse transformation method of biquadratic polynomial

If the selected basis functions $\varphi_i(x)$ take the form $\{1, x, x^2\}$ and $\psi_j(y)$ take the form $\{1, y, y^2\}$ of the product interpolation function (6.1.5), then we can obtain the expression for the relationship of a biquadratic direct transformation according to (6.1.5) as:

$$
\left.
\begin{aligned}
X = {}& a_{00} + a_{10}x + a_{01}y + a_{20}x^2 + a_{11}xy + a_{02}y^2 \\
& + a_{21}x^2y + a_{12}xy^2 + a_{22}x^2y^2 \\
Y = {}& b_{00} + b_{10}x + b_{01}y + b_{20}x^2 + b_{11}xy + b_{02}y^2 \\
& + b_{21}x^2y + b_{12}xy^2 + b_{22}x^2y^2
\end{aligned}
\right\}
\tag{6.1.15}
$$

The expression for the relationship of a biquadratic inverse transformation is as follows:

$$
\left.
\begin{aligned}
\varphi = {}& a_{00} + a_{10}x + a_{01}y + a_{20}x^2 + a_{11}xy + a_{02}y^2 \\
& + a_{21}x^2y + a_{12}xy^2 + a_{22}x^2y^2 \\
\lambda = {}& b_{00} + b_{10}x + b_{01}y + b_{20}x^2 + b_{11}xy + b_{02}y^2 \\
& + b_{21}x^2y + b_{12}xy^2 + b_{22}x^2y^2
\end{aligned}
\right\}
\tag{6.1.16}
$$

For calculating convenience, we introduce the following symbols:

$$A_{i0} = 1, \quad A_{i1} = x_i, \quad A_{i2} = y_i, \quad A_{i3} = x_i^2, \quad A_{i4} = x_i y_i,$$
$$A_{i5} = y_i^2, \quad A_{i6} = x_i^2 y_i, \quad A_{i7} = x_i y_i^2, \quad A_{i8} = x_i^2 y_i^2$$
$$c_i = X_i, \quad d_i = Y_i, \quad e_i = \varphi_i, \quad g_i = \lambda_i$$

Hence (6.1.15) and (6.1.16) can be rewritten as:

$$\sum_{j=0}^{8} A_{ij} a_j = c_i, \quad \sum_{j=0}^{8} A_{ij} b_j = d_i \quad (i = 0, 1, \ldots, 8) \tag{6.1.17}$$

$$\sum_{j=0}^{8} A_{ij} a_j = e_i, \quad \sum_{j=0}^{8} A_{ij} b_j = g_i \quad (i = 0, 1, \ldots, 8) \tag{6.1.18}$$

Formulae (6.1.17) and (6.1.18) provide the system of linear equations of order 9. For the same reason as described above, they can be written into a matrix similar to (6.1.13) and (6.1.14).

6.1.3 Least-squares method for determining the coefficients of an approximate polynomial

Direct and inverse transformation using bivariate cubic polynomials

We have discussed the conventional method of determining direct and inverse transformations using bivariate cubic polynomials in the last section. Its merit is keeping strict correspondence between 10 discrete points. Its demerit is probably producing substantial errors among the rest of the points. In order to reduce the errors of corresponding points in the transformation region and raise the stability of transformation polynomials, more than 10 points should be used to determine the bivariate cubic polynomial, that is to determine the transformation polynomials using the least-squares method.

Given the expansion of the bivariate cubic polynomial as (6.1.8), its expression using the least-squares condition is:

$$\varepsilon = \sum_{i=1}^{n} [X_i - X_i']^2 = \min, \quad \varepsilon' = \sum_{i=1}^{n} [Y_i - Y_i']^2 = \min \tag{6.1.19}$$

where $n > 10$, X_i', Y_i' are true values, X_i, Y_i are transformation values.

According to the extremum principle, we have:

$$\frac{\partial \varepsilon}{\partial a_{ij}} = 0, \quad \frac{\partial \varepsilon'}{\partial b_{ij}} = 0 \tag{6.1.20}$$

Considering (6.1.10), by substituting (6.1.7) into the above equation, and rearranging we have:

$$Aa = x, \quad Ab = y \tag{6.1.21}$$

where:

$$A = \begin{pmatrix} \sum A_{i0} A_{i0} & \sum A_{i0} A_{i1} & \sum A_{i0} A_{i2} & \cdots & \sum A_{i0} A_{i9} \\ \sum A_{i1} A_{i0} & \sum A_{i1} A_{i1} & \sum A_{i1} A_{i2} & \cdots & \sum A_{i1} A_{i9} \\ \cdots & \cdots & \cdots & \cdots & \cdots \\ \sum A_{i9} A_{i0} & \sum A_{i9} A_{i1} & \sum A_{i9} A_{i2} & \cdots & \sum A_{i9} A_{i9} \end{pmatrix}$$

$$a = (a_0, a_1, a_2, \ldots, a_9)^T, \quad b = (b_0, b_1, b_2, \ldots, b_9)^T$$

$$x = \left(\sum A_{i0} X_i', \sum A_{i1} X_i', \sum A_{i2} X_i', \ldots, \sum A_{i9} X_i' \right)^T$$

$$y = \left(\sum A_{i0} Y_i', \sum A_{i1} Y_i', \sum A_{i2} Y_i', \ldots, \sum A_{i9} Y_i' \right)^T$$

$$\sum \text{ denotes } \sum_{i=1}^{n}$$

and:

$$A_{i0} = 1, \quad A_{i1} = x_1, \quad A_{i2} = y_i, \quad A_{i3} = x_i^2, \quad A_{i4} = x_i y_i,$$
$$A_{i5} = y_i^2, \quad A_{i6} = x_i^3, \quad A_{i7} = x_i^2 y_i, \quad A_{i8} = x_i y_i^2, \quad A_{i9} = y_i^3$$

Similarly, the matrix form of the inverse transformation of (6.1.9) according to least-squares is:

$$Aa = \varphi, \quad Ab = \lambda \tag{6.1.22}$$

where:

$$\varphi = \left(\sum A_{i0} \varphi_i', \sum A_{i1} \varphi_i', \sum A_{i2} \varphi_i', \ldots, \sum A_{i9} \varphi_i' \right)^T$$

$$\lambda = \left(\sum A_{i0} \lambda_i', \sum A_{i1} \lambda_i', \sum A_{i2} \lambda_i', \ldots, \sum A_{i9} \lambda_i' \right)^T$$

Direct and inverse transformation method for biquadratic approximate polynomials

When the discrete points are more than 9, the matrix form for direct and inverse numerical transformation of biquadratic polynomials (6.1.15) and (6.1.16) can be obtained using the least-squares principle. That is,

$$Aa = x, \quad Ab = y \tag{6.1.23}$$

and:

$$Aa = \varphi, \quad Ab = \lambda \tag{6.1.24}$$

where:

$$A = \begin{pmatrix} \sum A_{i0} A_{i0} & \sum A_{i0} A_{i1} & \cdots & \sum A_{i0} A_{i8} \\ \sum A_{i1} A_{i0} & \sum A_{i1} A_{i1} & \cdots & \sum A_{i1} A_{i8} \\ \cdots & \cdots & \cdots & \cdots \\ \sum A_{i8} A_{i0} & \sum A_{i8} A_{i1} & \cdots & \sum A_{i8} A_{i8} \end{pmatrix}$$

$$a = (a_0, a_1, a_2, \ldots, a_8)^T, \quad b = (b_0, b_1, b_2, \ldots, b_8)^T$$

$$x = \left(\sum A_{i0} X_i', \sum A_{i1} X_i', \ldots, \sum A_{i8} X_i' \right)^T$$

$$y = \left(\sum A_{i0} Y_i', \sum A_{i1} Y_i', \ldots, \sum A_{i8} Y_i' \right)^T$$

$$\varphi = \left(\sum A_{i0} \varphi_i', \sum A_{i1} \varphi_i', \ldots, \sum A_{i8} \varphi_i' \right)^T$$

$$\lambda = \left(\sum A_{i0} \lambda_i', \sum A_{i1} \lambda_i', \ldots, \sum A_{i8} \lambda_i' \right)^T$$

$\sum$ denotes $\displaystyle\sum_{i=1}^{n}$

and:

$$A_{i0} = 1, \qquad A_{i1} = x_i, \qquad A_{i2} = y_i, \qquad A_{i3} = x_i^2,$$
$$A_{i4} = x_i y_i, \quad A_{i5} = y_i^2, \qquad A_{i6} = x_i^2 y_i,$$
$$A_{i7} = x_i y_i^2, \quad A_{i8} = x_i^2 y_i^2$$

The system matrix A, formed by the least-squares method, is generally a positive definite and symmetric matrix. It is also suitable for using the pivot elimination method to compute the coefficients a_i, b_i.

Examples of Section 6.1 will be presented in Appendix 3, Section A3.1.

6.2   REFINING PARALLELS AND MERIDIANS

6.2.1   Method of cubic splines

We have discussed the general method of numerical transformation by approximate polynomials. In order to guarantee the precision of transformation, the method of transformation by lower-order polynomial in divided regions is often used. If there are not enough discrete points, the coordinates of the discrete points need to be refined according to the coordinates of a few known points. In this section we will introduce the method of cubic splines. An example application can be found in Tobler (1977).

Representing cubic splines with function values and the first derivative

A 'spline' is a simple tool for a draftsperson to use to describe a smooth curve. Smooth interpolation is the mathematical analog of this kind of charting. This kind of interpolation is called spline interpolation.

In spline interpolation, cubic spline functions are most often used. The reason is that they have the merit of simple computation and guaranteed smooth results. Simultaneously they have better convergence and extremum properties, including minimum modulus and best approximation, etc., so cubic spline functions are widely used in practical applications.

Mathematically, cubic spline interpolation uses a group of nodes Δ: $a = x_0 < x_1 < \ldots < x_N = b$ and a group of coordinate values Y: $y_0, y_1, \ldots, y_N$ within an interval $[a, b]$. If the function $S_3(x)$ has the following properties (Zhu 1997):

1. $S_3(x)$ is a polynomial of degree less than 3 on each subinterval $[x_{i-1}, x_i]$ ($i = 0, 1, \ldots, N$);

2. $S_3(x) = y_i$ ($i = 0, 1, \ldots, N$);

3. $S_3(x) \in c^2[a, b]$; that is, $S_3(x)$ is continuously differentiable n times in the interval $[a, b]$.

then $S_3(x)$ is called a cubic spline function for which interpolations at nodes Δ are Y.

Next we represent the cubic spline function by its function values and the first derivative. Assume that the values of $S_3(x)$ and its first derivatives at the nodes are $S_3(x_i) = y_i$, $S_3'(x_i) = m_i$ ($i = 0, 1, \ldots, N$). Then $S_3(x)$ is a cubic polynomial in each subinterval $[x_{i-1}, x_i]$. According to the interpolation formula of Hermite we have:

$$S_2(x)_i = m_{i-1}\frac{(x_i - x)^2(x - x_{i-1})}{h_i^2} - m_i\frac{(x - x_{i-1})^2(x_i - x)}{h_i^2}$$

$$+ y_{i-1}\frac{(x_i - x)^2[2(x - x_{i-1}) + h_i]}{h_i^3}$$

$$+ y_i\frac{(x - x_{i-1})^2[2(x_i - x) + h_i]}{h_i^3}$$

(6.2.1)

where:

$$h_i = x_i - x_{i-1}, \quad x_{i-1} \le x \le x_i \quad (i = 1, 2, \ldots, N)$$

For the convenience of computer use, the above formula can be rewritten as follows:

$$S_3(x) = a_{i-1}[y_{i-1} + (x_{i-1} - x)(2b_{i-1}y_{i-1} - m_{i-1})]$$
$$+ a_i[y_i + (x_i - x)(2b_iy_i - m_i)]$$

(6.2.2)

where:

$$a_{i-1} = \left(\frac{x - x_i}{x_{i-1} - x_i}\right)^2, \quad b_{i-1} = \frac{1}{x_{i-1} - x_i}$$

$$a_i = \left(\frac{x - x_{i-1}}{x_{i-1} - x_i}\right)^2, \quad b_i = \frac{1}{x_i - x_{i-1}}$$

From (6.2.1) we know that the first and second derivatives of $S_3(x)$ are:

$$S_3'(x) = m_{i-1}\frac{(x_i - x)(2x_{i-1} + x_i - 3x)}{h_i^2} - m_i\frac{(x - x_{i-1})(2x_i + x_{i-1} - 3x)}{h_i^2}$$

$$+ 6\frac{y_i - y_{i-1}}{h_i^3}(x_i - x)(x - x_{i-1})$$

(6.2.3)

$$S_3''(x) = -2m_{i-1}\frac{x_{i-1} + 2x_i - 3x}{h_i^2} - 2m_i\frac{2x_{i-1} + x_i - 3x}{h_i^2}$$

$$+ 6\frac{y_i - y_{i-1}}{h_i^3}(x_i + x_{i-1} - 2x)$$

(6.2.4)

and the left and right limits at the respective points x_i are:

$$S_3''(x_i^-) = -2m_{i-1}\frac{x_{i-1} + 2x_i - 3x_i}{h_i^2} - 2m_i\frac{2x_{i-1} + x_i - 3x_i}{h_i^2}$$

$$+ 6\frac{y_i - y_{i-1}}{h_i^3}(x_i + x_{i-1} - 2x_i) = \frac{2m_i - 1}{h_i} + \frac{4m_i}{h_i} - 6\frac{y_i - y_{i-1}}{h_i^2}$$

$$S_3''(x_i^+) = -2m_i\frac{x_i + 2x_{i+1} - 3x_i}{h_{i+1}^2} - 2m_{i+1}\frac{2x_i + x_{i+1} - 3x_i}{h_{i+1}^2}$$

$$+ 6\frac{y_{i+1} - y_i}{h_{i+1}^3}(x_{i+1} + x_i - 2x_i)$$

$$= -\frac{4m_i}{h_{i+1}} - \frac{2m_{i+1}}{h_{i+1}} + 6\frac{y_{i+1} - y_i}{h_{i+1}^2}$$

Since $S_3''(x)$ is continuous at point x_i $(i = 1, 2, \ldots, N - 1)$, that is

$$S_3''(x_i^-) = S_3''(x_i^+)$$

then we obtain:

$$\frac{1}{h_i}m_{i-1} + 2\left(\frac{1}{h_i} + \frac{1}{h_{i+1}}\right)m_i + \frac{1}{h_{i+1}}m_{i+1} = 3\frac{y_i - y_{i-1}}{h_i^2} + 3\frac{y_{i+1} - y_i}{h_{i+1}^2} \tag{6.2.5}$$

Letting:

$$\alpha_i = \frac{h_i}{h_i + h_{i+1}}, \quad C_i = 3\left[(1 - \alpha_i)\frac{y_i - y_{i-1}}{h_i} + \alpha_i\frac{y_{i+1} - y_i}{h_{i+1}}\right]$$

then (6.2.5) can be rewritten in the following form:

$$(1 - \alpha_i)m_{i-1} + 2m_i + \alpha_i m_{i+1} = C_i \tag{6.2.6}$$

If the boundary conditions are given as:

$$2m_0 = 2v_0' = C_0, \quad 2m_N = 2y_N' = C_N \tag{6.2.7}$$

then we get the following system of equations:

$$Am = C \tag{6.2.8}$$

where:

$$A = \begin{pmatrix} 2 & 0 & & & & 0 \\ 1 - \alpha_1 & 2 & \alpha_1 & & & \\ & 1 - \alpha_2 & 2 & \alpha_2 & & \\ & & \ddots & \ddots & \ddots & \\ & & & 1 - \alpha_{N-1} & 2 & \alpha_{N+1} \\ 0 & & & & 0 & 2 \end{pmatrix}$$

$$m = (m_0, m_1, m_2, \ldots, m_{N-1}, m_N)^T, \quad C = (C_0, C_1, C_2, \ldots, C_{N-1}, C_N)^T$$

$\{m_i\}$ can be solved with the above system of equations by a 'speedup method'. After substituting them back into the piecewise representation (6.2.1), the spline function will be obtained.

The procedure for solving a cubic spline function can be summed up as follows:

1. Determining the boundary conditions.
2. Solving the system of equations (6.2.8) by the 'speedup method' to obtain the first derivatives $\{m_i\}$ at the nodes.
3. Substituting $\{m_i\}$ back into the piecewise representation (6.2.1) of the spline function.

Then the function values of the piecewise polynomial at any point are obtained.

As for the method of determining boundary conditions, it is just the method of determining m_0, m_N.

The precision of the given initial values will directly influence the precision of the region of interpolation near the boundary. If the required precision is not very high, a cubic Newton interpolation formula or a cubic Lagrange interpolation formula can be used; the result is the same. The following are the numerical derivative formulae at equidistant nodes.

For the Newton formula:

$$f'(x_0) = \frac{\left(\Delta y_0 - \frac{1}{2}\Delta^2 y_0 + \frac{1}{3}\Delta^3 y_0 \right)}{h}$$

$$f'(x_N) = \frac{\left(\nabla y_n - \frac{1}{2}\nabla^2 y_n + \frac{1}{3}\nabla^3 y_n \right)}{h}$$

(6.2.9)

where $h = x_i - x_{i-1}$ $(i = 1, 2, \ldots, N)$. Δ indicates a forward or downward difference, ∇ indicates a reverse or an upward difference. The Lagrange formula is as follows:

$$L'(x_0) = \frac{\left(-\frac{11}{6}y_0 + 3y_1 - \frac{3}{2}y_2 + \frac{1}{3}y_3 \right)}{h}$$

$$L'(x_N) = \frac{\left(-\frac{1}{3}y_{N-3} + \frac{3}{2}y_{N-2} - 3y_{N-1} + \frac{11}{6}y_N \right)}{h}$$

(6.2.10)

If the required precision is very high, the derivative formula for the 5th term of the interpolation needs to be obtained. The following are the numerical derivative formulae obtained for the 5th term Lagrange interpolation formula:

$$L'(x_0) = \frac{\left(-\frac{137}{60}y_0 + 5y_1 - 5y_2 + \frac{10}{3}y_3 - \frac{5}{4}y_4 + \frac{1}{5}y_5 \right)}{h}$$

$$L'(x_N) = \frac{\left(-\frac{1}{5}y_{N-5} + \frac{5}{4}y_{N-4} - \frac{10}{3}y_{N-3} + 5y_{N-2} - 5y_{N-1} + \frac{137}{60}y_N \right)}{h}$$

(6.2.11)

Speedup method

Given a system of equations of the following form:

$$\begin{aligned}
b_0 x_0 &+ c_0 x_1 & &= d_0 \\
a_1 x_0 &+ b_1 x_1 &+ c_1 x_2 &= d_1 \\
\cdots & \cdots & \cdots & \cdots \\
a_K x_{K-1} &+ b_K x_K &+ c_K x_{K+1} &= d_K \\
\cdots & \cdots & \cdots & \cdots \\
a_{N-1} x_{N-2} &+ b_{N-1} x_{N-1} &+ c_{N-1} x_N &= d_{N-1} \\
& a_N x_{N-1} &+ b_N x_N &= d_N
\end{aligned}$$

(6.2.12)

In matrix form it is as follows:

$$Ax = d$$

(6.2.13)

The coefficient matrix for this system of equations is:

$$A = \begin{pmatrix} b_0 & c_0 & & & & 0 \\ a_1 & b_1 & c_1 & & & \\ & \ddots & \ddots & \ddots & & \\ & & a_{N-1} & b_{N-1} & c_{N-1} \\ 0 & & & a_N & b_N \end{pmatrix}$$ (6.2.14)

The system of equations (6.2.13) is called the tridiagonal matrix. The solution of this kind of system is divided into two procedures of elimination and back substitution.

From the first expression of (6.2.12) we have:

$$x_0 + r_0 x_1 = p_0$$

where:

$$r_0 = \frac{c_0}{b_0}, \quad p_0 = \frac{d_0}{b_0}$$

Using the above equations, the second expression of the system of equations can be transformed into:

$$x_1 + r_1 x_2 = p_1$$

After the Nth step we get:

$$x_{N-1} + r_{N-1} x_N = p_{N-1}$$

where:

$$r_{N-1} = \frac{c_{N-1}}{b_{N-1} - r_{N-2} a_{N-1}}, \quad p_{N-1} = \frac{d_{N-1} - p_{N-2} a_{N-1}}{b_{N-1} - r_{N-2} a_{N-1}}$$

From the simultaneous solution of this and the final equation we get:

$$x_N = p_N = \frac{d_N - p_{N-1} a_N}{b_N - r_{N-1} a_N}$$

Therefore, through the elimination procedure, (6.2.12) can be presented in the following form:

$$\left. \begin{aligned} x_0 &+ r_0 x_1 &= p_0 \\ \cdots & \cdots & \cdots \\ x_K &+ r_K x_{K+1} &= p_K \\ \cdots & \cdots & \cdots \\ x_{N-1} &+ r_{N-1} x_N &= p_{N-1} \\ & x_N &= p_N \end{aligned} \right\}$$ (6.2.15)

where:

$$r_K = \frac{c_K}{b_K - r_{K-1} a_K}, \quad p_K = \frac{d_K - p_{K-1} a_K}{b_K - r_{K-1} a_K}, \quad r_0 = \frac{c_0}{b_0}, \quad p_0 = \frac{d_0}{b_0}$$

The system of equations (6.2.15) is called bidiagonal. $x_N, x_{N-1}, \ldots, x_0$, can be obtained by back substituting stepwise from bottom to top in (6.2.15). The formula for computation is:

$$\left.\begin{aligned} x_N &= p_N \\ x_K &= p_K - r_K x_{K+1} \ (K = N - 1, N - 2, \ldots, 0) \end{aligned}\right\} \tag{6.2.16}$$

The last algorithm is called the 'speedup method'. This elimination procedure and back-substitution procedure are called the pursuit procedure and the recall procedure, respectively.

General method for numerical transformation in refining parallels and meridians on a map by cubic spline interpolation

This method can be summed up as follows:

- Step 1. Partitioning the transformation region and measuring or choosing coordinates of points of intersection of parallels and meridians in the map region to be transformed.
- Step 2. Refining the coordinates of the points of intersection of the parallels and meridians in the transformation region.

   Knowing a few coordinates of nodes, the coordinates can be refined along parallels and meridians respectively by cubic spline interpolation. The corresponding new coordinates of the points of intersection of parallels and meridians in the new projection can be obtained by analytic computation.

- Step 3. Making numerical transformation between the two projections by approximate polynomials. A detailed introduction of this method is referenced in Sections 6.1.2 and 6.1.3.

6.2.2 Lagrange interpolation

Lagrange interpolation – general

Given projection coordinates (x_i, y_i) of the points of intersection of the geographical graticule with a constant difference in longitude and latitude. They are represented respectively by $x = f_1(B, l_i)$, $y = f_2(B, l_i)$ for meridian $l = l_i$; that is, (x, y) are functions of latitude B; or by $x = f_3(B_i, l)$, $y = f_4(B_i, l)$ for parallel $B = B_i$; that is, (x, y) is a function of longitude difference l. We want to refine the coordinates of the points of intersection of parallels and meridians with a constant difference in longitude and latitude by (x_i, y_i).

Mathematically, the above problem consists of the known node values x_i ($i = 0, 1, 2, \ldots, n$) of the function $f(x)$ and the corresponding function values of y_i ($i = 0, 1, 2, \ldots, n$). The function value $f(x)$ corresponding to x, which is given and is not a node point, must be computed using a unary Lagrange interpolation formula of $n + 1$ points.

The Lagrange interpolation formula is obtained from (3.5.9) as follows:

$$f(x) = \sum_{i=0}^{n} \prod_{\substack{i=0 \\ i \neq j}}^{n} \left(\frac{x - x_i}{x_j - x_i} \right) y_i \tag{6.2.17}$$

Unary four-point Lagrange interpolation

A higher-degree interpolation function $f(x)$ is not guaranteed to approximate a primitive function $F(x)$ very well between two interpolation nodes, and sometimes the difference is quite large and the Runge phenomenon appears. From the theory of algebraic interpolation we know that high-degree interpolation, for example a degree more than 8, is rarely used.

For the precision of the coordinates of parallels and meridians refined by a Lagrange interpolation discussed above, a four-point interpolation may meet the requirements for refining coordinates. Therefore we discuss the unary four-point Lagrange interpolation. Mathematically, it involves the node values x_i ($i = 0, 1, 2, \ldots, n$) of function $f(x)$, the corresponding function value $f(x_i) = y_i$ ($i = 0, 1, 2, \ldots, n$) for a given x which is not a node point, and four selected interpolated points that are closest to it; then the corresponding function value $f(x)$ can be computed with a unary four-point Lagrange interpolation.

Unary four-point Lagrange interpolation formulae can be obtained from (6.2.17) as follows:

$$f(x) = \sum_{i=K}^{K+3} \prod_{\substack{i=K \\ i \neq j}}^{K+3} \left(\frac{x - x_i}{x_j - x_i} \right) y_i \qquad (6.2.18)$$

where:

$$K = \begin{cases} 0 & x \leq x_2 \\ s - 1 & x_s \leq x \leq x_{s+1} \\ N - 3 & x \geq x_{N-2} \end{cases}$$

General method of numerical transformation in refining parallels and meridians on a map using Lagrange interpolation

This method is similar to that discussed in Section 6.2.1; that is:

- Step 1. Partitioning the transformation region and measuring or choosing the coordinates of points of intersection of parallels and meridians in the transformed region.

- Step 2. Refining coordinates of the points of intersection of parallels and meridians using Lagrange interpolation.

- Step 3. Making a numerical transformation between the two projections using approximate polynomials.

Bivariate four-point Lagrange interpolation

If given a group of rectangular interpolation points (x_s, y_t), $s = 1, 2, \ldots, N$; $t = 1, 2, \ldots, M$ on $[a, b] \times [c, d]$, then a Lagrange interpolation surface (5.1.6) can be obtained.

Mathematically, a bivariate four-point Lagrange interpolation consists of the node values x_i ($i = 0, 1, 2, \ldots, n$; equidistance is not necessarily required) of the first variable (x) of the function $F(x, y)$ and the node values y_j ($j = 0, 1, 2, \ldots, m$; equidistance is not necessarily required) of the second variable (y). The function values of the corresponding nodes are F_{ij} ($i = 0, 1, 2, \ldots, n$; $j = 0, 1, 2, \ldots, m$). For a given point (x, y) which is not a node, four points $(x_q, x_{q+1}, x_{q+2}, x_{q+3})$ which are closest to x and four points $(y_p, y_{p+1}, y_{p+2}, y_{p+3})$ which are closest to y are selected, respectively. Then the corresponding function value $f(x, y)$ can be computed with a bivariate Lagrange interpolation.

The formula for a bivariate four-point Lagrange interpolation is:

$$f(x, y) = \sum_{i=q}^{q+3} \sum_{j=p}^{p+3} \left(\prod_{\substack{s=q \\ s \neq i}}^{q+3} \frac{x - x_s}{x_i - x_s} \right) \left(\prod_{\substack{l=p \\ l \neq j}}^{p+3} \frac{y - y_l}{y_j - y_l} \right) F_{ij}$$

(6.2.19)

where:

$$q = \begin{cases} 0 & x \leq x_2 \\ t - 1 & x_t \leq x \leq x_{t+1}, \\ N - 3 & x \geq x_{N-2} \end{cases} \quad p = \begin{cases} 0 & y \leq y_2 \\ t - 1 & y_t \leq y \leq y_{t+1} \\ M - 3 & y \geq y_{M-2} \end{cases}$$

Examples for Section 6.2 will be presented in Appendix 3, Section A3.2; see also Brandenberger (1985).

6.3   ORTHOGONAL POLYNOMIAL APPROXIMATION

We have discussed the numerical transformation method for refining coordinates of points of intersection of a geographical graticule with a cubic spline and Lagrange interpolation, respectively. They are approximate interpolations. However, the information for a given point on a map, obtained by applying a digital map and computer cartography, always has some errors. Thus an approximate function is not required to agree strictly with the coordinates of a digital point; approximate agreement is sufficient. This kind of curve or surface approximation passing near the given point is called curve or surface fitting. We have discussed least-squares surface fitting of bivariate cubic polynomials and biquadratic polynomials in the above sections. Determining the coefficients a_i, b_i of polynomials by the least-squares method can be reduced to solving a system of normal equations. This system of equations often appears to be ill-conditioned, especially with an increase in the order p of the equation. This creates trouble during the solution. If an orthogonal poly-nomial is used, the coefficients of the matrix A of the system of normal equations may be transformed into a diagonal matrix, thus solving an ill-conditioned equation can be avoided, and the precision of solution can be improved. Meanwhile the system of normal equa-tions becomes one of p linear equations, and its solution becomes rather simple.

6.3.1   Orthogonal polynomials

Mathematically, orthogonal polynomials include n different values x_i ($i = 0, 1, 2, \ldots, n$) of argument x and a group of polynomials $\psi_i(x)$ ($j = 0, 1, 2, \ldots, p$) which take x as their argument. If they satisfy the following orthogonal condition (Yu 1980):

$$\left. \begin{aligned} \sum_{i=1}^{n} \psi_i(x_i) &= 0, \quad j \neq 0 \\ \sum_{i=1}^{n} \psi_i(x_i)\psi_k(x_i) &= 0, \quad j \neq k \end{aligned} \right\}$$

(6.3.1)

then polynomials $\psi_j(x)$ are called orthogonal polynomials.

Next we introduce a method for constructing orthogonal polynomials. Suppose a group of orthogonal polynomials are:

$$\left.\begin{aligned}
\psi_0(x) &= 1 \\
\psi_1(x) &= x^1 + a_{10}\psi_0(x) \\
\psi_2(x) &= x^2 + a_{20}\psi_0(x) + a_{21}\psi_1(x) \\
\psi_3(x) &= x^3 + a_{30}\psi_0(x) + a_{31}\psi_1(x) + a_{32}\psi_2(x) \\
&\cdots \\
\psi_m(x) &= x^m + a_{m,0}\psi_0(x) + a_{m,1}\psi_1(x) + \ldots + a_{m,m-1}\psi_{m-1}(x)
\end{aligned}\right\} \tag{6.3.2}$$

Using the orthogonality conditions, we have:

$$\sum_i \psi_0(x)\psi_1(x) = \sum_i x\psi_0(x) + a_{10}\sum_i \psi_0(x) = 0$$

$$\therefore \quad a_{10} = -\frac{1}{n}\sum x = -\bar{x}$$

Similarly there is:

$$\sum_i \psi_0(x)\psi_m(x) = \sum_i x^m\psi_0(x) + a_{m0}\sum_i \psi_0(x) = 0$$

$$\therefore \quad a_{m0} = -\frac{1}{n}\sum_i x^m = -\bar{x}^m$$

and:

$$\sum_i \psi_1(x)\psi_m(x) = \sum_i x^m\psi_1(x) + a_{m1}\sum_i \psi_1^2(x) = 0$$

$$\therefore \quad a_{m1} = -\frac{\sum x^m\psi_1(x)}{\sum \psi_1^2(x)}$$

By analogy, for any (k, j), there exists:

$$a_{jk} = -\frac{\sum x^j\psi_k(x)}{\sum \psi_k^2(x)} = -\frac{\sum x^j\psi_k}{\sum \psi_k^2} \tag{6.3.3}$$

where:

$$j = 1, 2, \ldots, p; \; k = 0, 1, \ldots, p-1; \; j > k$$

Substituting (6.3.3) into (6.3.2) yields:

$$\left.\begin{aligned}
\psi_0(x) &= 1 \\
\psi_1(x) &= x - \frac{\sum x}{n} \\
\psi_2(x) &= x^2 - \frac{\sum x^3 - \dfrac{\sum x \sum x^2}{n}}{\sum x^2 - \dfrac{\left(\sum x\right)^2}{n}}\left(x - \frac{\sum x}{n}\right) - \frac{\sum x^2}{n} \\
&\cdots
\end{aligned}\right\} \tag{6.3.4}$$

Especially, when each x_i is equidistant, without losing generality, we can assume that $x_1 = 1$, $x_2 = 2, \ldots , x_n = n$. Then from (6.3.4) we obtain:

$$
\left.
\begin{aligned}
&\psi_0(x) = 1 \\[4pt]
&\psi_1(x) = x - \bar{x} \\[4pt]
&\psi_2(x) = (x - \bar{x})^2 - \frac{n^2 - 1}{12} \\[4pt]
&\psi_3(x) = (x - \bar{x})^3 - \frac{3n^2 - 7}{20}(x - \bar{x}) \\[4pt]
&\psi_4(x) = (x - \bar{x})^4 - \frac{3n^2 - 13}{14}(x - \bar{x})^2 + \frac{3(n^2 - 1)(n^2 - 9)}{560} \\[4pt]
&\psi_5(x) = (x - \bar{x})^5 - \frac{5(n^2 - 7)}{18}(x - \bar{x})^3 + \frac{15n^4 - 230n^2 + 407}{1\,008}(x - \bar{x}) \\[4pt]
&\cdots \\[4pt]
&\psi_{K+1}(x) = \psi_1(x)\psi_K(x) - \frac{K^2(n^2 - k^2)}{4(4k^2 - 1)}\psi_{K-1}(x)
\end{aligned}
\right\}
\qquad (6.3.5)
$$

where:

$$
\bar{x} = \frac{1}{n}\sum_{i=1}^{n} x_i
$$

$\psi_K(x)$ are polynomials of degree k in x in the above formula, and they satisfy the orthogonality condition (6.3.1).

Formulae (6.3.5) are the orthogonal polynomials we need.

Since $\psi_j(x)$ are not certain to be integers equal to the integer value for x, they must be multiplied by the proper numbers q_j in order to let

$$
X_j(x) = q_j\psi_j(x) \qquad (6.3.6)
$$

become integers for which the absolute values are as small as possible at n integral points. Meanwhile:

$$
\sum_{x=1}^{n} X_j(x)X_k(x) =
\begin{cases}
0, & j \neq k \\
S_j, & j = k
\end{cases}
\qquad (6.3.7)
$$

6.3.2   Curve fitting

Given a general curve to be fitted:

$$
y = a_0 + a_1x + a_2x^2 + \ldots + a_px^p \qquad (6.3.8)
$$

construct the following form:

$$
y = b_0\psi_0(x) + b_1\psi_1(x) + \ldots + b_p\psi_p(x) = \sum_{j=0}^{p} b_j\psi_j(x) \qquad (6.3.9)
$$

If $\psi_j(x)$ satisfies the orthogonality condition, then:

$$\sum_{i=1}^{m} \psi_j(x_i) = 0, \quad j = 1, 2, \ldots, p; \ m \gg p$$

$$\sum_{i=1}^{m} \psi_j(x_i)\psi_k(x_i) = 0, \quad j \neq k$$

(6.3.10)

is called an orthogonal polynomial. The system of equations:

$$\begin{pmatrix} \sum \psi_0^2(x) & & & 0 \\ & \sum \psi_1^2(x) & & \\ & & \ddots & \\ 0 & & & \sum \psi_p^2(x) \end{pmatrix} \begin{pmatrix} b_0 \\ b_1 \\ \vdots \\ b_p \end{pmatrix} = \begin{pmatrix} \sum y'\psi_0(x) \\ \sum y'\psi_1(x) \\ \vdots \\ \sum y'\psi_p(x) \end{pmatrix}$$

(6.3.11)

is developed from (6.3.9). Here the system of normal equations has changed into p linear equations, with the undetermined coefficients:

$$b_j = \frac{\sum\limits_{i=1}^{m} y_i\psi_j(x_i)}{\sum\limits_{i=1}^{m} \psi_j^2(x_i)} \quad (j = 0, 1, 2, \ldots, p)$$

(6.3.12)

Example of application

Given the coordinates of some nodes along the meridian of the Mercator projection, we shall try to use an orthogonal polynomial of degree $p = 3$ to fit the curve.

Transform the coordinates first. Letting

$$x' = \frac{B - a}{h}$$

when $B = 10°$, $a = 5°$, $h = 5°$, $x' = 1$, then the results are as in Table 6.1.

Table 6.1 Coordinates for a Mercator projection

$B \rightarrow$	10°	15°	20°	25°	30°	35°
Coord. $\downarrow$						
x	1	2	3	4	5	6
y'	111.150	167.818	225.846	285.774	348.225	413.945

Using orthogonal polynomials, when $n = 6$,

$$\psi_0(x) = 1$$
$$\psi_1(x) = x - \bar{x}$$
$$\psi_2(x) = (x - \bar{x})^2 - \frac{n^2 - 1}{12}$$
$$\psi_3(x) = (x - \bar{x})^3 - \frac{3n^2 - 7}{20}(x - \bar{x})$$

then the curve can be fitted using (6.3.9) and (6.3.12).

6.3.3 Surface fitting

Given a bivariate quadratic fitting surface,

$$z = a_{00} + a_{10}x + a_{01}y + a_{20}x^2 + a_{02}y^2 + a_{11}xy \tag{6.3.13}$$

using orthogonal polynomials

$$\psi_{10}(x) = x - \bar{x}, \quad \psi_{20}(x) = (x - \bar{x})^2 - \frac{n^2 - 1}{12}$$

$$\psi_{01}(y) = y - \bar{y}, \quad \psi_{02}(y) = (y - \bar{y})^2 - \frac{m^2 - 1}{12}$$

(6.3.13) can be rewritten as:

$$z = b_{00} + b_{10}\psi_{10}(x) + b_{01}\psi_{10}(y) + b_{20}\psi_{20}(x) + b_{02}\psi_{02}(y) + b_{11}\psi_{10}(x)\psi_{01}(y) \tag{6.3.14}$$

As the same method, we can use an orthogonal polynomial to fit the surface. Here the system of normal equations is:

$$
\begin{pmatrix}
\sum_i \sum_j 1 & & & & 0 \\
& \sum_i \sum_j \psi_{10}^2 & & & \\
& & \sum_i \sum_j \psi_{01}^2 & & \\
& & & \ddots & \\
0 & & & & \sum_i \sum_j \psi_{10}^2 \psi_{01}^2
\end{pmatrix}
\begin{pmatrix}
b_{00} \\
b_{10} \\
b_{01} \\
\vdots \\
b_{11}
\end{pmatrix}
=
\begin{pmatrix}
\sum_i \sum_j Z'_{ij} \\
\sum_i \sum_j Z'_{ij}\psi_{10} \\
\sum_i \sum_j Z'_{ij}\psi_{01} \\
\vdots \\
\sum_i \sum_j Z'_{ij}\psi_{10}\psi_{01}
\end{pmatrix}
\tag{6.3.15}
$$

and the coefficients b_{ij} can be obtained as follows:

$$b_{00} = \frac{\sum_{i=1}^{n}\sum_{j=1}^{m} Z'_{ij}}{nm}, \quad b_{10} = \frac{\sum_i \sum_j \psi_{10}(x_i)Z'_{ij}}{m\sum_i \psi_{10}^2(x_i)}$$

$$b_{20} = \frac{\sum_i \sum_j \psi_{20}(x_i)Z'_{ij}}{m\sum_i \psi_{20}^2(x_i)}, \quad b_{01} = \frac{\sum_i \sum_j \psi_{01}(y_i)Z'_{ij}}{m\sum_i \psi_{01}^2(y_i)} \tag{6.3.16}$$

$$b_{02} = \frac{\sum_i \sum_j \psi_{02}(y_i)Z'_{ij}}{m\sum_i \psi_{02}^2(y_i)}, \quad b_{11} = \frac{\sum_i \sum_j \psi_{10}(x_i)\psi_{01}(y_i)Z'_{ij}}{\sum_i \sum_j [\psi_{10}(x_i)\psi_{01}(y_i)]^2}$$

Similarly we can obtain a bivariate cubic fitting surface:

$$Z = b_{00} + b_{10}\psi_{10}(x) + b_{01}\psi_{01}(y) + b_{20}\psi_{20}(x) + b_{02}\psi_{02}(y) + b_{11}\psi_{10}(x)\psi_{01}(y)$$
$$+ b_{30}\psi_{30}(x) + b_{21}\psi_{20}(x)\psi_{01}(y) + b_{12}\psi_{10}(x)\psi_{02}(y) + b_{03}\psi_{03}(y) \tag{6.3.17}$$

Besides (5.3.14) there exist other orthogonals, such as:

$$\psi_{30}(x) = (x - \bar{x})^3 - \frac{3n^2 - 7}{20}(x - \bar{x})$$

$$\psi_{03}(x) = (y - \bar{y})^3 - \frac{3m^2 - 7}{20}(y - \bar{y})$$

Similarly we can obtain the other coefficients such as:

$$\left. \begin{array}{l} b_{30} = \dfrac{\displaystyle\sum_i \sum_j \psi_{30}(x_i)Z'_{ij}}{m\displaystyle\sum_j \psi_{30}^2(x_i)}, \quad b_{03} = \dfrac{\displaystyle\sum_i \sum_j \psi_{03}(y_i)Z'_{ij}}{n\displaystyle\sum_j \psi_{03}^2(y_i)} \\[3em] b_{21} = \dfrac{\displaystyle\sum_i \sum_j \psi_{20}(x_i)\psi_{01}(y_i)Z'_{ij}}{\displaystyle\sum_i \sum_j \psi_{20}^2(x_i)\psi_{01}^2(y_i)}, \quad b_{12} = \dfrac{\displaystyle\sum_i \sum_j \psi_{10}(x_i)\psi_{02}(y_i)Z'_{ij}}{\displaystyle\sum_i \sum_j \psi_{10}^2(x_i)\psi_{02}^2(y_i)} \end{array} \right\} \qquad (6.3.18)$$

6.4 ACCURACY CONSIDERATIONS IN NUMERICAL TRANSFORMATION

Numerical transformation of a map projection belongs to the approximation problem of a bivariate function. In the above sections we have discussed several common methods, including the bivariate cubic polynomial approximation, the biquadratic polynomial approximation and the orthogonal polynomial approximation that are used in numerical transformation of map projections.

The theory and method of approximation of a bivariate function are important research subjects of numerical mathematics that face a series of theoretical and practical problems. For example, is the data distribution regular or not? The selection of numerical methods will depend on the accuracy of the original data, the selection of a global approximation, a local approximation, or a two-step approximation. The selection of the basis function is a most important and difficult problem; determination of the number of basis functions is also a practical problem.

In this section we can only make a preliminary analysis of some factors that affect the precision of a polynomial approximation of a map projection.

6.4.1 Estimate of error of the approximating polynomial

We have introduced the remainder term for several kinds of polynomials used in the numerical approximation of curves and surfaces, respectively. Next we take linear functions as an example to illustrate the estimation of the error of polynomial approximations. For example, we know that the remainder term using a Lagrange interpolation is:

$$R_n(x) = f(x) - P_n(x) = \frac{f^{(n+1)}(\xi)}{(n+1)!}\omega_{n+1}(x), \quad (a \le \xi \le b) \qquad (6.4.1)$$

where:

$$\omega_{n+1}(x) = (x - x_0)(x - x_1)\dots(x - x_n)$$

If we denote

$$M_{n+1} = \max_{a \le x \le b} |f^{(n+1)}(x)|$$

then we have:

$$|R_n(x)| \le M_{n+1} \frac{|\omega_{n+1}(x)|}{(n+1)!}$$

When $n = 1$, then:

$$R_1(x) = \frac{1}{2}\omega_2(x)f''(\xi), \quad a \le \xi \le b$$

$$\omega_2(x) = (x - x_0)(x - x_1)$$

Let $x_1 - x_0 = h$, $x = x_0 + \theta h$, $0 < \theta < 1$, then:

$$\omega_2(x) = \theta(\theta - 1)h^2 = -\theta(1 - \theta)h^2$$

When $0 < \theta < 1$, the maximum of $\theta(\theta - 1)$ is 1/4, and the estimate of error in linear interpolation is:

$$|R_1(x)| \le \frac{1}{8}h^2|f''(\xi)| \tag{6.4.2}$$

When $n = 3$, then:

$$R_3(x) = \frac{f^{(4)}(\xi)}{4!}\omega_4(x)$$

$$\omega_4(x) = (x - x_0)(x - x_1)(x - x_2)(x - x_3)$$

For equidistant nodes, we have:

$$x_1 - x_0 = x_2 - x_1 = x_3 - x_2 = h$$

Let $x = x_1 + t \cdot h$, then:

$$\omega_4(x) = \omega_4(x_1 + th) = (t + 1)t(t - 1)(t - 2)h^2$$

When $t = -1, 0, 1, 2$, the values of function $(t + 1)t(t - 1)(t - 2)$ are zero. It reaches the minimum -1 at $t = \frac{1}{2}(1 + \sqrt{5})$ and the maximum 9/16 at $t = 1/2$.

If we denote by

$$M = |f^{(4)}(\xi)|$$

then:

$$|R_3(x)| \le \frac{h^4}{4!}M|(t + 1)t(t - 1)(t - 2)|$$

From this we get:

$$|R_3(x)| \le \begin{cases} \dfrac{9}{16} \cdot \dfrac{h^4}{4!}M & (0 \le t \le h) \\[2mm] \dfrac{h^4}{4!}M & (-h \le t \le 0, \quad h \le t \le 2h) \end{cases} \tag{6.4.3}$$

The above formula indicates that using interpolation in the middle of interval $0 \leq t \leq h$ is more accurate than that at either end.

Using another example, the remainders term for a Hermite interpolation is:

$$R_n(x) = \frac{f^{(2n+2)}(\xi)}{(2n+2)!}\omega_{n+1}^2(x) \tag{6.4.4}$$

When

$$n = 1, \quad R_1(x) = \frac{f^{(4)}(\xi)}{4!}\omega_2^2(x), \quad \omega_2(x) = (x - x_0)(x - x_1)$$

then we have:

$$|R_1(x)| \leq \frac{1}{4!}\left(\frac{1}{4}h^2\right)^2 |f^{(4)}(\xi)|$$

that is:

$$|R_1(x)| \leq \frac{1}{384}h^4|f^{(4)}(\xi)| \tag{6.4.5}$$

Formula (6.4.5) is also suitable for the estimate of the error of a cubic spline function for equidistant nodes.

Next we give an example to show the error estimate method for refining coordinates in the latitude interval $[30°–35°]$ of the Mercator projection by using the cubic spline.

Given the coordinates formula for the Mercator projection

$$y = aq \tag{6.4.6}$$

where $a = 6\ 378\ 245$ m, q is equilatitude (or isometric latitude).

From (6.4.6) we have:

$$\frac{d^4y}{dB^4} = \frac{a}{\cos B}\tan B\,(5 + 6\tan^2 B - \eta^2) \tag{6.4.7}$$

where $\eta^2 = e'^2\cos^2 B$, e' is the second eccentricity.

If the map scale $\mu_0 = 1:1\ 000\ 000$, $a = 637.8245$ cm.

When $B = 35°$, from (6.4.7) we have:

$$\frac{d^4y}{dB^4} \leq 4\ 330$$

Noticing that $h = $ arc $5° = 0.0873$, from (6.4.5) we get:

$$|R_1(B)| \leq 0.0007 \text{ cm}$$

When estimating the error of the remainder term of polynomials, the factor $f^{(n+1)}(\xi)$ of the remainder term often has a great influence upon the error, where ξ is an indeterminate point in $[a, b]$. Now suppose $f^{(n+1)}(y)$ is a bounded function in the interval $[a, b]$, and let

$$M_{n+1} = \max_{a<y<b}|f^{(n+1)}(y)|$$

then:

$$|R_1(y)| \leq M_{n+1}\frac{|\omega_{n+1}(y)|}{(n + 1)!} \tag{6.4.8}$$

In a general sense, a higher derivative increases very rapidly except for integer functions. This indicates that the effect of higher interpolation is not ideal; in some cases the result will be worse and the Runge phenomenon appears.

The precision of the polynomial approximation of a numerical transformation of a map projection is a stability problem, a problem that needs to be probed further. Generally, the precision of the approximation has something to do with the construction and order of the approximate polynomials as well as the size of the region of transformation, the distribution of nodes, the projection property, etc. The factors mentioned above must be restricted in concrete numerical transformations; therefore, the advantage and generality of some transformation methods are relative. For example, some factors affecting the approximate precision analyzed in the following are discussed under the condition of restricting the order of the approximating polynomials.

6.4.2   The size of the transformation region and the precision of the polynomial approximation

The size of the transformation region directly influences the precision of the polynomial approximation. In the following we illustrate this with bivariate cubic polynomials.

In direct transformation of directly determined approximate polynomials, for a transformation region $B = 20°$–$35°$, $L = 15°$–$30°$, the deviation of coordinates for the transformation from the Mercator projection to a conical conformal projection with bivariate cubic polynomials is 0.273 cm, but for the transformation region $B = 20°$–$30°$, $L = 15°$–$25°$, the deviation of coordinates is less than 0.002 cm. This means that the precision of bivariate cubic polynomial approximation is much worse when the Mercator projection is transformed into a conformal conic projection in a transformation region $15° \times 15°$, and the map scale is 1:1 000 000. But in a transformation region $10° \times 10°$, the precision would perfectly satisfy mapping requirements.

6.4.3   The distribution of common points and the precision of polynomial approximation

The transformation of directly determined polynomials using a cubic polynomial requires the selection of 10 common or control points.

It has been proven in practice that the distribution of common points in the same transformation region directly influences the precision of polynomial approximation. To illustrate this, we have computed the results of six different distributions of common points in a region $B = 20°$–$30°$, $L = 15°$–$25°$ using the transformation from the Mercator projection to a conformal conic projection with cubic polynomials.

It can be seen from comparing the computation results for the six schemes that the distribution of the common points has a great influence on the computation results.

Depending on the distribution of the common points, the errors of solution are very large, or even unusable when three points lie along the same meridian. A suitable form of distribution of common points is random or even a crisscross distribution over the region of transformation.

6.4.4 Computation method and the precision of polynomial approximation

The computation method directly influences the precision of polynomial approximation. When solving by the least-squares method, the coefficient matrix is positive definite and symmetric. With an increasing degree of the polynomial, the coefficient matrix is generally ill-conditioned. At this time we should select a proper method of computation such as solution by using an improved square-root method, conjugate gradient method, or orthogonalization method, etc.; for example, direct coordinate transformation by using the orthogonal polynomials introduced in Section 6.3 to fit the surface is a very simple computing method.

6.4.5 Property of the map projection and selection of basis functions

The theory and method of a general polynomial such as a bivariate cubic polynomial, bivariate quadratic polynomial, product polynomial, biquadratic polynomial, etc., used in the numerical transformation of map projections and their programming have been generally discussed in this chapter. Comparing each kind of transformation method indicates that the precision of transformation is affected by many factors. The stability of numerical transformation is a problem needing further study.

In order to improve the stability of the numerical transformation of map projections, selecting a different basis function, that is different polynomials, for different map projections, is also a key factor besides noting the factors that have influence upon the precision of polynomial approximation mentioned above.

For this purpose, we will study in detail the theory and method of numerical transformation between two conformal projections by conformal or harmonic polynomials in the next chapter.

CHAPTER SEVEN

Numerical Transformation for Conformal Projection

▌

7.1 CONFORMAL POLYNOMIAL APPROXIMATION

7.1.1 Mathematical principle

A method for numerical transformation between conformal projections will be discussed in this chapter. Its mathematical principle is that only a unique power series of $(z - z_0)$ may be summed in a neighborhood of point Z_0 according to the theory of functions of a complex variable (Yang 1982b).

The method of undetermined coefficients is helped by the uniqueness of a series developed at $(z - z_0)$. That is, if there are two series of $(z - z_0)$:

$$A_0 + A_1(z - z_0) + A_2(z - z_0)^2 + \ldots + A_n(z - z_0)^n + \ldots$$
$$B_0 + B_1(z - z_0) + B_2(z - z_0)^2 + \ldots + B_n(z - z_0)^n + \ldots$$

and if their sums are equal at infinity, for example at set E of which z_0 is a summation, then their corresponding coefficients are equal, so that:

$$A_0 = B_0, \quad A_1 = B_1, \quad A_2 = B_2, \quad \ldots, \quad A_n = B_n, \quad \ldots$$

For a fixed transformation region, the coordinates (x_0, y_0) and (X_0, Y_0) of the expansion point or auxiliary point 0 and the coordinates (x_i, y_i) and (X_i, Y_i) $(i = 1, 2, 3, 4)$ of common points in the two coordinate systems can be selected and the following system of linear equations can be formed from (5.1.36):

$$\left. \begin{array}{l} \displaystyle\sum_{K=1}^{4}(a_K P_K^{(i)} - b_K Q_K^{(i)}) = \Delta Y^{(i)} \\[4mm] \displaystyle\sum_{K=1}^{4}(b_K P_K^{(i)} + a_K Q_K^{(i)}) = \Delta X^{(i)} \end{array} \right\} \tag{7.1.1}$$

where $\Delta X = X - X_0$, $\Delta Y = Y - Y_0$, $i = 1, 2, 3, 4$.

Formula (7.1.1) is a system of linear equations with eight unknowns. The coefficients a_K, b_K $(K = 1, 2, 3, 4)$ can be obtained by solving this system of equations.

The method mentioned above is the general method for the numerical transformation for conformal projections.

If the series are equal at finite points, then the polynomial $F(z)$ obtained is a polynomial of finite degree, but it is not the sum function $f(z)$ in the transformation region. So long as the value of $(F(z) - f(z))$ fits the defined permitted limitation, then $F(z)$ is a uniformly convergent analytic function that has the stability of an analytic function in the transformation region. As a result, $F(z)$ can replace the required sum function $f(z)$ completely.

Equation (5.1.36) is referred to as a conformal polynomial. It is quite different from the bivariate polynomial of degree n and bivariate-nth polynomial discussed previously.

As a conformal polynomial satisfies the conformal condition, that is:

$$\frac{\partial X}{\partial x} = -\frac{\partial Y}{\partial y}, \quad \frac{\partial X}{\partial y} = \frac{\partial Y}{\partial x} \tag{7.1.2}$$

so (5.1.36) is called a conformal polynomial.

Similarly, approximate interpolation and surface fitting can be achieved in a numerical transformation with conformal polynomials.

In the following, we take a conformal polynomial of 4th degree as an example to show the steps of numerical transformation between two conformal projections:

- Step 1: Select coordinates (x_0, y_0) and (X_0, Y_0) of an auxiliary point 0 in the two coordinate systems.

- Step 2: Select coordinates (x_i, y_i) and (X_i, Y_i) $(i = 1, 2, 3, 4)$ of four common points in the two coordinate systems.

- Step 3: Form a system of linear equations using (7.1.1) and solve it with a Gaussian elimination method to obtain coefficients (a_i, b_i).

- Step 4: Make a coordinate transformation using (5.1.36) in the region of transformation.

The method mentioned above is an algebraic interpolation of a directly determined conformal polynomial.

When the number of common points, except for the auxiliary point, is more than four, then the least-squares method can be used to determine the conformal polynomial. This is then a method of surface fitting.

Numerical transformation for functions of complex variables can be achieved with Lagrange interpolation similar to that for a real variable.

Otherwise, from (5.1.8) and (5.1.9) we know that transformation between two conformal projections can be reduced to the following Dirichlet problem:

$$\frac{\partial^2 X}{\partial x^2} + \frac{\partial^2 X}{\partial y^2} = 0 \quad (x, y) \in \Omega$$

$$X|_\Gamma = D'(x, y) \tag{7.1.3}$$

$$\frac{\partial^2 Y}{\partial x^2} + \frac{\partial^2 Y}{\partial y^2} = 0 \quad (x, y) \in \Omega$$

$$Y|_\Gamma = D(x, y) \tag{7.1.4}$$

where Ω is the interior of the region of transformation, Γ is its boundary; and D and D' are functions given on the boundary.

If the boundary of the transformation region is simple, then the expression of the relationship for an analytic transformation using the Dirichlet problem can be obtained.

Otherwise it is generally very difficult to solve a Dirichlet problem in the transformation between two projections. We often use a numerical method to solve it. There are many numerical methods for solving Dirichlet problems, such as the finite difference method and the finite element method. Next we introduce each of these methods, respectively.

7.1.2 Determining coefficients for a conformal polynomial by a conventional method

Direct solution for transformation between two conformal projections

The case of the expansion point 0 in an arbitrary position Numerical transformation between conformal projections involves forming a system of linear equations using co-ordinates of some common points of the two conformal projections and determining the coefficients (a_k, b_k) of formula (5.1.36) by an 'undetermined coefficient method'.

In the following we use a quartic conformal polynomial as an example to show this transformation method. Given coordinates (x_0, y_0) and (X_0, Y_0) of the expansion point 0 and coordinates (x_i, y_i) and (X_i, Y_i) $(i = 1, 2, 3, 4)$ of four common points in the trans-formation region, (7.1.1) can be rewritten as:

$$Ay = b \tag{7.1.5}$$

where:

$$A = \begin{pmatrix} P_{11} & P_{12} & P_{13} & P_{14} & -Q_{11} & -Q_{12} & -Q_{13} & -Q_{14} \\ Q_{11} & Q_{12} & Q_{13} & Q_{14} & P_{11} & P_{12} & P_{13} & P_{14} \\ \cdots & & \cdots & & \cdots & & \cdots & \\ P_{41} & P_{42} & P_{43} & P_{44} & -Q_{41} & -Q_{42} & -Q_{43} & -Q_{44} \\ Q_{41} & Q_{42} & Q_{43} & Q_{44} & P_{41} & P_{42} & P_{43} & P_{44} \end{pmatrix}$$

$$y = [a_1, \quad a_2, \quad a_3, \quad a_4, \quad b_1, \quad b_2, \quad b_3, \quad b_4]^T$$
$$b = [Y_1 - Y_0, \quad X_1 - X_0, \quad Y_2 - Y_0, \quad X_2 - X_0, \quad \ldots, \quad Y_4 - Y_0, \quad X_4 - X_0]^T$$
$$P_{i, K+1} = \Delta y P_{iK} - \Delta x Q_{iK}, \quad Q_{i, K+1} = \Delta x P_{iK} + \Delta y Q_{iK},$$
$$P_{i1} = \Delta x, \quad Q_{i1} = \Delta y, \quad \Delta y = y_i - y_0, \quad \Delta x = x_i = x_0$$
$$i = 1, 2, 3, 4; \quad K = 1, 2, 3$$

Through solution of these linear equations, coefficients a_K, b_K can be found. Then the coordinate transformation can be conducted using formula (5.1.36).

The case of the expansion point 0 along the central meridian When the expansion point 0 is along the central meridian, we have formula (5.1.37).

The expression of the relationship of conformal projection coordinates along the cen-tral meridian $(x = 0, X = 0)$ is:

$$a_1 \Delta y + a_2 \Delta y^2 + a_3 \Delta y^3 + \ldots = \Delta Y \tag{7.1.6}$$

In the following we also use a quartic conformal polynomial as an example to show a numerical transformation when the expansion point is along the central meridian.

Given coordinates y_0 and Y_0 of the expansion point 0 along the central meridian and coordinates y_i and Y_i of four common points along the central meridian $(i = 1, 2, 3, 4)$, we can form a system of linear equations of four unknown numbers according to (7.1.6):

$$\sum_{K=1}^{4} a_K \Delta y_i^K = \Delta Y_i \quad (i = 1, 2, 3, 4) \tag{7.1.7}$$

Coefficients a_K can be obtained by solving this system of linear equations using a Gaussian elimination method, and then the transformation of coordinates can be made according to (5.1.37).

Inverse solution transformation of a conformal projection

The case of the expansion point 0 at an arbitrary position We know that the inverse transformation formulae for conformal projections are as given in formulae (5.1.46).

Once again we take a quartic conformal polynomial as an example to illustrate the numerical method for the inverse transformation.

Given coordinates (x_0, y_0) and (l_0, q_0) of expansion point 0 and coordinates (x_i, y_i) and (l_i, B_i) $(i = 1, 2, 3, 4)$ of four common points in the transformation region, we can obtain a matrix similar to (7.1.5):

$$Ay = c \tag{7.1.8}$$

where the matrices A and y are the same as those in (7.1.5). However,

$$c = [q_1 - q_0, \quad l_1 - l_0, \quad q_2 - q_0, \quad l_2 - l_0, \quad \ldots, \quad q_4 - q_0, \quad l_4 - l_0]^T \tag{7.1.9}$$

where:

$$q_i = \ln \left[\tan \left(\frac{\pi}{4} + \frac{B_i}{2} \right) \left(\frac{1 - e \sin B_i}{1 + e \sin B_i} \right)^{e/2} \right]$$

Through solution of these linear equations, coefficients (a_K, b_K) can be found. Then the coordinate transformation can be conducted with formulae (5.1.46) and (3.7.18).

The case of the expansion point 0 along the central meridian Again taking a quartic conformal polynomial as an example to illustrate its numerical method, and selecting coordinates y_0 and q_0 of the expansion point 0 and coordinates y_i and B_i $(i = 1, 2, 3, 4)$ of four common points in the transformation region, with equations similar to (7.1.7) we can get a system of linear equations with four unknown numbers.

$$\sum_{K=1}^{4} a_K \Delta y_i^K = \Delta q_i \quad (i = 1, 2, 3, 4) \tag{7.1.10}$$

Coefficients a_K can be obtained by solving this system of linear equations; then the transformation of coordinates can be made according to (5.1.47) and (3.7.18).

Direct transformation for conformal projections

For the case of the expansion point 0 located at an arbitrary position, the formula for direct transformation for conformal projections is the same as formula (5.1.42). If given coordinates (B_i, l_i) and (y_i, x_i) $(i = 1, 2, 3, 4)$ of the four common points, we can get a matrix form similar to (7.1.5):

$$Ay = d \qquad (7.1.11)$$

where matrices A and y are the same as those in (7.1.5). However,

$$d = [y_1 - y_0, \quad x_1 - x_0, \quad y_2 - y_0, \quad x_2 - x_0, \quad \ldots, \quad y_4 - y_0, \quad x_4 - x_0]^T \qquad (7.1.12)$$

Through solution of these linear equations, coefficients (a_K, b_K) can be found. Then the coordinate transformation can be conducted with formulae (5.1.36).

In the case of the expansion point 0 along the central meridian, the formula for direct transformation for conformal projections is as in (5.1.43), similar to (7.1.7), and we can obtain a system of linear equations as follows:

$$\sum_{K=1}^{n} a_K \Delta y_i^K = \Delta y_i \quad (i = 1, 2, \ldots, n) \qquad (7.1.13)$$

Through the solution of these linear equations, coefficients (a_K, b_K) can be found. Then coordinate transformation can be conducted using formula (5.1.43).

7.1.3 Determining coefficients for a conformal polynomial by least squares

Direct transformation between two conformal projections

The case of the expansion point 0 at an arbitrary position   In the following we take a quartic conformal polynomial as an example to illustrate this method of transformation. Given coordinates (x_0, y_0) and (X_0, Y_0) of expansion point 0 and coordinates (x_i, y_i) and (X_i, Y_i) $(i = 1, 2, 3, 4, \ldots m)$ of $m(> 4)$ common points in the transformation region. From (5.4.1) we can get the least-squares condition:

$$\varepsilon = \sum_{i=1}^{m}[(\Delta X_i - \Delta X_i')^2 + (\Delta Y_i - \Delta Y_i')^2] = \min \qquad (7.1.14)$$

where $\Delta X_i, \Delta Y_i$ are the coordinates after transformation; and $\Delta X_i', \Delta Y_i'$ are the theoretical coordinates.

Taking the extremum condition of a multivariate function we have:

$$\frac{\partial \varepsilon}{\partial a_k} = 0, \quad \frac{\partial \varepsilon}{\partial b_k} = 0 \quad (k = 1, 2, 3, 4) \qquad (7.1.15)$$

Considering (7.1.1) and substituting (7.1.14) into (7.1.15), after rearranging we obtain the following matrix equation:

$$Ay = b \qquad (7.1.16)$$

where:

$$A = \begin{pmatrix} \sum P_{i1}P_{i1} + \sum Q_{i1}Q_{i1} & \sum P_{i1}P_{i2} + \sum Q_{i1}Q_{i2} & \cdots & -\sum P_{i1}Q_{i4} + \sum Q_{i1}P_{i4} \\ \sum P_{i2}P_{i1} + \sum Q_{i2}Q_{i1} & \sum P_{i2}P_{i2} + \sum Q_{i2}Q_{i2} & \cdots & -\sum P_{i2}Q_{i4} + \sum Q_{i2}P_{i4} \\ \cdots & \cdots & & \cdots \\ -\sum Q_{i3}P_{i1} + \sum P_{i3}Q_{i1} & -\sum Q_{i3}P_{i2} + \sum P_{i3}Q_{i2} & \cdots & \sum Q_{i3}Q_{i4} + \sum P_{i3}P_{i4} \\ -\sum Q_{i4}P_{i1} + \sum P_{i4}Q_{i1} & -\sum Q_{i4}P_{i2} + \sum P_{i4}Q_{i4} & \cdots & \sum Q_{i4}Q_{i4} + \sum P_{i4}P_{i4} \end{pmatrix}$$

and:

$$y = \begin{pmatrix} a_1 \\ a_2 \\ a_3 \\ a_4 \\ b_1 \\ b_2 \\ b_3 \\ b_4 \end{pmatrix} \qquad b = \begin{pmatrix} \sum P_{i1}\Delta Y_i + \sum Q_{i1}\Delta X_i \\ \sum P_{i2}\Delta Y_i + \sum Q_{i2}\Delta X_i \\ -\sum Q_{i3}\Delta Y_i + \sum P_{i3}\Delta X_i \\ -\sum Q_{i4}\Delta Y_i + \sum P_{i4}\Delta X_i \end{pmatrix}$$

Coefficients (a_K, b_K) can be obtained from the above formula by Gaussian elimination. Then the transformation between two conformal projections may be computed with (5.1.36).

The case of the expansion point 0 along the central meridian   Once again we take a quartic conformal polynomial as an example to illustrate this transformation method.

Given the coordinates y_0 and Y_0 of expansion point 0 and coordinates y_i and Y_i ($i = 1$, 2, 3, 4, ... m) of m (> 4) common points along the central meridian, they are substituted into (7.1.6), yielding:

$$\Delta Y_i = a_1\Delta y + a_2\Delta y^2 + a_3\Delta y^3 + a_4\Delta y^4 \qquad (7.1.17)$$

where $i = 1, 2, 3, 4, \ldots m$.

According to the least-squares principle, the condition is:

$$\varepsilon = \sum_{i=1}^{m}(\Delta Y_i - \Delta Y_i')^2 = \min \qquad (7.1.18)$$

From the extremum principle we get the following matrix form:

$$\begin{pmatrix} \sum \Delta y_i^2 & \sum \Delta y_i^3 & \sum \Delta y_i^4 & \sum \Delta y_i^5 \\ \sum \Delta y_i^3 & \sum \Delta y_i^4 & \sum \Delta y_i^5 & \sum \Delta y_i^6 \\ \sum \Delta y_i^4 & \sum \Delta y_i^5 & \sum \Delta y_i^6 & \sum \Delta y_i^7 \\ \sum \Delta y_i^5 & \sum \Delta y_i^6 & \sum \Delta y_i^7 & \sum \Delta y_i^8 \end{pmatrix} \cdot \begin{pmatrix} a_1 \\ a_2 \\ a_3 \\ a_4 \end{pmatrix} = \begin{pmatrix} \sum \Delta y_i \cdot \Delta Y_i \\ \sum \Delta y_i^2 \cdot \Delta Y_i' \\ \sum \Delta y_i^3 \cdot \Delta Y_i' \\ \sum \Delta y_i^4 \cdot \Delta Y_i' \end{pmatrix} \qquad (7.1.19)$$

The coordinates a_K can be obtained by the principal-element elimination method; then the coordinate transformation can proceed using (5.1.43).

Inverse transformation of conformal projections

The case of the expansion point 0 at an arbitrary position   We use a quartic conformal polynomial as an example to illustrate this transformation method as well. Given coordinates (x_0, y_0) and (l_0, q_0) of the expansion point 0 and the coordinates (x_i, y_i) and (l_i, B_i) ($i = 1, 2, 3, 4, \ldots m$) of m (> 4) common points in the transformation region, in a manner similar to (7.1.16) we can get the following matrix equation:

$$Ay = c \qquad (7.1.20)$$

where the matrices A and y are identical with those in (7.1.16). However,

$$c = \left[\sum P_{i1}\Delta q_i' + \sum Q_{i1}\Delta l_i', \quad \sum P_{i2}\Delta q_i' + \sum Q_{i2}\Delta l_i', \quad \dots, \quad -\sum Q_{i4}\Delta q_i' + \sum P_{i4}\Delta l_i',\right]^T$$

$$(7.1.21)$$

Coefficients (a_K, b_K) can be obtained by solving the above system of equations with a principal-element elimination; then the latitude B and the difference of longitude l can be computed from (5.1.46) and (3.7.18).

The case of the expansion point 0 along the central meridian   Continuing to take a quartic conformal polynomial as an example to illustrate this transformation method, we select the coordinates y_0 and q_0 of the expansion point 0 and coordinates y_i and B_i ($i = 1, 2, 3, 4, \dots, m$) of m (> 4) common points along the central meridian.

Similar to (7.1.19), we obtain a matrix equation in which a different term is on the right side as in the following:

$$\left[\sum \Delta y_i \cdot \Delta q_i', \quad \sum \Delta y_i^2 \cdot \Delta q_i', \quad \sum \Delta y_i^3 \cdot \Delta q_i', \quad \sum \Delta y_i^4 \cdot \Delta q_i'\right]^T \qquad (7.1.22)$$

Coordinates a_K can be obtained with the method of principal-element elimination, and then the coordinate transformation can proceed using (5.1.47) and (3.7.18).

Examples of Sections 7.1.2 and 7.1.3 will be presented in Appendix 3, Section A3.3.

7.2   INTERPOLATION

7.2.1   Difference quotient method

We know that the formula for coordinate transformation between two conformal projections is:

$$Y + iX = f(y + ix) \quad \text{or} \quad Z = f(z)$$

If the coordinates of common points in the regions of two conformal projections are respectively given as $z_1, z_2, \dots, z_n$ and $Z_1, Z_2, \dots, Z_n$, the transformation from point z on one conformal projection to point Z on another can be obtained by the difference quotient method following an example of a real variable function (Liu 1985).

The definition of a first-order difference quotient is:

$$[Z_1 Z_2] = \frac{Z_1 - Z_2}{z_1 - z_2}, \quad \dots, \quad [Z_{n-1} Z_n] = \frac{Z_{n-1} - Z_n}{z_{n-1} - z_n}$$

The definitions of second- and higher-order difference quotients are, respectively:

$$[Z_1 Z_2 Z_3] = \frac{[Z_1 Z_2] - [Z_2 Z_3]}{z_1 - z_3}$$

$$\dots$$

$$[Z_1 Z_2 Z_3 \dots Z_n] = \frac{[Z Z_1 Z_2 \dots Z_{n-1}] - [Z_2 Z_3 \dots Z_n]}{z_1 - z_n}$$

Assume the value of Z corresponding to z is known, then the corresponding difference quotients are:

$$[ZZ_1] = \frac{Z - Z_1}{z - z_1}$$

$$[ZZ_1Z_2] = \frac{[ZZ_1] - [Z_1Z_2]}{z - z_2}$$

. . .

$$[ZZ_1Z_2 \ldots Z_n] = \frac{[ZZ_1Z_2 \ldots Z_{n-1}] - [Z_1Z_2 \ldots Z_n]}{z - z_n}$$

From the above formulae we get:

$$Z = Z_1 + (z - z_1)[ZZ_1]$$
$$Z = Z_1 + (z - z_1)[Z_1Z_2] + (z - z_1)(z - z_2)[Z_1Z_2Z_3]$$

. . .

$$Z = Z_1 + (z - z_1)[Z_1Z_2] + (z - z_1)(z - z_2)[Z_1Z_2Z_3] + \ldots$$
$$+ (z - z_1)(z - z_2) \ldots (z - z_{n-1})[Z_1Z_2 \ldots Z_{n-1}] + R(Z)$$

where:

$$R(Z) = (z - z_1)(z - z_2) \ldots (z - z_n)[ZZ_1Z_2 \ldots Z_n]$$

Next, we write out formulae for 5th-order difference quotients with six common points:

$$\begin{aligned}
Z = Z_1 &+ (z - z_1)[Z_1Z_2] + (z - z_1)(z - z_2)[Z_1Z_2Z_3] \\
&+ (z - z_1)(z - z_2)(z - z_3)[Z_1Z_2Z_3Z_4] \\
&+ (z - z_1)(z - z_2)(z - z_3)(z - z_4)[Z_1Z_2Z_3Z_4Z_5] \\
&+ (z - z_1)(z - z_2)(z - z_3)(z - z_4)(z - z_5)[Z_1Z_2Z_3Z_4Z_5Z_6] + R(Z)
\end{aligned} \tag{7.2.1}$$

where:

$$[Z_1Z_2Z_3Z_4Z_5Z_6] = \frac{[Z_1Z_2Z_3Z_4Z_5] - [Z_2Z_3Z_4Z_5Z_6]}{z_1 - z_6}$$

$$[Z_1Z_2Z_3Z_4Z_5] = \frac{[Z_1Z_2Z_3Z_4] - [Z_2Z_3Z_4Z_5]}{z_1 - z_5}$$

$$[Z_2Z_3Z_4Z_5Z_6] = \frac{[Z_2Z_3Z_4Z_5] - [Z_3Z_4Z_5Z_6]}{z_2 - z_6}$$

$$[Z_1Z_2Z_3Z_4] = \frac{[Z_1Z_2Z_3] - [Z_2Z_3Z_4]}{z_1 - z_4}$$

$$[Z_2Z_3Z_4Z_5] = \frac{[Z_2Z_3Z_4] - [Z_3Z_4Z_5]}{z_2 - z_5}$$

$$[Z_3Z_4Z_5Z_6] = \frac{[Z_3Z_4Z_5] - [Z_4Z_5Z_6]}{z_3 - z_6}$$

$$[Z_1Z_2Z_3] = \frac{[Z_1Z_2] - [Z_2Z_3]}{z_1 - z_3}, \quad [Z_2Z_3Z_4] = \frac{[Z_2Z_3] - [Z_3Z_4]}{z_2 - z_4}$$

$$[Z_3Z_4Z_5] = \frac{[Z_3Z_4] - [Z_4Z_5]}{z_3 - z_5}, \quad [Z_4Z_5Z_6] = \frac{[Z_4Z_5] - [Z_5Z_6]}{z_4 - z_6}$$

$$[Z_1Z_2] = \frac{Z_1 - Z_2}{z_1 - z_2}, \quad [Z_2Z_3] = \frac{Z_2 - Z_3}{z_2 - z_3}, \quad [Z_3Z_4] = \frac{Z_3 - Z_4}{z_3 - z_4}$$

$$[Z_4Z_5] = \frac{Z_4 - Z_5}{z_4 - z_5}, \quad [Z_5Z_6] = \frac{Z_5 - Z_6}{z_5 - z_6}$$

$$R(Z) = (z - z_1)(z - z_2)(z - z_3)(z - z_4)(z - z_5)(z - z_6) \cdot [ZZ_1Z_2Z_3Z_4Z_5Z_6]$$

7.2.2 Interpolation

Imitating the Lagrange formula for a real variable, the interpolation formula for conformal polynomials can be given as follows:

$$Z = Z_0 \frac{(z - z_1)(z - z_2) \ldots (z - z_n)}{(z_0 - z_1)(z_0 - z_2) \ldots (z_0 - z_n)} + Z_1 \frac{(z - z_1)(z - z_2) \ldots (z - z_n)}{(z_1 - z_1)(z_1 - z_2) \ldots (z_1 - z_n)}$$

$$+ \ldots + Z_n \frac{(z - z_1)(z - z_2) \ldots (z - z_{n-1})}{(z_n - z_1)(z_n - z_2) \ldots (z_n - z_{n-1})} + R(Z) \tag{7.2.2}$$

where:

$$R(Z) = (z - z_1)(z - z_2) \ldots (z - z_n)[Z_0Z_1Z_2 \ldots Z_nZ]$$

The interpolation formula for conformal polynomials can also be written as:

$$Z = \sum_{j=0}^{n} \prod_{\substack{i=0 \\ i \neq j}}^{n} \left(\frac{z - z_i}{z_j - z_i} \right) Z_j \tag{7.2.3}$$

The interpolation formula for a six-point conformal polynomial is:

$$Z = \sum_{j=0}^{5} \prod_{\substack{i=0 \\ i \neq j}}^{5} \left(\frac{z - z_i}{z_j - z_i} \right) Z_j \tag{7.2.4}$$

In the following we use a 5th-order difference as an example to illustrate this computation method.

For convenience in computer programming, formula (7.2.1) needs to be transformed properly. Thus let:

$$(z - z_1) = P(1) + iQ(1)$$

$$(z - z_1)(z - z_2) = P(2) + iQ(2)$$

$$\ldots$$

$$(z - z_1)(z - z_2) \ldots (z - z_5) = P(5) + iQ(5)$$

$$(z - z_1)^{-1} = T(1, 1) + iR(1, 1)$$

$$(z_2 - z_3)^{-1} = T(1, 2) + iR(1, 2)$$

$$\ldots$$

$$(z_5 - z_6)^{-1} = T(1, 5) + iR(1, 5)$$

In the general case the equations consist of:

$$(z_i - z_{i+k})^{-1} = T(k, i) + iR(k, i)$$
$$(k = 1, 2, \ldots, 5; \quad i = 1, 2, \ldots, 5; \quad i + k \le 5)$$
$$[Z_1 Z_2] = T(6, 1) + iR(6, 1)$$
$$[Z_2 Z_3] = T(6, 2) + iR(6, 2)$$

$\cdots$

$$[Z_5 Z_6] = T(6, 5) + iR(6, 5)$$
$$[Z_1 \ldots Z_3] = T(5, 2) + iR(5, 2)$$

$\cdots$

$$[Z_4 \ldots Z_6] = T(5, 5) + iR(5, 5)$$
$$[Z_1 \ldots Z_4] = T(4, 3) + iR(4, 3)$$

$\cdots$

$$[Z_3 \ldots Z_6] = T(4, 5) + iR(4, 5)$$
$$[Z_1 \ldots Z_5] = T(3, 4) + iR(3, 4)$$
$$[Z_2 \ldots Z_6] = T(3, 5) + iR(3, 5)$$
$$[Z_1 \ldots Z_6] = T(2, 5) + iR(2, 5)$$

The stored units of the above variable are allocated in Table 7.1.

Table 7.1   *The stored units of the complex variable*

$(z_1 - z_2)^{-1}$	$(z_2 - z_3)^{-1}$	$(z_3 - z_4)^{-1}$	$(z_4 - z_5)^{-1}$	$(z_5 - z_6)^{-1}$
$(z_1 - z_3)^{-1}$	$(z_2 - z_4)^{-1}$	$(z_3 - z_5)^{-1}$	$(z_4 - z_6)^{-1}$	$[z_1 \ldots z_6]$
$\cdots$	$\cdots$	$\cdots$	$\cdots$	$\cdots$
$(z_1 - z_6)^{-1}$	$[z_1 \ldots z_3]$	$[z_2 \ldots z_4]$	$[z_3 \ldots z_5]$	$[z_4 \ldots z_6]$
$[z_1 z_2]$	$[z_2 z_3]$	$[z_3 z_4]$	$[z_4 z_5]$	$[z_5 z_6]$

Hence formula (7.2.1) can be rewritten as follows:

$$Y + iX = YT(1) + iXT(1) + \sum_{i=1}^{5}(P(i)T(7 - i, i) - Q(i)R(7 - i, i))$$

$$+ i\sum_{i=1}^{5}(Q(i)T(7 - i, i) + P(i)R(7 - i, i)) \tag{7.2.5}$$

The computing speed of the difference quotient method is faster than that of interpolation, but the programming of interpolation is simpler.

7.3   DIFFERENCE METHOD

From (7.1.3) and (7.1.4) we know that transformation between two conformal projections can be reduced to solving the following Dirichlet problem.

If we are given the projection region Ω surrounded by meridians and parallels in the (x, y) conformal projection plane and the coordinates (x_i, y_i) of nodes on the boundary Γ; and the conformal projection region Ω' corresponding to (X, Y) and the coordinates (X_i, Y_i) of nodes on boundary Γ_3, let us obtain the coordinates (X, Y) of the inner point of Ω_3 corresponding to (x, y) of the inner point of Ω. Generally the exact solution of the boundary value problem of the Laplace equation is difficult to obtain, but a numerical solution can be approximated by the difference method.

7.3.1 Constructing difference equations

Adopting two groups of parallel lines on the (x, y) projection plane, that is,

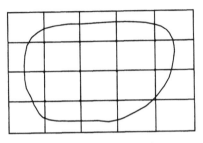

Figure 7.1 Rectangular grid

$$y = y_0 + ih, \quad i = 0, \pm 1, \pm 2, \ldots$$
$$x = x_0 + jk, \quad j = 0, \pm 1, \pm 2, \ldots$$

to form a rectangular grid, see Figure 7.1.

Let us plan only to compute the approximate value F_{ij} of solutions $F(x_j, y_i)$ at intersection points of the grid in Ω. The intersection points $(x_j, y_i) = (y_0 + ih, x_0 + jk)$ of the grid are called nodes, simply denoted by (i, j).

Assume the total number of nodes in region Ω is N. Obviously, in order to determine the approximate value F_{ij} of the solution at each node, N systems of algebraic equations must be formed, one algebraic equation for each node. In the following we introduce the method of forming difference equations.

The Taylor expansion of a bivariate function $F(x, y)$ is:

$$F(y + h, x + k) = F(x, y) + \frac{\partial F}{\partial y}h + \frac{\partial F}{\partial x}k$$

$$+ \frac{1}{2}\left[\frac{\partial^2 F}{\partial y^2}h^2 + 2\frac{\partial^2 F}{\partial y\partial x}hk + \frac{\partial^2 F}{\partial x^2}k^2\right]$$

$$+ \frac{1}{6}\left[\frac{\partial^3 F}{\partial y^3}h^3 + 3\frac{\partial^3 F}{\partial y^2\partial x}h^2 k + 3\frac{\partial^3 F}{\partial y\partial x^2}hk^2 + \frac{\partial^3 F}{\partial x^3}k^3\right] \qquad (7.3.1)$$

$$+ \frac{1}{24}\left[\frac{\partial^4 F}{\partial y^4}h^4 + 4\frac{\partial^4 F}{\partial y^3\partial x}h^3 k + 6\frac{\partial^4 F}{\partial y^2\partial x^2}h^2 k^2\right.$$

$$\left. + 4\frac{\partial^4 F}{\partial y\partial x^3}hk^3 + \frac{\partial^4 F}{\partial x^4}k^4\right] + \ldots$$

For any node (i, j) and its four neighbor nodes 1, 2, 3, 4, see Figure 7.2.

From (7.3.1) we can respectively obtain the function values of four nodes 1, 2, 3, 4, that is,

$$F(y - h, x) = F(x, y) - \frac{\partial F}{\partial y}h + \frac{1}{2}\frac{\partial^2 F}{\partial y^2}h^2 - \frac{1}{6}\frac{\partial^3 F}{\partial y^3}h^3 + \frac{1}{24}\frac{\partial^4 F}{\partial y^4}h^4 + \ldots \qquad (7.3.2)$$

$$F(y, x - k) = F(x, y) - \frac{\partial F}{\partial x}k + \frac{1}{2}\frac{\partial^2 F}{\partial x^2}k^2 - \frac{1}{6}\frac{\partial^3 F}{\partial x^3}k^3 + \frac{1}{24}\frac{\partial^4 F}{\partial x^4}k^4 - \ldots \qquad (7.3.3)$$

$$F(y + h, x) = F(x, y) + \frac{\partial F}{\partial y}h + \frac{1}{2}\frac{\partial^2 F}{\partial y^2}h^2 + \frac{1}{6}\frac{\partial^3 F}{\partial y^3}h^3 + \frac{1}{24}\frac{\partial^4 F}{\partial y^4}h^4 + \ldots \qquad (7.3.4)$$

$$F(y, x + k) = F(x, y) + \frac{\partial F}{\partial x}k + \frac{1}{2}\frac{\partial^2 F}{\partial x^2}k^2 + \frac{1}{6}\frac{\partial^3 F}{\partial x^3}k^3 + \frac{1}{24}\frac{\partial^4 F}{\partial x^4}k^4 + \ldots \qquad (7.3.5)$$

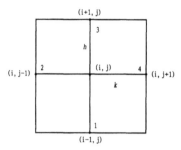

Figure 7.2 Relation of neighbor nodes

Adding (7.3.2) to (7.3.4) and adding (7.3.3) to (7.3.5), and then introducing the number of a point provides:

$$F_1 + F_3 = 2F_0 + \frac{\partial^2 F}{\partial y^2}h^2 + \frac{1}{12}\frac{\partial^4 F}{\partial y^4}h^4 + \cdots$$

$$F_2 + F_4 = 2F_0 + \frac{\partial^2 F}{\partial x^2}k^2 + \frac{1}{12}\frac{\partial^4 F}{\partial x^4}k^4 + \cdots$$

(7.3.6)

From (7.3.6) we get:

$$\frac{\partial^2 F}{\partial y^2} = \frac{F_1 - 2F_0 + F_3}{h^2} - \frac{h^2}{12}\frac{\partial^4 F}{\partial y^4}$$

(7.3.7)

$$\frac{\partial^2 F}{\partial x^2} = \frac{F_2 - 2F_0 + F_4}{k^2} - \frac{k^2}{12}\frac{\partial^4 F}{\partial x^4}$$

(7.3.8)

Considering $(X, Y) = F(x, y)$ in (7.1.3) and (7.1.4), by substituting (7.3.7) and (7.3.8) into (7.1.3) or (7.1.4) yields:

$$\frac{1}{h^2}(F_1 - 2F_0 + F_3) + \frac{1}{k^2}(F_2 - 2F_0 + F_4) = \frac{h^2}{12}\frac{\partial^4 F}{\partial x^4} + \frac{k^2}{12}\frac{\partial^4 F}{\partial y^4}$$

(7.3.9)

Omitting the error term of the right side we can get the difference equation for (7.1.3) or (7.1.4):

$$\frac{1}{h^2}(F_1 - 2F_0 + F_3) + \frac{1}{k^2}(F_2 - 2F_0 + F_4) = 0$$

(7.3.10)

The term omitted in (7.3.9) is:

$$R(y_i, x_j) = \frac{h^2}{12}\frac{\partial^4 F(\zeta_i, x_j)}{\partial y^4} + \frac{k^2}{12}\frac{\partial^4 F(y_i, \eta_j)}{\partial x^4}$$

(7.3.11)

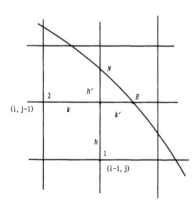

Figure 7.3 The circumstance when nodes are not equidistant

This is called a truncation error or error of approximation of the difference equation (7.3.10) approximating the differential equation (7.1.3) or (7.1.4). Obviously, the approximate error is $O(h^2) + O(k^2)$.

In the case of $h = k$, equation (5.4.37) can be reduced to:

$$F_1 + F_2 + F_3 + F_4 - 4F_0 = 0$$

(7.3.12)

Nodes are often not equidistant in practice. For example, points E and N on boundary Γ are not equidistant with respect to node (i, j) (see Figure 7.3). Next we find from the difference equations that nodes are not equidistant.

Applying (7.3.1) to the nodes in Figure 7.3, then:

$$F(N) = F_0 + h'\frac{\partial F}{\partial y} + \frac{h'^2}{2}\frac{\partial^2 F^2}{\partial y^2} + \frac{h'^3}{6}\frac{\partial^3 F}{\partial y^3} + \frac{h'^4}{24}\frac{\partial^4 F}{\partial y^4} + \cdots \tag{7.3.13}$$

$$F_1 = F_0 - h\frac{\partial F}{\partial y} + \frac{h^2}{2}\frac{\partial^2 F^2}{\partial y^2} - \frac{h^3}{6}\frac{\partial^3 F}{\partial y^3} + \frac{h^4}{24}\frac{\partial^4 F}{\partial y^4} - \cdots \tag{7.3.14}$$

Multiplying (7.3.13) and (7.3.14) respectively by h and h' yields:

$$hF(N) = hF_0 + hh'\frac{\partial F}{\partial y} + \frac{hh'^2}{2}\frac{\partial^2 F}{\partial y^2} + \frac{hh'^3}{6}\frac{\partial^3 F}{\partial y^3} + \frac{hh'^4}{24}\frac{\partial^4 F}{\partial y^4} + \cdots \tag{7.3.15}$$

$$h'F_1 = h'F_0 - h'h\frac{\partial F}{\partial y} + \frac{h'h^2}{2}\frac{\partial^2 F^2}{\partial y^2} - \frac{h'h^3}{6}\frac{\partial^3 F}{\partial y^3} + \frac{h'h^4}{24}\frac{\partial^4 F}{\partial y^4} - \cdots \tag{7.3.16}$$

Adding the two foregoing formulae yields:

$$hF(N) + h'F_1 = (h + h')F_0 + \frac{hh'}{2}(h + h')\frac{\partial^2 F}{\partial y^2}$$
$$+ \frac{1}{2}hh'(h'^2 - h^2)\frac{\partial^3 F}{\partial y^3} + \frac{1}{24}hh'(h^3 + h'^3)\frac{\partial^4 F}{\partial y^4}$$

then:

$$\frac{\partial^2 F}{\partial y^2} = \frac{2}{h + h'}\left(\frac{F(N) - F_0}{h'} - \frac{F_0 - F_1}{h}\right) + \frac{1}{3}(h - h')\frac{\partial^3 F}{\partial y^3} + \cdots$$

Thus we obtain the difference equation:

$$\frac{2}{h + h'}\left(\frac{F_N - F_0}{h'} - \frac{F_0 - F_1}{h}\right) + \frac{2}{k + k'}\left(\frac{F_E - F_0}{k'} - \frac{F_0 - F_2}{k}\right) = 0 \tag{7.3.17}$$

Its approximate error is $O(h - h') + O(k - k') + \cdots$.

7.3.2 Application example

To use the difference method to achieve a numerical transformation from the Mercator projection to a conformal conical projection, the regions of a Mercator and a conformal conical projection are shown in Figure 7.4.

Considering that the meridian intervals are equidistant and that the parallel intervals are gradually extended; by renumbering the nodes from top to bottom and from left to right, we can rewrite (7.3.17) as:

$$\frac{2}{(h + h')h'}F_{i-1,j} + \frac{1}{k^2}F_{i,j-1} - \left(\frac{2}{hh'} + \frac{2}{k^2}\right)F_{ij} + \frac{1}{k^2}F_{i,j+1} + \frac{2}{(h + h')h}F_{i+1,j} = 0 \tag{7.3.18}$$

Applying the above equation to nodes 1, 2, 3, 4, 5, 6 in Figure 7.4 we can obtain the system of linear equations:

$$TF = C \tag{7.3.19}$$

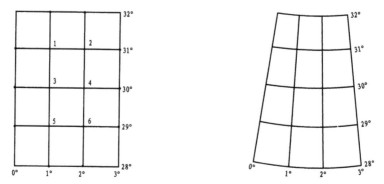

Figure 7.4 The regions of a Mercator and a conformal conical projection

where:

$$
T = \begin{pmatrix}
T_{00} & T_{01} & T_{02} & 0 & 0 & 0 \\
T_{10} & T_{11} & 0 & T_{13} & 0 & 0 \\
T_{20} & 0 & T_{22} & T_{23} & T_{24} & 0 \\
0 & T_{31} & T_{32} & T_{33} & 0 & T_{35} \\
0 & 0 & T_{42} & 0 & T_{44} & T_{45} \\
0 & 0 & 0 & T_{53} & T_{54} & T_{55}
\end{pmatrix}
\quad
F = \begin{pmatrix}
F_1 \\
F_2 \\
F_3 \\
F_4 \\
F_5 \\
F_6
\end{pmatrix}
\quad
C = \begin{pmatrix}
T_{06} \\
T_{16} \\
T_{26} \\
T_{36} \\
T_{46} \\
T_{56}
\end{pmatrix}
$$

7.4 FINITE ELEMENT METHOD

In this section we introduce another numerical method, the finite element method, for solving the Dirichlet problem (7.1.3) or (7.1.4). The foundation of the theory of the finite element method is the principle of variation; that is, solving a Dirichlet problem (7.1.3) is equivalent to obtaining the minimum of the functional (Li 1985):

$$
J[Y(x, y)] = \frac{1}{2} \iint_{\Omega} \left[\left(\frac{\partial Y}{\partial y} \right)^2 + \left(\frac{\partial Y}{\partial x} \right)^2 \right] dydx
\tag{7.4.1}
$$

under the boundary condition:

$$
Y|_{\Gamma} = D(x, y)
$$

This is a variational problem.

In fact, from the variational principle we know the Euler equation for the functional:

$$
J = \iint_{\Omega} F(y, x, Y, p, q)dydx
$$

i.e.,

$$
F_y - \frac{\partial}{\partial y} F_p - \frac{\partial}{\partial x} F_q = 0
\tag{7.4.2}
$$

where:

$$p = \frac{\partial Y}{\partial y}, \quad q = \frac{\partial Y}{\partial x}$$

If the functional $J = \iint_{\Omega} F(y, x, Y, p, q)dydx$ reaches an extremum at a function $Y(x, y)$,

then $Y(x, y)$ should satisfy the Euler equation (7.4.2).

In the functional (7.4.1), $D(x, y)$ is a continuous function on a given boundary. If the functional J reaches its extremum at the function $Y(x, y)$, then what kind of equation should $Y(x, y)$ satisfy?

Because:

$$F(y, x, Y, p, q) = p^2 + q^2$$

then:

$$F_p = 2p, \quad F_q = 2q, \quad F_x = 0$$

Substituting these into the Euler equation (7.4.2) yields:

$$\frac{\partial^2 Y}{\partial y^2} + \frac{\partial^2 Y}{\partial x^2} = 0 \quad (x, y) \in \Omega$$

$$Y|_\Gamma = D(x, y)$$

This is the Dirichlet problem of the Laplace equation.

This shows that to obtain the extremum of the functional

$$J = \iint_{\Omega} \left[\left(\frac{\partial Y}{\partial y} \right)^2 + \left(\frac{\partial Y}{\partial x} \right)^2 \right] dydx$$

is equivalent to solving the Dirichlet problem. Below we show how to obtain the numerical solution of a variational problem (7.4.1).

7.4.1 Principle of finite element method

Building a finite element net (dissection)

Suppose boundary Γ of the transformation region Ω is formed by meridians and parallels of the projection (see Figure 7.5). If new and old coordinates of the L boundary points are given, then we get the new coordinates of N interior points in the region of transformation. For this purpose, we take all the boundary points and all the transformation points (i.e., interior points) as vertices to link up finite non-overlapping triangular units (called surface units), and they cover the entire transformation region Ω.

The triangular units in Figure 7.5 are numbered (1)–(k_0). 1–N are the numbers of the transformation points, $N + 1$–$N + L$ are the numbers of boundary points, and i, j, m are the numbers given to the three vertices of each triangular unit (they should be arranged counterclockwise).

The purpose of the network partition is to turn the finding of a continuous solution of the Dirichlet problem into one of obtaining a numerical solution for the problem of

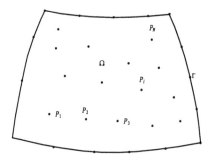 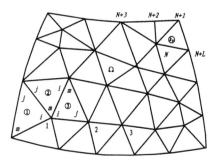

Figure 7.5   The transformation region and the construction of triangular units

variation (7.1.3). This involves obtaining approximate values $Y_1, Y_2, \ldots, Y_n$ for the true solution $Y(x, y)$ of the problem of variation (7.1.3) at each interior point.

After dissecting Ω as mentioned above, the functional $J[Y(x, y)]$ in (7.1.3) can be approximately written:

$$J(Y) \approx \sum_\Delta J_\Delta \tag{7.4.3}$$

where:

$$J_\Delta = \frac{1}{2} \iint_\Delta \left[\left(\frac{\partial Y}{\partial y} \right)^2 + \left(\frac{\partial Y}{\partial x} \right)^2 \right] dydx \tag{7.4.4}$$

In order to compute these integrals approximately, we assume that Y is a linear function of x and y in each triangular unit, that is, $Y(x, y)$ can be approximately replaced by a linear interpolation function.

Linear interpolation formula for the triangular unit

Suppose the vertices of the triangular units are, A_i, A_j, A_m, and p is an arbitrary point in Δ (see Figure 7.6). If there is a linear interpolation function:

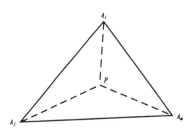

$$Y(P) \approx u = \alpha + \beta y + x \tag{7.4.5}$$

then there should be:

$$\left. \begin{array}{l} Y_i = \alpha + \beta y_i + \gamma x_i \\ Y_j = \alpha + \beta y_j + \gamma x_j \\ Y_m = \alpha + \beta y_m + \gamma x_m \end{array} \right\} \tag{7.4.6}$$

Figure 7.6   The triangular units

Regarding (7.4.5) and (7.4.6) as homogeneous systems of linear equations of unknown numbers $1, \alpha, \beta, \gamma$, since their solutions are not all zero, the determinant of the system should be zero according to the theory of systems of linear algebraic equations; that is:

$$\begin{vmatrix} u & 1 & y & x \\ Y_i & 1 & y_i & x_i \\ Y_j & 1 & y_j & x_j \\ Y_m & 1 & y_m & x_m \end{vmatrix} = 0$$

Expanding, beginning with the first line, we have:

$$u\begin{vmatrix}1 & y_i & x_i \\ 1 & y_j & x_j \\ 1 & y_m & x_m\end{vmatrix} - Y_i\begin{vmatrix}1 & y & x \\ 1 & y_j & x_j \\ 1 & y_m & x_m\end{vmatrix} + Y_j\begin{vmatrix}1 & y & x \\ 1 & y_i & x_i \\ 1 & y_m & x_m\end{vmatrix} - Y_m\begin{vmatrix}1 & y & x \\ 1 & y_i & x_i \\ 1 & y_j & x_j\end{vmatrix} = 0$$

Moving the last three terms of the above determinant to the right side and dividing simultaneously by the first determinant into the two sides, while taking account of the fact that these determinants respectively denote twofold coverage of the areas of the corresponding triangles, we obtain the linear interpolation formula for the triangular units as follows:

$$Y(P) \approx u = N_i Y_i + N_j Y_j + N_m Y_m \tag{7.4.7}$$

where:

$$N_i = \frac{2\Delta PA_j A_m}{2\Delta A_i A_j A_m} = \frac{1}{2\Delta}(a_i + b_i y + c_i x)$$

$$N_j = \frac{2\Delta PA_m A_j}{2\Delta A_i A_j A_m} = \frac{1}{2\Delta}(a_j + b_j y + c_j x) \tag{7.4.8}$$

$$N_m = \frac{2\Delta PA_i A_j}{2\Delta A_i A_j A_m} = \frac{1}{2\Delta}(a_m + b_m y + c_m x)$$

and:

$$\begin{aligned}
a_i &= y_j x_m - y_m x_j, & b_i &= x_j - x_m, & c_i &= y_m - y_j \\
a_j &= y_m x_i - y_i x_m, & b_j &= x_m - x_i, & c_j &= y_i - y_m \\
a_m &= y_i x_j - y_j x_i, & b_m &= x_i - x_j, & c_m &= y_j - y_i
\end{aligned} \tag{7.4.9}$$

In the linear interpolation formula (7.4.7), because N_i, N_j, N_m are the ratios of the triangular areas, they are called the area coordinates of point p, and:

$$N_i + N_j + N_m = 1 \tag{7.4.10}$$

Analysis of the triangular unit

The integral J_Δ in (7.4.4) can be computed if the linear interpolation formulae for the triangular units are given:

From (7.4.7) we have:

$$\frac{\partial Y}{\partial y} = \frac{\partial N_i}{\partial y}Y_i + \frac{\partial N_j}{\partial y}Y_j + \frac{\partial N_m}{\partial y}Y_m$$

$$= \frac{1}{2\Delta}(b_i Y_i + b_j Y_j + b_m Y_m) = \frac{1}{2\Delta}B_\Delta^T Y_\Delta$$

where:

$$Y_\Delta = (Y_i, Y_j, Y_m)^T, \quad B_\Delta = (b_i, b_j, b_m)^T$$

Thus,

$$\iint_\Delta \left(\frac{\partial Y}{\partial y}\right)^2 dydx = \iint_\Delta \frac{1}{4\Delta^2}(B_\Delta^T Y_\Delta)^2 dydx$$

Noting that B_Δ^T and Y_Δ are constant, and that the integral $\iint\limits_\Delta dy dx$ equals the area Δ of the triangular unit,

$$\iint\limits_\Delta \left(\frac{\partial Y}{\partial y}\right)^2 dy dx = Y_\Delta^T \frac{1}{4\Delta} B_\Delta B_\Delta^T Y_\Delta \tag{7.4.11}$$

where:

$$\Delta = \frac{1}{2}\begin{vmatrix} 1 & y_i & x_i \\ 1 & y_j & x_j \\ 1 & y_m & x_m \end{vmatrix} = \frac{1}{2}|c_i b_j - c_j b_i|$$

In fact,

$$B_\Delta^T Y_\Delta = (b_i b_j b_m)\begin{pmatrix} Y_i \\ Y_j \\ Y_m \end{pmatrix} = b_i Y_i + b_j Y_j + b_m Y_m$$

$$(B_\Delta^T Y_\Delta)^2 = Y_\Delta^T B_\Delta B_\Delta^T Y_\Delta = (Y_i Y_j Y_m)\begin{pmatrix} b_i \\ b_j \\ b_m \end{pmatrix}$$

$$(b_i b_j b_m)\begin{pmatrix} Y_i \\ Y_j \\ Y_m \end{pmatrix} = (b_i Y_i + b_j Y_j + b_m Y_m)^2 = (Y_i Y_j Y_m)\begin{pmatrix} b_i^2 & b_i b_j & b_i b_m \\ b_i b_j & b_j^2 & b_j b_m \\ b_i b_m & b_j b_m & b_m^2 \end{pmatrix}\begin{pmatrix} Y_i \\ Y_j \\ Y_m \end{pmatrix}$$

Similarly we have:

$$\frac{\partial Y}{\partial x} = \frac{\partial N_i}{\partial x} Y_i + \frac{\partial N_j}{\partial x} Y_j + \frac{\partial N_m}{\partial x} Y_m$$

$$= \frac{1}{2\Delta}(c_i Y_i + c_j Y_j + c_m Y_m) = \frac{1}{2\Delta} C_\Delta^T Y_\Delta$$

where:

$$Y_\Delta = (Y_i, Y_j, Y_m)^T, \quad C_\Delta = (c_i, c_j, c_m)^T$$

Thus,

$$\iint\limits_\Delta \left(\frac{\partial Y}{\partial x}\right)^2 dy dx = \iint\limits_\Delta \frac{1}{4\Delta^2}(C_\Delta^T Y_\Delta)^2 dy dx = Y_\Delta^T \frac{1}{4\Delta} C_\Delta C_\Delta^T Y_\Delta \tag{7.4.12}$$

Substituting (7.4.11) and (7.4.12) into (7.4.4) yields:

$$J_\Delta \approx \frac{1}{2} Y_\Delta^T \left(\frac{1}{4\Delta} B_\Delta B_\Delta^T + \frac{1}{4\Delta} C_\Delta C_\Delta^T\right) Y_\Delta$$

$$\therefore \quad J_\Delta \approx \frac{1}{2} Y_\Delta^T k_\Delta Y_\Delta \tag{7.4.13}$$

where $Y_\Delta^T = (Y_i Y_j Y_m)$, $Y_\Delta = (Y_i Y_j Y_m)^T$, K_Δ is called the stiffness matrix of a triangular unit, and its representative is:

$$K_\Delta = \begin{pmatrix} K_{ii}^{(\Delta)} & K_{ij}^{(\Delta)} & K_{im}^{(\Delta)} \\ K_{ji}^{(\Delta)} & K_{jj}^{(\Delta)} & K_{jm}^{(\Delta)} \\ K_{mi}^{(\Delta)} & K_{mj}^{(\Delta)} & K_{mm}^{(\Delta)} \end{pmatrix} \qquad (7.4.14)$$

and may be written:

$$K_\Delta = \frac{1}{4\Delta} \begin{pmatrix} b_i b_i & b_i b_j & b_i b_m \\ b_j b_i & b_j b_j & b_j b_m \\ b_m b_i & b_m b_j & b_m b_m \end{pmatrix} + \frac{1}{4\Delta} \begin{pmatrix} c_i c_i & c_i c_j & c_i c_m \\ c_j c_i & c_j c_j & c_j c_m \\ c_m c_i & c_m c_j & c_m c_m \end{pmatrix}$$

Then:

$$K_\Delta = \frac{1}{4\Delta} \begin{pmatrix} b_i^2 + c_i^2 & b_i b_j + c_i c_j & b_i b_m + c_i c_m \\ b_j b_i + c_j c_i & b_j^2 + c_j^2 & b_j b_m + c_j c_m \\ b_m b_i + c_m c_i & b_m b_j + c_m c_j & b_m^2 + c_m^2 \end{pmatrix} \qquad (7.4.15)$$

where:

$$\Delta = \frac{1}{2}|c_i b_j - c_j b_i|$$

Population analysis of the region

We have obtained the representative of the functional J_Δ of the triangular unit. Now we want to obtain the functional representative $J(Y)$ of the whole region. Thus we let:

$$Y = (Y_1, Y_2, \ldots, Y_N, Y_{N+1}, \ldots, Y_{N+L})^T$$

$$P_\Delta = \begin{pmatrix} 1 & & 0 & & 0 & \\ 0 & 0 & 0 & 1 & 0 & 0 & 0 \\ 0 & & 0 & & 1 & \end{pmatrix}$$

$$\underbrace{\hspace{2cm}}_{i \text{ column}} \quad \underbrace{\hspace{2cm}}_{j \text{ column}} \quad \underbrace{\hspace{2cm}}_{m \text{ column}}$$

P_Δ is a rectangular matrix of order $3 \times (N + L)$, and only three elements are 1, the rest being 0. Therefore:

$$Y_\Delta = P_\Delta Y$$

According to this formula, (5.4.59) can be written:

$$J_\Delta \approx \frac{1}{2} Y_\Delta^T P_\Delta^T K_\Delta P_\Delta Y$$

Substituting into (7.4.3) yields:

$$J(Y) \approx \frac{1}{2} Y^T K Y \qquad (7.4.16)$$

where Y^T is a row vector of dimension $N + L$, Y is a column vector of dimension $N + L$, and K is a matrix of order $(N + L) \times (N + L)$, called the total stiffness matrix.

$$K_\Delta = \sum_\Delta P_\Delta^T K_\Delta P_\Delta = \sum_\Delta \begin{pmatrix} \cdots & & \cdots & & \cdots & \\ \cdots & K_{i1}^{(\Delta)} & \cdots & K_{ij}^{(\Delta)} & \cdots & K_{im}^{(\Delta)} & \cdots \\ \cdots & & & & & \\ & K_{ji}^{(\Delta)} & \cdots & K_{jj}^{(\Delta)} & \cdots & K_{jm}^{(\Delta)} & \cdots \\ \cdots & & & & & \\ \cdots & K_{mi}^{(\Delta)} & \cdots & K_{mj}^{(\Delta)} & \cdots & K_{mm}^{(\Delta)} & \cdots \\ \cdots & & & \cdots & & \cdots & \end{pmatrix}$$

$$\qquad\qquad\qquad\qquad i \qquad\qquad j \qquad\qquad m$$

(7.4.17)

It can be seen from the above formula that the total stiffness matrix is a square matrix of order $(N + L)$, its elements $K_{\alpha\beta}$ being the sum of the elements $K_{\alpha\beta}^{(1)}$ and $K_{\alpha\beta}^{(N_0)}$ of each unit stiffness matrix.

Forming a system of equations with finite elements

Formula (7.4.16) shows that $J(Y)$ is a multivariate quadratic function of $Y_1, Y_2, \ldots,$ $Y_N, \ldots, Y_{N+L}$. According to the necessary condition for an extreme's existence, we have:

$$\frac{\partial}{\partial Y_s}\left(\frac{1}{2}Y^T KY\right) = 0, \quad s = 1, 2, \ldots, N + L \tag{7.4.18}$$

Noting that the total stiffness matrix K is positive definite symmetric, after derivation of the quadratic form (7.4.16) and substituting it into (7.4.18), we obtain the system of algebraic equations of $Y_1, Y_2, \ldots, Y_{N+L}$

$$KY = 0 \tag{7.4.19}$$

or writing them as:

$$\sum_{j=1}^{N+L} K_{ij}Y_j = 0, \quad i = 1, 2, \ldots, N + L \tag{7.4.20}$$

In order to form a system of equations with finite elements, boundary conditions must be forced on (7.4.20). Because the coordinates of each of the boundary points for the new projection are known, that is at boundary points:

$$Y_s = D(y_s, x_s), \quad s = N + 1, \ldots, N + L$$

by substituting the above formula into (7.4.20) and crossing off the last L superfluous equations, since the last L points are boundary points, we obtain the finite element system of equations for computing the new projection coordinates for every transformation point:

$$\sum_{j=1}^{N} K_{ij}Y_j = G_i, \quad i = 1, 2, \ldots, N \tag{7.4.21}$$

where:

$$G_i = -\sum_{j=N+1}^{N+L} K_{ij}D_j, \quad i = 1, 2, \ldots, N$$

Characteristics of the finite element method and an estimate of its error

The coefficient matrix K for the system of algebraic equations $KY = G$ obtained by finite element discretization has the following characteristics.

The positive definite coefficient matrix is symmetric and sparse; that is, most of its elements are zero. Every equation in the system of equations has at most only a certain number of elements that are not zero; that is K is a band matrix. It can be proved that problem (7.4.1) is positive definite and has a unique minimum solution. Based on this it can be furthermore proved that a system of equations $KY = G$ has a unique solution.

The properties of positive definite, symmetry, and sparseness of the coefficient matrix substantially benefit the actual solving. Generally the convenient method involving a Gaussian elimination and iteration can be used to obtain a solution. Next we explain simply the error of the numerical solution obtained from a finite element method.

According to the variational principle, the function $u(x, y)$ at which the functional $J(v)$ reaches its minimum can be regarded as the solution of the boundary problem, and is called a generalized solution. Suppose the maximum of the line element length after dissection is h, and the finite element solution is $u_h(x, y)$ then, under a condition of no roundoff, the error of the numerical solution is:

$$\varepsilon(x, y) = u_h(x, y) - u(x, y)$$

After computation we obtain the formula for estimating the error of a finite element solution:

$$\|u_h - u\| \le \frac{\sqrt{K_1}}{\sin\theta} M_2 h \tag{7.4.22}$$

where:

$$K_1 = 32 \iint_\Omega dy dx, \quad \left|\frac{\partial^2 u}{\partial y^2}\right| \le M_2, \quad \left|\frac{\partial^2 u}{\partial x^2}\right| \le M_2$$

and $\sin\theta$ is the minimum value of the sine of an interior angle of a surface element triangle. It can be seen from (7.4.22) that in order to guarantee the precision of a finite element solution, the maximum length h should be sufficiently small and the value of $|\sin\theta|$, and therefore the interior angle of a surface element triangle, should not be very small.

7.4.2 Programming of the finite element method

In order to solve the boundary problem (7.4.1) using a finite element method, we summarize the procedure as follows:

- Step 1: Dissecting the transformation region into some triangular units, then individually coding the triangular units and the vertices of each triangle and points for transformation as well as the respective boundary points.
- Step 2: Computing the stiffness matrix of each triangular unit using formulae (7.4.15), (7.4.9), and (7.4.17), and gradually accumulating the results to form a total stiffness matrix.
- Step 3: Forming a finite element system of equations using (7.4.21).

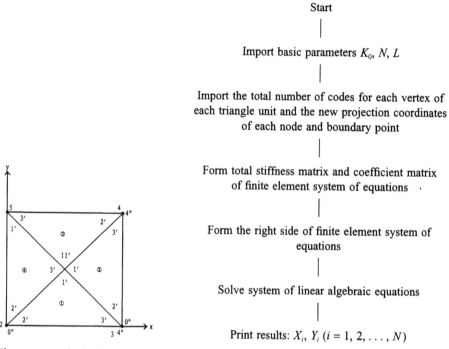

Start

|

Import basic parameters K_0, N, L

|

Import the total number of codes for each vertex of each triangle unit and the new projection coordinates of each node and boundary point

|

Form total stiffness matrix and coefficient matrix of finite element system of equations ·

|

Form the right side of finite element system of equations

|

Solve system of linear algebraic equations

|

Print results: X_i, Y_i ($i = 1, 2, \ldots, N$)

Figure 7.7   Block diagram for finite element transformation of conformal projections

- Step 4: Solving the system of linear algebraic equations to obtain the coordinates for the transformation points of the new projection.

A total block diagram for finite element transformation of conformal projections is as shown in Figure 7.7.

The key to finite element programming is to form the total stiffness matrix. The total stiffness matrix $[K]$ is formed by the superposition of unit stiffness matrices $[K]^e$.

Note that the elements of the unit stiffness matrix are arranged as (i, j, m) by the local coding of nodes, but the elements of the total stiffness matrix are arranged as $(1, 2, \ldots, N, \ldots, N + L)$ in the total coding of nodes. The order of these two kinds of matrices is different. The unit $[K]^e$ matrices are 3rd order, but $[K]$ is of order $(N + L)$ by $(N + L)$. Therefore, the unit stiffness matrix must be remolded before superimposition.

First, expand $[K]^e$ from an order 3 by 3 to an order $(N + L)$ by $(N + L)$. This kind of expanded unit stiffness matrix is called a unit contribution matrix $[K]^{(e)}$.

Second, move the nine elements which are arranged with local coding in $[K]^e$ and rearrange them in an expanded matrix according to the sequence of the total coding, meanwhile filling the gaps with zero elements.

The unit contribution matrix can be involved only after remolding as above and forming a total stiffness matrix.

In order to save storage, forming the unit contribution matrix should be combined with superimposition in programming; that is, by being integrated with a method of 'sit in the right seat, superimpose, and move at the same time.' Examples of Section 7.4 will be presented in Appendix 3, Section A3.4.

Comparing the finite element method and the difference method for solving the same boundary problem, it seems that the finite element method is more troublesome, but the precision of computation is almost the same as that using the difference method. However, the steps in the solution using the finite element method are rather standard and invariant, although the regions to be transformed are not divided into equal parts. This is convenient for computerized calculation. Based on the variational principle, the finite element method is a further development of the Riesz method. Using it to solve a boundary problem in a complicated region is a very good discretization method. It can overcome some shortcomings of the difference method and realize the numerical transformation between any two conformal projections and the direct and inverse numerical transformation for a single conformal projection.

CHAPTER EIGHT

The Third Type of Coordinate Transformation

8.1 PRINCIPLE AND APPLICATIONS

In computer-assisted mapping, the computation of the coordinates of a point proceeds according to the following three cases. For the first, given the geographical coordinates (B_i, l_i) of points in the mapping area, we can calculate their plane rectangular coordinates (x_i, y_i) from map projection equations $x = f_1(B, l), y = f_2(B, l)$. For the second, if the plane rectangular coordinates (x_i, y_i) on the original base map are known, we can obtain their geographical coordinates (B_i, l_i) using the inverse formulae for the projection, and then obtain the values (x_i, y_i) of the newly edited map from its forward projection equations. For the third case, as in the second case, the plane rectangular coordinates (x_i, y_i) of the original base map are given. First, from formulae for map projection transformation, we can directly compute the plane rectangular coordinates (x_i, y_i) on the corresponding newly edited map. All these cases belong to the problem of transformation of the coordinates of the points. They can be summed up as three types according to nature of the research. The first type, $(B_i, l_i) \Leftrightarrow (x_i, y_i)$, is the goal of map projection study, and the second type, i.e. $(x_i, y_i) \to (X_i, Y_i)$, is the aim of map projection transformation.

Now we consider the problem of the third type of coordinate transformation. It can be described as finding coordinates (x, y) of the point of intersection between a curve $y = f(x)$ on the plane of projection and a meridian $x = f_1(B, l_0), y = f_2(B, l_0)$ or parallel of latitude $x = f_1(B_0, l), y = f_2(B_0, l)$ (Yang 1984b).

For the general solution:

$$f_2(B, l_0) = f[f_1(B, l_0)] \tag{8.1.1}$$

or:

$$f_2(B_0, l) = f[f_1(B_0, l)] \tag{8.1.2}$$

Let:

$$F(B) = f_2(B, l_0) - f[f_1(B, l_0)] \tag{8.1.3}$$

or:

$$F(l) = f_2(B_0, l) - f[f_1(B_0, l)] \tag{8.1.4}$$

We can calculate B or l by Newton's iteration formula, i.e.,

$$B_{i+1} = B_i - \frac{F(B_i)}{F'(B_i)} \quad |B_{i+1} - B_i| < \varepsilon \tag{8.1.5}$$

or:

$$l_{i+1} = l_i - \frac{F(l_i)}{F'(l_i)} \quad |l_{i+1} - l_i| < \varepsilon \tag{8.1.6}$$

Since (B, l_0) or (B_0, l) are usually known, the coordinates (x, y) of the intersecting point can be obtained.

In particular, assuming that x is constant, and the value of l_0 is given, we can get (B, y) from l:

$$f_1(B, l_0) = x$$

Let:

$$F(B) = f_1(B, l_0) - x \tag{8.1.7}$$

or, assuming that x is constant and the value of B_0 is given, we can calculate (l, y) by letting:

$$F(l) = f_1(B_0, l) - x \tag{8.1.8}$$

Assuming that y is constant and the value of l_0 is known, we can calculate (B, x) by letting:

$$F(B) = f_2(B, l_0) - y \tag{8.1.9}$$

or assuming that y is constant and the value of B_0 is known, we can calculate (l, x) by letting:

$$F(l) = f_2(B_0, l) - y \tag{8.1.10}$$

All these cases can be summed up in two expressions:

$$\{B, y\} \Leftrightarrow \{l, x\}; \quad \{B, x\} \Leftrightarrow \{l, y\}$$

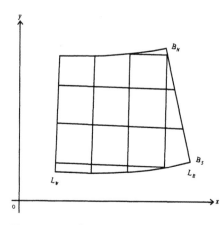

Figure 8.1 Plotting coordinate grid at latitude/longitude along the margin

They belong to the problem of coordinate transformation for points on a marginal line. This is the so-called third type of coordinate transformation or mixed coordinate transformation.

As is shown in Figure 8.1, if we want to plot an ordinate line at a northern or southern parallel which defines the margin, we must find (l, y) from B_S and x or find (l, y) from B_N and x. To plot an abscissa line at an east or west limiting meridian, we are required to find (B, x) from l_W and y, or find (B, x) from l_E and y, or find (l, x) from (B_S, y), or find (l, x) from (B_N, y).

In Figure 8.2, in order to show the intersections for the graticule of latitude/longitude along the neatline or map border, coordinate

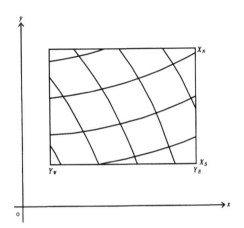

Figure 8.2 Plotting the graticule at rectangular margin

transformation must be performed just as described above.

In conventional cartography, whether we draw a square grid with the Chinese unit of length 'li' along the margin of the graticule of latitude/longitude or show border divisions of the latitude/longitude graticule along a rectangular neatline, the lines or tic marks are determined directly by graphical methods. As a result, the problem of the third type of coordinate transformation was not obvious in the past and has rarely been studied. It is a critical problem, however, in computer-assisted mapping. We must give considerable attention to such research; otherwise coordinate transformation for points along the marginal lines cannot be achieved.

In this chapter, we present an analytical method, a numerical method, and a method of linear interpolation for the third type of coordinate transformation, illustrating this with some of the map projections most commonly used. We will take conic projection as an example to illustrate the method of the third type of coordinate transformation.

First, we will apply the third type of coordinate transformation to conic projections. From equation (2.4.1), coordinate formulae for conic projections have the form:

$$\left. \begin{array}{ll} y = \rho_s - \rho\cos\delta, & x = \rho\sin\delta \\ \rho = f(B), & \delta = \alpha l \end{array} \right\} \tag{8.1.11}$$

They can also take the form:

$$l = \frac{1}{\alpha}\delta, \quad \delta = \arcsin\frac{x}{\rho} = \arccos\frac{\rho_s - y}{\rho} \tag{8.1.12}$$

or:

$$\rho = \frac{x}{\sin\delta} \tag{8.1.13}$$

or:

$$\rho = \frac{\rho_s - y}{\cos\delta} \tag{8.1.14}$$

8.1.1 The third type of coordinate transformation for the conformal conic projection

The equation for the conformal conic projection has the form:

$$\rho = \frac{C}{U^\alpha} \tag{8.1.15}$$

Let us consider the following conditions:

- Calculating (l, y) if the values of (B, x) are known:
  Since (B, x) is given, we can find these unknowns with equations (8.1.15) and (8.1.12)

$$\rho = \frac{C}{U^\alpha}, \quad \delta = \arcsin\frac{x}{\rho} \tag{8.1.16}$$

Hence, (l, y) can be calculated with the formulae

or:
$$\left. \begin{array}{l} l = \dfrac{1}{\alpha}\delta, \quad y = \rho_s - \rho\cos\delta \\[2mm] y = \rho_s - \sqrt{(\rho+x)(\rho-x)} \end{array} \right\} \tag{8.1.17}$$

- Calculating (B, x) if values of (l, y) are known:
  Since (l, y) is given, we can calculate (B, x) with equations (8.1.11) and (8.1.14):

$$\delta = \alpha l, \quad \rho = \frac{\rho_s - y}{\cos\delta} \tag{8.1.18}$$

From equation (8.1.15), we obtain:

$$\ln\rho = \ln C - \alpha \ln U$$

Then:

$$q = \frac{1}{\alpha}(\ln C - \ln\rho) \tag{8.1.19}$$

where q is isometric latitude. Using equation (3.7.18), we can calculate latitude B. But the value of x can be calculated from equation (8.1.11):

$$x = \rho\sin\delta \tag{8.1.20}$$

- Given values of (B, y), to obtain (l, x):
  Since values of (B, y) are known, equations (8.1.15) and (8.1.12) can be written in the form:

$$\rho = \frac{C}{U^\alpha}, \quad \delta = \arccos\frac{\rho_S - y}{\rho} \tag{8.1.21}$$

Then we obtain:

$$l = \frac{1}{\alpha}\delta, \quad x = \rho\sin\delta \tag{8.1.22}$$

- Given values of (l, x), to calculate (B, y):
  Since the values of (l, x) are known, from equations (8.1.11) and (8.1.13), we have:

$$\delta = \alpha l, \quad \rho = \frac{x}{\sin\delta} \tag{8.1.23}$$

Then we obtain:

$$\left. \begin{array}{l} q = \dfrac{1}{\alpha}(\ln C - \ln\rho) \\[2mm] y = \rho_s - \rho\cos\delta \end{array} \right\} \tag{8.1.24}$$

where q is isometric latitude. Latitude B can be found with equation (3.7.18).

8.1.2 The third type of coordinate transformation for the equivalent conic projection

The equivalent or equal-area conic projection takes the following form:

$$\rho^2 = \frac{2}{\alpha}(C - F) \tag{8.1.25}$$

Hence, we get:

$$F = C - \frac{\alpha\rho^2}{2} \tag{8.1.26}$$

where:

$$F = \int_0^B MrdB$$

- Using values of (B, x) to find (l, y):
  Since the (B, x) are known, from equations (8.1.25) and (8.1.12), we have:

$$\rho^2 = \frac{2}{\alpha}(C - F), \quad \delta = \arcsin\frac{x}{\rho} \tag{8.1.27}$$

Then we get:

$$\left. \begin{array}{l} l = \dfrac{1}{\alpha}\delta, \quad y = \rho_s - \rho\cos\delta \\[2mm] \text{or:} \\[1mm] y = \rho_s - \sqrt{(\rho + x)(\rho - x)} \end{array} \right\} \tag{8.1.28}$$

- Using values of (l, y) to get (B, x):
  From equations (8.1.11) and (8.1.26) we have:

$$x = \rho\sin\delta, \quad F = C - \frac{\alpha\rho^2}{2} \tag{8.1.29}$$

where:

$$\delta = \alpha l, \quad \rho = \frac{\rho_s - y}{\cos\delta}$$

Now that equivalent isometric latitude F is known, we can obtain latitude B with equation (3.7.22).

- Using values of (B, y) to find (l, x):
  Incorporating equations (8.1.25) and (8.1.12), we have:

$$\rho^2 = \frac{2}{\alpha}(C - F), \quad \delta = \arccos\frac{\rho_s - y}{\rho} \tag{8.1.30}$$

Then, we obtain:

$$l = \frac{1}{\alpha}\delta, \quad x = \rho\sin\delta \tag{8.1.31}$$

- Using values of (l, x) to find (B, y):
  Incorporating equations (8.1.11) and (8.1.26) we have:

$$y = \rho_s - \rho\cos\delta, \quad F = C - \frac{\alpha\rho^2}{2} \tag{8.1.32}$$

where:

$$\delta = \alpha l, \quad \rho = \frac{x}{\sin \delta}$$

Now that equivalent isometric latitude F is given, we can obtain latitude B using equation (3.7.22).

8.1.3 The third type of coordinate transformation for the equidistant conic projection

The equations for the conical equidistant projection take the form:

$$\rho = C - S \qquad\qquad (8.1.33)$$

Hence,

$$S = C - \rho \qquad\qquad (8.1.34)$$

where:

$$S = \int_0^B M dB$$

- Using given values of (B, x) to find (l, y):
  Taking into account equations (8.1.33) and (8.1.12), we have:

$$\rho = C - S, \quad \delta = \arcsin\frac{x}{\rho} \qquad\qquad (8.1.35)$$

  Substituting from (8.1.35) into (8.1.28), we can get values of (l, y).
- Using given values of (l, y) to find (B, x):
  Combining equations (8.1.11) and (8.1.34) we obtain the formula for calculating y:

$$x = \rho \sin \delta, \quad S = C - \rho \qquad\qquad (8.1.36)$$

where:

$$\delta = \alpha l, \quad \rho = \frac{\rho_s - y}{\cos \delta}$$

Given the value of equivalent isometric latitude S, we can calculate the latitude B from the formula (3.7.20).

- Using given values of (B, y) to find (l, x):
  Taking into account equations (8.1.33) and (8.1.12), we have:

$$\rho = C - S, \quad \delta = \arccos\frac{\rho_s - y}{\rho} \qquad\qquad (8.1.37)$$

Applying these values to equation (8.1.31), we can obtain values of (l, x).
- Using values of (l, x) to find (B, y):
  Taking into account equations (8.1.11) and (8.1.34) we have:

$$y = \rho_s - \rho \cos \delta, \quad S = C - \rho \qquad\qquad (8.1.38)$$

where:

$$\delta = \alpha l, \quad \rho = \frac{x}{\sin\delta}$$

As equidistant isometric latitude S is given, we can find latitude B from equation (3.7.20).

8.1.4 Applications

Coordinate transformation for the points on the rectangular map border

Knowing that the map has adopted the conformal conic projection with two standard parallels, $B_1 = 24°$, $B_2 = 29°$. With central meridian $L_0 = 117°$, map scale $\mu_0 = 1:500\ 000$, for the coordinate computing formula of the projection see (8.1.11) and (8.1.15).

Projection constants are $\rho_s = 13\ 394\ 552$ m, $a = 0.4463428$, $C = 15\ 818\ 197$ m.

Now we will solve for the coordinates (x, y) of the intersection between the rectangle margin line and the parallels and meridians.

1. To find the coordinates of different intersections with the south–north margin lines, that is, knowing l, y, find B, y.

    For example, given $l = 119°30'$, $y = 874\ 000$, then the computed result is: $B = 28°52'00''.1683$, $x = 243\ 873.3246$.

2. To find the coordinate intersections on the east–west line, that is given B, x, find l, y.

    For example, known are $B = 29°$, $x = 222\ 000$, then the computed result is: $l = 2°16'42''.3558$, $y = 888\ 352.2875$.

8.2   TRANSVERSE MERCATOR (OR GAUSS–KRÜGER) PROJECTION

From (2.3.17) and (2.3.20), the direct and inverse coordinate formulae of the Gauss–Krüger projection with variable coefficients, respectively, are known. Now we deduce analytical expressions for the third type of coordinate transformation of the Gauss–Krüger projection.

8.2.1   Calculating (*l*, *y*) when values of (*B*, *x*) are given

Inverting series (A1.6) from Appendix 1 and the second expression of equation (2.3.17), we can obtain:

$$
\begin{aligned}
l = {}& \frac{1}{N\cos B}x + \frac{1}{6N^3\cos B}(-1 + t^2 - \eta^2)x^3 \\
& + \frac{1}{120N^5\cos B}(5 - 2t^2 + 9t^4 + 6\eta^2 + 38t^2\eta^2)x^5 \\
& + \frac{1}{5040N^7\cos B}(-61 + 31t^2 + 45t^4 + 225t^6)x^7
\end{aligned} \tag{8.2.1}
$$

Then, introducing values of (B, l) into the first expression of equation (2.3.17), we can obtain the value of y.

8.2.2 Calculating (B, x) if values of (l, y) are known

Since y is known, we can obtain $S = y$. Then the latitude of footpoint latitude B_f can be calculated from equation (3.7.20).

Using the inversion series (A1.6) from Appendix 1 and the second expression of equation (2.3.20), we obtain:

$$x = N_f \cos B_f \cdot l + \frac{1}{6} N_f \cos^3 B_f (1 + 2t_f^2 + \eta_f^2) l^3$$

$$+ \frac{1}{120} N_f \cos^5 B_f (5 + 12t_f^2 + 16t_f^4 + 14\eta_f^2 + 32t_f^2\eta_f^2) l^5 \qquad (8.2.2)$$

$$+ \frac{1}{5040} N_f \cos^7 B_f (61 + 214t_f^2 + 200t_f^4 + 272t_f^6) l^7$$

Finally, we can obtain the latitude B by introducing values of x and B_f into the first expression of equation (2.3.20).

8.2.3 Calculating (l, x) when values of (B, y) are given

Modify the first expression of equation (2.3.17) in the following form:

$$y - S = a_1 l' + a_2 l'^2 + a_3 l'^3 + \dots \qquad (8.2.3)$$

where:

$$l' = l^2, \quad a_1 = \frac{1}{2} Nt \cos^2 B$$

$$a_2 = \frac{1}{24} Nt (5 - t^2 + 9\eta^2 + 4\eta^4) \cos^4 B$$

$$a_3 = \frac{1}{720} Nt (61 - 58t^2 + 270\eta^2 - 330t^2\eta^2) \cos^6 B$$

From the inverse series (A1.6) in Appendix 1 we get:

$$l' = b_1(y - S) + b_2(y - S)^2 + b_3(y - S)^3 + \dots \qquad (8.2.4)$$

where:

$$b_1 = \frac{1}{a_1}, \quad b_2 = -\frac{a_2}{a_1^3}, \quad b_3 = \frac{2a_2^2}{a_1^5} - \frac{a_3}{a_1^4}$$

Then, we obtain:

$$l = \sqrt{l'} \qquad (8.2.5)$$

Introducing (l, B) into the second expression of equation (2.3.17) gives the value of x.

8.2.4 Calculating (B, y) when values of (l, x) are given

Using Newton's method of iteration, we can calculate the value of latitude B from the second expression of equation (2.3.17). It takes the form:

$$B_{i+1} = B_i - \frac{f(B_i)}{f'(B_i)} \quad (i = 0, 1, 2, \ldots) \Bigg\}$$

$$B_0 = \arccos\frac{x}{al}, \qquad |B_{i+1} - B_i| < \varepsilon$$

(8.2.6)

where:

$$f(B) = N\cos B \cdot l + \frac{1}{6}N(1 - t^2 + \eta^2)\cos^2 B \cdot l^3$$

$$+ \frac{1}{120}N(5 - 18t^2 + t^4 + 14\eta^2 - 58t^2\eta^2)\cos^5 B \cdot l^5 - x$$

$$f'(B) = -N(1 - \eta^2 + \eta^4)\sin B \cdot l - \frac{1}{6}Nt(5 - t^2 + 4\eta^2 + t^2\eta^2)\cos^3 B \cdot l^3$$

$$- \frac{1}{120}Nt(61 - 58t^2 + 209\eta^2 - 272t^2\eta^2)\cos^5 B \cdot l^5$$

and a is the semimajor axis of the earth-ellipsoid, ε is the given precision. Latitude B can be calculated by iteration using the above equation.

Substituting values of (B, l) into the first expression of equation (2.3.17) will give the value of y.

8.3   CONFORMAL PROJECTIONS

Above we dealt with the problem of the third type of coordinate transformation for the transverse Mercator projection and conic projections by an analytical method. In this section we will derive numerical formulae to calculate the third type of coordinate transformation for conformal projections.

According to equations (5.1.42) and (5.1.46), we can respectively write expansions of direct and inverse coordinate transformation formulae for conformal projections with constant coefficients:

$$\begin{aligned}
y = {}& y_0 + a_1\Delta q - b_1 l + a_2(\Delta q^2 - l^2) - b_2 \cdot 2\Delta ql + a_3(\Delta q^3 - 3\Delta ql^2) \\
& - b_3(3\Delta q^2 l - l^3) + a_4(\Delta q^4 - 6\Delta q^2 l^2 + l^4) - b_4(4\Delta q^3 l - 4\Delta ql^3) \\
x = {}& x_0 + b_1\Delta q + a_1 l + b_2(\Delta q^2 - l^2) + a_2 \cdot 2\Delta ql + b_3(\Delta q^3 - 3\Delta ql^2) \\
& + a_3(3\Delta q^2 l - l^3) + b_4(\Delta q^4 - 6\Delta q^2 l^2 + l^4) + a_4(4\Delta q^3 l - 4\Delta ql^3)
\end{aligned} \Bigg\}$$

(8.3.1)

where l is the difference in longitude and Δq is the difference in isometric latitude relative to origin (x_0, y_0) for the series expansion.

$$\begin{aligned}
q = {}& q_0 + a_1\Delta y - b_1\Delta x + a_2(\Delta y^2 - \Delta x^2) - b_2 \cdot 2\Delta y\Delta x + a_3(\Delta y^3 - 3\Delta y\Delta x^2) \\
& - b_3(3\Delta y^2\Delta x - \Delta x^3) + a_4(\Delta y^4 - 6\Delta y^2\Delta x^2 + \Delta x^4) - b_4(4\Delta y^3\Delta x - 4\Delta y\Delta x^3) \\
l = {}& l_0 + b_1\Delta y + a_1\Delta x + b_2(\Delta y^2 - \Delta x^2) + a_2 \cdot 2\Delta y\Delta x + b_3(\Delta y^3 - 3\Delta y\Delta x^2) \\
& + a_3(3\Delta y^2\Delta x - \Delta x^3) + b_4(\Delta y^4 - 6\Delta y^2\Delta x^2 + \Delta x^4) + a_4(4\Delta y^3\Delta x - 4\Delta y\Delta x^3)
\end{aligned} \Bigg\}$$

(8.3.2)

where $\Delta x = x - x_0$, $\Delta y = y - y_0$, and (l_0, q_0) are the longitude and isometric latitude at (x_0, y_0). Coefficients in equations (8.3.1) and (8.3.2) can be calculated by a method of numerical transformation for conformal projections.

8.3.1 Calculating (*l*, *y*) if values of (*B*, *x*) are given

The second expression of equation (8.3.1) can also be expressed in the form:

$$x = x_0 + (b_1\Delta q + b_2\Delta q^2 + b_3\Delta q^3 + b_4\Delta q^4) + (a_1 + 2a_2\Delta q + 3a_3\Delta q^2 + 4a_4\Delta q^3) \cdot l$$
$$+ (-b_2 - 3b_3\Delta q - 6b_4\Delta q^2)l^2 + (-a_3 - 4a_4\Delta q)l^3 + b_4 l^4 \tag{8.3.3}$$

Introducing the following symbols,

$$\left.\begin{aligned}
X &= x - x_0 - (b_1\Delta q + b_2\Delta q^2 + b_3\Delta q^3 + b_4\Delta q^4) \\
A_1 &= a_1 + 2a_2\Delta q + 3a_3\Delta q^2 + 4a_4\Delta q^3 \\
A_2 &= -b_2 - 3b_3\Delta q - 6b_4\Delta q^2 \\
A_3 &= -a_3 - 4a_4\Delta q \\
A_4 &= b_4
\end{aligned}\right\} \tag{8.3.4}$$

equation (8.3.3) takes the form:

$$X = A_1 l + A_2 l^2 + A_3 l^3 + A_4 l^4 \tag{8.3.5}$$

From the inverse series we can obtain:

$$l = l_0 + B_1 X + B_2 X^2 + B_3 X^3 + B_4 X^4 \tag{8.3.6}$$

where:

$$B_1 = \frac{1}{A_1}, \quad B_2 = \frac{A_2}{A_1^3}, \quad B_3 = \frac{2A_2^2}{A_1^5} - \frac{A_3}{A_1^4},$$

$$B_4 = -\frac{5A_2^3}{A_1^7} + \frac{5A_2 A_3}{A_1^6} - \frac{A_4}{A_1^5}$$

After *l* is calculated from equation (8.3.6), we can obtain *y* by introducing (Δ*q*, *l*) into the first expression of equation (8.3.1).

8.3.2 Calculating (*B*, *x*) if values of (*l*, *y*) are given

The second expression of equation (8.3.2) can be written as:

$$l = l_0 + (b_1\Delta y + b_2\Delta y^2 + b_3\Delta y^3 + b_4\Delta y^4) + (a_1 + 2a_2\Delta y + 3a_3\Delta y^2 + 4a_4\Delta y^3)\Delta x$$
$$+ (-b_2 - 3b_3\Delta y - 6b_4\Delta y^2)\Delta x^2 + (-a_3 - 4a_4\Delta y)\Delta x^3 + b_4\Delta x^4 \tag{8.3.7}$$

Introducing the following symbols,

$$\left.\begin{aligned}
L &= l - l_0 - (b_1\Delta y + b_2\Delta y^2 + b_3\Delta y^3 + b_4\Delta y^4) \\
A_1 &= a_1 + 2a_2\Delta y + 3a_3\Delta y^2 + 4a_4\Delta y^3 \\
A_2 &= -b_2 - 3b_3\Delta y - 6b_4\Delta y^2 \\
A_3 &= -a_3 - 4a_4\Delta y \\
A_4 &= b_4
\end{aligned}\right\} \tag{8.3.8}$$

we get:

$$L = A_1 \Delta x + A_2 \Delta x^2 + A_3 \Delta x^3 + A_4 \Delta x^4 \tag{8.3.9}$$

Using the inverse series (A1.6) in Appendix 1, we can obtain:

$$\Delta x = B_1 L + B_2 L^2 + B_3 L^3 + B_4 L^4 \tag{8.3.10}$$

where:

$$B_1 = \frac{1}{A_1}, \quad B_2 = \frac{A_2}{A_1^3}, \quad B_3 = \frac{2A_2^2}{A_1^5} - \frac{A_3}{A_1^4},$$

$$B_4 = -\frac{5A_2^3}{A_1^7} + \frac{5A_2 A_3}{A_1^6} - \frac{A_4}{A_1^5}$$

But,

$$x = x_0 + \Delta x \tag{8.3.11}$$

Introducing values of $(\Delta x, \Delta y)$ into the first expression of equation (8.3.2), we can obtain the isometric latitude q. Further, from equation (3.7.18) we can calculate the latitude B.

8.3.3 Calculating (l, x) if values of (B, y) are given

The first expression of equation (8.3.1) can be expressed as:

$$y = y_0 + (a_1 \Delta q + a_2 \Delta q^2 + a_3 \Delta q^3 + a_4 \Delta q^4) + (-b_1 - 2b_2 \Delta q - 3b_3 \Delta q^2 - 4b_4 \Delta q^3)l$$
$$+ (-a_2 - 3a_3 \Delta q - 6a_4 \Delta q^2)l^2 + (b_3 + 4b_4 \Delta q)l^3 + a_4 l^4 \tag{8.3.12}$$

Introducing the following symbols,

$$\left. \begin{array}{l} Y = y - y_0 - (a_1 \Delta q + a_2 \Delta q^2 + a_3 \Delta q^3 + a_4 \Delta q^4) \\ A_1 = -b_1 - 2b_2 \Delta q - 3b_3 \Delta q^2 - 4b_4 \Delta q^3 \\ A_2 = -a_2 - 3a_3 \Delta q - 6a_4 \Delta q^2 \\ A_3 = b_3 + 4b_4 \Delta q \\ A_4 = a_4 \end{array} \right\} \tag{8.3.13}$$

we get:

$$Y = A_1 l + A_2 l^2 + A_3 l^3 + A_4 l^4 \tag{8.3.14}$$

From inversion of the series (A1.6) in Appendix 1, we obtain:

$$l = l_0 + B_1 Y + B_2 Y^2 + B_3 Y^3 + B_4 Y^4 \tag{8.3.15}$$

where:

$$B_1 = \frac{1}{A_1}, \quad B_2 = -\frac{A_2}{A_1^3}, \quad B_3 = \frac{2A_2^2}{A_1^5} - \frac{A_3}{A_1^4},$$

$$B_4 = -\frac{5A_2^3}{A_1^7} + \frac{5A_2 A_3}{A_1^6} - \frac{A_4}{A_1^5}$$

After calculating λ from equation (8.3.15), we can obtain x by substituting values of (B, l) into the second expression of equation (8.3.1).

8.3.4 Calculating (B, y) if the values of (l, x) are given

The second expression of equation (8.3.2) can be expressed as:

$$l = l_0 + (a_1\Delta x - b_2\Delta x^2 - a_3\Delta x^3 + b_4\Delta x^4) + (b_1 + 2a_2\Delta x - 3b_3\Delta x^2 - 4a_4\Delta x^3)\Delta y$$
$$+ (b_2 + 3a_3\Delta x - 6b_4\Delta x^2)\Delta y^2 + (b_3 + 4a_4\Delta x)\Delta y^3 + b_4\Delta y^4 \tag{8.3.16}$$

Introducing the following symbols,

$$\left.\begin{array}{l} L' = l - l_0 - (a_1\Delta x - b_2\Delta x^2 - a_3\Delta x^3 + b_4\Delta x^4) \\ A_1 = b_1 + 2a_2\Delta x - 3b_3\Delta x^2 - 4a_4\Delta x^3 \\ A_2 = b_2 + 3a_3\Delta x - 6b_4\Delta x^2 \\ A_3 = b_3 + 4a_4\Delta x \\ A_4 = b_4 \end{array}\right\}$$

we get:

$$L' = A_1\Delta y + A_2\Delta y^2 + A_3\Delta y^3 + A_4\Delta y^4 \tag{8.3.17}$$

Hence, using inversion of the series (A1.6) in Appendix 1 we can obtain:

$$\Delta y = B_1 L' + B_2 L'^2 + B_3 L'^3 + B_4 L'^4 \tag{8.3.18}$$

where:

$$B_1 = \frac{1}{A_1}, \quad B_2 = \frac{A_2}{A_1^3}, \quad B_3 = \frac{2A_2^2}{A_1^5} - \frac{A_3}{A_1^4},$$

$$B_4 = -\frac{5A_2^3}{A_1^7} + \frac{5A_2 A_3}{A_1^6} - \frac{A_4}{A_1^5}$$

But:

$$y = y_0 + \Delta y \tag{8.3.19}$$

Introducing $(\Delta x, \Delta y)$ into the second expression of equation (8.3.2), we can obtain isometric latitude q. We can then calculate latitude B from equation (3.7.18).

For computing examples see Appendix 3, Section A3.5.

8.4 LINEAR INTERPOLATION

Above we discussed how to obtain geographic coordinates and rectangular coordinates of a point on the margin of the map by both an analytical method and a numerical method. The methods and formulae mentioned are not only important in theory, but they are also of practical value.

If analytical expressions for the third type of coordinate transformation are difficult to find, or those formulae are too complicated to be used, it is best to select and use linear interpolation.

As we know, when a computer is used to assist in drawing meridians or parallels, we must first calculate at the map scale the coordinates of points using their latitudes and longitudes. Then we join segments of lines one by one to obtain the corresponding meridians or parallels. For that reason, the coordinate transformation of points on the

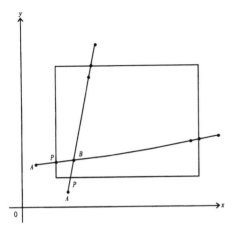

Figure 8.3 The coordinate transformation of a point at map border

margin (or on a grid line) is, in the final analysis, the problem of finding coordinates for the intersection of two lines.

As is shown in Figure 8.3, finding the coordinates of a point P at the intersection of a meridian and an abscissa line can be treated as finding the coordinates of the point of intersection of the line AB and a line of constant x.

When the coordinates of point A (x_A, y_A) and B (x_B, y_B) are given, we can obtain the following linear equation by means of the two-point formula in analytic geometry:

$$x = \frac{x_B - x_A}{y_B - y_A}(y - y_A) + x_A \qquad (8.4.1)$$

If coordinate x_P of point P is known from the above equation, we can obtain:

$$y_P = \frac{y_B - y_A}{x_B - x_A}(x_P - x_A) + y_A \qquad (8.4.2)$$

In the same manner we can find the coordinates of a point P which is the intersection of a latitude line and an ordinate line. Since y_P is given, equation (8.4.1) takes the form:

$$x_P = \frac{x_B - x_A}{y_B - y_A}(y_P - y_A) + x_A \qquad (8.4.3)$$

This linear interpolation is far from a rigorous solution. But it can satisfy the precision required by computer-assisted mapping within a sufficiently small interval, and its computation is rather simple.

8.5   SPHERICAL COORDINATES

In Section 3.4.2 we discussed the transformation among spherical coordinate systems, i.e., the transformation from geographical coordinates (φ, λ) to polar spherical coordinates (Z, a), and the transformation from polar spherical coordinate (Z, a) to geographical coordinates (φ, λ).

In computer-assisted mathematics for transformation on topographic maps and for map projection transformation, sometimes we need to use the third type of transformation of spherical coordinates, such as $(Z, \lambda) \Leftrightarrow (a, \varphi)$; $(Z, \varphi) \Leftrightarrow (a, \lambda)$. Here we introduce mathematical models for the third type of transformation of spherical coordinates.

8.5.1   Calculating (a, φ) using (Z, λ)

Figure 8.4 shows the elements in spherical coordinates. Let:

$$Z_0 = 90° - \varphi_0, \quad Z_1 = 90° - \varphi$$

Since the two sides Z_0, Z_1 and the opposite angle $\Delta\lambda$ of the spherical triangle are given, the side Z_1 $(= 90° - \varphi)$ and its opposite angle a can be calculated.

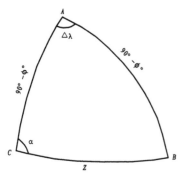

Figure 8.4 Spherical triangle

Referring to the literature (Compile Group 1979), the solution of a spherical triangle takes the form:

$$\sin B = \frac{\sin Z_0 \sin \Delta\lambda}{\sin Z} \tag{8.5.1}$$

$$\tan \frac{a}{2} = \frac{\cos \dfrac{Z - Z_0}{2} \cot \dfrac{\Delta\lambda + B}{2}}{\cos \dfrac{Z + Z_0}{2}} \tag{8.5.2}$$

$$\tan \frac{Z_1}{2} = \frac{\cos \dfrac{\Delta\lambda + B}{2} \tan \dfrac{Z + Z_0}{2}}{\cos \dfrac{\Delta\lambda - B}{2}} \tag{8.5.3}$$

Combining equation (8.5.2) with (8.5.3) we can obtain:

$$a = 2\arctan\left(\frac{\cos \dfrac{Z - Z_0}{2} \cot \dfrac{\Delta\lambda + B}{2}}{\cos \dfrac{Z + Z_0}{2}}\right) \tag{8.5.4}$$

$$\varphi = \frac{\pi}{2} - 2\arctan\left(\frac{\cos \dfrac{\Delta\lambda + B}{2} \tan \dfrac{Z + Z_0}{2}}{\cos \dfrac{\Delta\lambda - B}{2}}\right) \tag{8.5.5}$$

where $B = \arcsin\left(\dfrac{\sin Z_0 \sin \Delta\lambda}{\sin Z}\right)$, $Z_0 = 90° - \varphi_0$, $\Delta\lambda = \lambda - \lambda_0$.

8.5.2 Calculating (Z, λ) if values of (a, φ) are given

When two sides Z_0, Z_1 and one opposite angle a of the spherical triangle are known, we can calculate the side Z and its opposite angle $\Delta\lambda$ (= $\lambda - \lambda_0$).

Analogous to equation (8.5.1), the solution of a spherical triangle takes the form:

$$\sin B = \frac{\sin Z_0 \sin a}{\sin Z_1} \tag{8.5.6}$$

$$\tan \frac{\Delta\lambda}{2} = \frac{\cos \dfrac{Z_1 - Z_0}{2} \cot \dfrac{a + B}{2}}{\cos \dfrac{Z_1 + Z_0}{2}} \tag{8.5.7}$$

$$\tan \frac{Z}{2} = \frac{\cos \dfrac{a + B}{2} \tan \dfrac{Z_1 + Z_0}{2}}{\cos \dfrac{a - B}{2}} \tag{8.5.8}$$

From equations (8.5.7) and (8.5.8) we get:

$$\lambda = \lambda_0 + 2\arctan\left(\dfrac{\cos\dfrac{Z_1 - Z_0}{2}\cot\dfrac{a + B}{2}}{\cos\dfrac{Z_1 + Z_0}{2}}\right) \tag{8.5.9}$$

$$Z = 2\arctan\left(\dfrac{\cos\dfrac{a + B}{2}\tan\dfrac{Z_1 + Z_0}{2}}{\cos\dfrac{a - B}{2}}\right) \tag{8.5.10}$$

where $B = \arcsin\left(\dfrac{\sin Z_0 \sin a}{\sin Z_1}\right)$, $Z_0 = 90° - \varphi_0$, $Z_1 = 90° - \varphi$.

8.5.3 Calculating (a, λ) if values of (Z, φ) are given

When three sides Z_0, Z_1, Z of the spherical triangle are given, we can calculate two of its angles a, $\Delta\lambda$.

According to the literature, solution of the spherical triangle takes the form:

$$\tan\dfrac{a}{2} = \dfrac{m}{\sin(P - Z_1)} \tag{8.5.11}$$

$$\tan\dfrac{\Delta\lambda}{2} = \dfrac{m}{\sin(P - Z)} \tag{8.5.12}$$

$$m = \sqrt{\dfrac{\sin(P - Z)\sin(P - Z_1)\sin(P - Z_0)}{\sin P}} \tag{8.5.13}$$

where $P = \dfrac{1}{2}(Z + Z_1 + Z_0)$.

From the above we get:

$$\lambda = \lambda_0 + 2\arctan(m/\sin(P - Z)) \tag{8.5.14}$$

$$a = 2\arctan(m/\sin(P - Z_1)) \tag{8.5.15}$$

where $P = \dfrac{1}{2}(Z + Z_1 + Z_0)$, $Z_1 = 90° - \varphi$, $Z_0 = 90° - \varphi_0$.

8.5.4 Calculating (Z, φ) if (a, λ) are given

When two angles of a spherical triangle a, $\Delta\lambda$ and their adjacent side Z_0 are given, we can calculate the other two sides Z, Z_1 ($= 90° - \varphi$).

Again from the literature, the solution of the spherical triangle takes the form:

$$\tan\frac{Z+Z_1}{2} = \frac{\cos\dfrac{\Delta\lambda - a}{2}\tan\dfrac{Z_0}{2}}{\cos\dfrac{\Delta\lambda + a}{2}} = t_1 \tag{8.5.16}$$

$$\tan\frac{Z-Z_1}{2} = \frac{\sin\dfrac{\Delta\lambda - a}{2}\tan\dfrac{Z_0}{2}}{\sin\dfrac{\Delta\lambda + a}{2}} = t_2 \tag{8.5.17}$$

From the above we get:

$$Z + Z_1 = 2\arctan t_1, \quad Z - Z_1 = 2\arctan t_2$$

Solving them simultaneously we obtain:

$$Z = \arctan t_1 + \arctan t_2 \tag{8.5.18}$$

$$Z_1 = \arctan t_1 - \arctan t_2 \tag{8.5.19}$$

$$\varphi = \frac{\pi}{2} - Z_1 \tag{8.5.20}$$

8.5.5 Applications

Example 1

Suppose $\varphi_0 = 30°$, $\lambda_0 = 105°$, given Z, λ, to find a, φ.

From (8.5.1), there are two solutions to spherical triangle B, i.e. B (<90°), 180°−B when $\Delta\lambda > 0$. When $\varphi > \varphi_0$ angle B > 90°, and $\varphi < \varphi_0$ angle B < 90°.

Computing results: given $\lambda = 120°$, $Z = 15°48'11''.7$, to obtain $\varphi = 39°59'59''.898$, $a = 46°43'10''.3329$.

Example 2

Given a, φ, to find Z, λ.

Using (8.5.9), (8.5.10) to compute. In designing a program we should consider $\varphi > \varphi_0$ angle B > 90°.

Computing results: given $a = 46°43'10''$, $\varphi = 48°$, to obtain $\lambda = 119°59'59''.984$, $Z = 15°48'11''.7477$.

Example 3

Given Z, φ, to find a, λ.

Computing results: given $Z = 15°48'11''.7$, $\varphi = 40°$, to obtain $a = 46°43'15''.8781$, $\lambda = 119°59'59''.908$.

Example 4

Given a, λ, to find Z, φ.

Computing results: given $a = 46°43'16''$, $\lambda = 120°$, to obtain $Z = 15°48'11''.7624$, $\varphi = 40°00'00''.0081$.

CHAPTER NINE

Zone Transformation for the Transverse Mercator (Gauss–Krüger) Projection

9.1 THE GENERAL METHOD

The Gauss–Krüger projection is a map projection widely used in surveying and mapping. Due to the manner in which zones are projected and the fact that the origin of coordinate points of each zone always start at the point of intersection of the central meridian and equator of the zone, the coordinate systems of each projection zone are independent of each other. In order to relate the neighboring zones, it is defined that the neighboring zones should overlap to some extent.

It is specified in surveying and mapping that map sheets within a difference of longitude of 30' from the western border of each zone, as well as within a difference of longitude of 7.5' (for 1:25 000) and of 15' (for 1:50 000) from the eastern border of each zone should overprint the grid of the adjacent zone. This means that a control point situated in the overlap area should have coordinates of both the east and west zones. As a result, we should study the problem of Gauss–Krüger coordinate transformation from one zone into its adjacent zones.

When mapping occurs at the edge of a projection zone, it is usually necessary to use triangulation points in the neighboring zone as control points. Thus they should be transformed into a unified coordinate system.

In large-scale mapping, zones of 3° are usually used or else the projection zone has an arbitrary meridian for its central meridian. On the other hand the coordinates of national control points use a zone of 6°. For the sake of the relationship to each other and for the application of results, we should explore the Gauss–Krüger coordinate transformation from a zone of 6° to a zone of 3° and from a zone of 3° to a certain local system.

In geodetic surveying the triangulation network frequently extends across different zones. In order to implement the adjustment and calculation of coordinates, we need to transfer part of the coordinates from a neighboring zone. This can be concluded as the problems of Gauss–Krüger coordinate transformation between 6° zones.

In other words, zone transformation on the Gauss–Krüger is a useful method encountered frequently in surveying and mapping. To solve zone coordinate transformation, many scholars from China and other countries have presented several calculating methods over a long period. They can be summed up as the following two types of methods.

1. Indirect transformation

   Given Gauss–Krüger coordinates (x_1, y_1) in the first zone, we can compute their geodetic coordinates using inverse equations for the Gauss–Krüger projection. Considering the results we can obtain the Gauss–Krüger coordinates (x_2, y_2) based on the central meridian of the second zone as a datum.

   This is a precision method and also a laborious one because of its numerous calculations, so it is not suitable for the transformation of a large number of points.

2. Direct transformation

   Here we derive the relationship between the coordinates of a point that is in two adjacent zones. Then, using a special transformation table, we can calculate the coordinates of the point in an adjacent zone according to its coordinates in the first zone. There exist many such equations and conversion tables, but these tables are only compatible with conventional calculations. Nowadays personal computers are widely used in surveying and mapping. As a result, it is necessary to explore more suitable transformations for computers.

   Programming with the widely used indirect transformation with variable coefficients is uneconomical in calculating time. Hence we introduce a numerical method of indirect transformation with constant coefficients and direct zone transformation for the Gauss–Krüger projection. These are new calculating methods suitable for automatic transformation.

9.2   INDIRECT TRANSFORMATION

9.2.1 Variable coefficient method

Given the Gauss–Krüger coordinates (x_G, y_G) of a point in west zone (or east zone), how can we obtain its Gauss–Krüger coordinates (x'_G, y'_G) in the adjacent east zone (or west zone)? The following steps can calculate this:

- First, transform the Gauss–Krüger coordinates into geographical coordinates, i.e., $(x_G, y_G) \rightarrow (B, l')$, then we can get the latitude B and difference in longitude l' by using a variable-coefficient inverse solution equation.

- Second, changing the difference of longitude l' to the difference of longitude l of the adjacent zone, the calculating equation is $l = l' \pm \Delta l$, where Δl is the difference of central longitude between two adjacent zones. The difference Δl equals $6°$ for the two adjacent $6°$ zones and equals $3°$ for the two adjacent $3°$ zones (or $6°$ zone $\leftrightarrow 3°$ zone). Also, Δl takes a negative value when the changes are from west to east and takes a positive value when the changes are from east to west.

- Third, finding Gauss–Krüger coordinates from B, l, i.e., $(B, l) \rightarrow (x'_G, y'_G)$. According to the variable-coefficient direct solution equation (2.3.20) we can obtain the new Gauss–Krüger coordinates. Now we realize $(x_G, y_G) \rightarrow (x'_G, y'_G)$.

9.2.2 One of the constant coefficient methods

The calculating step is as follows:

- First, we convert the Gauss–Krüger coordinates into isometric coordinates, i.e., $(x_G, y_G) \rightarrow (q, l')$. Using the constant-coefficient inverse solution equation (5.4.1) and

(5.4.3) we can get the values of the isometric latitude q and the difference in longitude l'. Where the latitude of the expansion point 0 is $B_0 = y_G/6\,378\,245$. Or taking the latitude of the central parallel of the conversion point's map sheet on the millionth scale as B_0.

- Second, we convert the difference of longitude l' as the adjacent zone's difference of longitude l. The calculating method is as above.

- Third, we compute the Gauss–Krüger coordinates from q, l, that is, $(q, l) \rightarrow (x'_G, y'_G)$.

According to the constant-coefficient direct solution equation (5.3.1) and (5.3.3) we can obtain the Gauss–Krüger coordinates x'_G, y'_G. Now we form $(x_G, y_G) \rightarrow (x'_G, y'_G)$.

9.2.3   Another constant coefficient method

Now we introduce the constant coefficient method of Gauss–Krüger projection zone transformation based on a Mercator coordinate transformation. The following are the computing steps.

- First, transform the Gauss–Krüger coordinates into Mercator coordinates, i.e., $(x_G, y_G) \rightarrow (x'_M, y'_M)$. The transformation can be realized by using equations (5.2.1) and (5.2.13).

- Second, transform the Mercator coordinates into the Mercator coordinates of the adjacent zone's coordinate system, i.e., $(x'_M, y'_M) \rightarrow (x_M, y_M)$.

$$y_M = y'_M, \quad x_M = x'_M \pm x_0$$

where $x_0 = r_0 \text{ arc } \Delta l$

- Third, transform the Mercator coordinates of the adjacent zone's coordinate system into Gauss–Krüger coordinates, that is, $(x_M, y_M) \rightarrow (x'_G, y'_G)$.
  This is effected using equations (5.2.1) and (5.2.7). Then we now obtain $(x_G, y_G) \rightarrow (x'_G, y'_G)$.

Example

Given 6° zones, Gauss–Krüger coordinates $x_G = 327\,982.5946$ m, $y_G = 3\,102\,979.1909$ m in the western zone at $B = 28°$, $l = 3°20'$, we now compute the point's Gauss–Krüger coordinates in the adjacent eastern zone (6° zone).

Results: The Mercator coordinates in the adjacent zone are $x_M = -257\,301.0744$, $y_M = 2\,798\,717.9679$, and the Gauss–Krüger coordinates in the adjacent zone are $x'_G = -262\,356.0720$, $y'_G = 3\,101\,364.4732$.

9.3   DIRECT TRANSFORMATION

Zone transformation of Gauss–Krüger coordinates can be taken as a coordinate transformation between two conformal projections, i.e., transformation from the Gauss–Krüger coordinates (x, y) in one coordinate system to the Gauss–Krüger coordinates (X, Y) in another system. From equations (5.1.36) we have the general equations for the zone transformation of the Gauss–Krüger projection as follows:

$$Y = Y_0 + \sum_{K=1}^{4}(a_K P_K - b_K Q_K) \left.\vphantom{\sum_{K=1}^{4}}\right\}$$

$$X = X_0 + \sum_{K=1}^{4}(b_K P_K + a_K Q_K) \left.\vphantom{\sum_{K=1}^{4}}\right\}$$

$$(9.3.1)$$

where:

$$P_{K+1} = \Delta y P_K - \Delta x Q_K, \quad Q_{K+1} = \Delta x P_{K+1} + \Delta y Q_K$$
$$P_1 = \Delta y, \quad Q_1 = \Delta x, \quad \Delta y = y - y_0, \quad \Delta x = x - x_0$$

It is clear from (9.3.1) that if the two Gauss–Krüger coordinate systems (x_0, y_0) and (X_0, Y_0) with auxiliary point 0 as well as coefficients (a_i, b_i) are known, then the specific coordinate transformation equations fitting that area can be determined. Below we discuss the analytical method to derive coefficients (a_i, b_i).

9.3.1 The analytical method

The mathematical principle of analytical transformation is given in Heristov (1957).

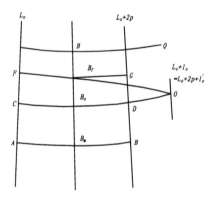

Figure 9.1 Two adjacent coordinate systems for Gauss–Krüger projection

According to (9.3.1) we conclude that to solve the problem of coordinate transformation on the Gauss–Krüger is to find the values of coefficients (a_i, b_i).

Assuming that the central meridians of two Gauss–Krüger coordinate systems are respectively L_0 and $(L_0 + 2p)$, the coordinates of an arbitrary point $Q(B, L)$ in the two systems are respectively (x, y) and (X, Y).

As seen in Figure 9.1, the parallel $B = B_m$ intersects the central meridian L_0 at point A and the central meridian $(L_0 + 2p)$ at point B. Considering equations (2.2.21) and (2.3.13), we obtain series expansions for the Gauss–Krüger projection respectively at points A and B:

$$z = z_A + N_m \cos B_m \Delta\omega + \frac{1}{2}N_m \cos^2 B_m t_m(-1)\Delta\omega^2$$

$$+ \frac{1}{6}N_m \cos^3 B_m(-1 + t_m^2 - \eta_m^2)\Delta\omega^3 + \ldots$$

$$(9.3.2)$$

where:

$$\Delta\omega = \omega - \omega_A$$

and:

$$Z = Z_B + N_m \cos B_m \Delta'\omega + \frac{1}{2}N_m \cos^2 B_m t_m(-1)\Delta'\omega^2$$

$$+ \frac{1}{6}N_m \cos^3 B_m(-1 + t_m^2 - \eta_m^2)\Delta'\omega^3 + \ldots$$

$$(9.3.3)$$

where:

$$\Delta'\omega = \omega - \omega_B$$

The following algorithm exists for the derivation of ω:

$$\frac{d}{d\omega} = \frac{d}{d\Delta\omega} = \frac{d}{d\Delta'\omega} \qquad (9.3.4)$$

Differentiating (9.3.2) and (9.3.3) separately with respect to ω yields:

$$\frac{dz}{d\omega} = N_m \cos B_m + N_m t_m \cos^2 B_m (-1)\Delta\omega$$

$$+ \frac{1}{2} N_m \cos^3 B_m (-1 + t_m^2 - \eta_m^2)\Delta\omega^2 + \ldots \qquad (9.3.5)$$

$$\frac{dZ}{d\omega} = N_m \cos B_m + N_m t_m \cos^2 B_m (-1)\Delta'\omega$$

$$+ \frac{1}{2} N_m \cos^3 B_m (-1 + t_m^2 - \eta_m^2)\Delta'\omega^2 + \ldots \qquad (9.3.6)$$

From (9.3.5) we obtain:

$$\frac{d\omega}{dz} = 1 \bigg/ \frac{dz}{d\omega} = \frac{1}{N_m \cos B_m} + \frac{1}{N_m} t_m \Delta\omega$$

$$+ \frac{1}{2N_m} \cos B_m (1 + t_m^2 + \eta_m^2)\Delta\omega^2 + \ldots \qquad (9.3.7)$$

and (9.3.6) multiplied by (9.3.7) yields:

$$\frac{dZ}{dz} = \frac{dZ}{d\omega} \cdot \frac{d\omega}{dz} = 1 + t_m \cos B_m \Delta\omega + t_m \cos B_m (-1)\Delta'\omega$$

$$+ \frac{1}{2} \cos^2 B_m (1 + t_m^2 + \eta_m^2)\Delta\omega^2 + \cos^2 B_m \cdot (-t_m^2)\Delta\omega\Delta'\omega + \ldots \qquad (9.3.8)$$

From (9.3.8) we obtain:

$$\frac{d}{d\omega}\left(\frac{dZ}{dz}\right) = \cos^2 B_m (1 + \eta_m^2)\Delta\omega + \cos^2 B_m (-1 - \eta_m^2)\Delta'\omega + \ldots \qquad (9.3.9)$$

$$\frac{d^2 Z}{dz^2} = \frac{d}{d\omega}\left(\frac{dZ}{dz}\right) \cdot \frac{d\omega}{dz} = \frac{1}{N_m} \cos B_m (1 + \eta_m^2)\Delta\omega$$

$$+ \frac{1}{N_m} \cos B_m (-1 - \eta_m^2)\Delta'\omega + \ldots \qquad (9.3.10)$$

Following this method we obtain the analytical expressions of all higher derivatives

$$\frac{d^3 Z}{dz^3}, \quad \frac{d^4 Z}{dz^4}, \quad \ldots$$

In order to obtain the values of the coefficients of a series expansion about the expansion point 0, it is necessary to place the value ω_0 of the point 0 into $(\Delta\omega, \Delta'\omega)$, from which we can derive the analytical equations for calculating the coefficients (a_i, b_i) in (9.3.1).

In the computerized calculation of the zone transformation for Gauss–Krüger projections it is unnecessary to derive analytical expressions for coefficients (a_i, b_i) in (9.3.1) beforehand and to find their values, since it is very complicated. We can select an auxiliary point 0 and coordinates for several common points in the two coordinate systems from the transformation area and form a linear equation system in which (a_i, b_i) are undetermined elements, but from which we can obtain the values of each coefficient (a_i, b_i). A polynomial of 3rd degree needs three common points besides the auxiliary point 0, of 4th degree needs four, and of 5th degree needs five.

The method using coordinate values of several common points from the transformation area to create the zone transformation equations of (9.3.1) is called the numerical method of zone transformation.

9.3.2 The numerical method

For the mathematical fundamentals of the numerical method see Section 7.1.1. Now we discuss the applications of the numerical method for Gauss–Krüger zone transformation (Yang 1982a).

Applications and analysis

Calculating procedure   To meet the requirements of surveying and mapping, we take a polynomial of 4th degree as an example. The calculating procedure follows.

First, we choose the coordinates (x_0, y_0) and (X_0, Y_0) of auxiliary point 0 in the two coordinate systems.

Second, we choose coordinates (x_i, y_i) and (X_i, Y_i) of several common points in the two coordinate systems.

For the coordinate equations for the Gauss–Krüger projection see equation (2.3.17), in which S_m is the arc length of the meridian from the equator, and:

$$N = \frac{c}{\sqrt{1 + \eta^2}}, \quad c = \frac{a}{\sqrt{1 - e^2}} \tag{9.3.11}$$

For the Krasovskiy ellipsoid,

$$e'^2 = 0.0067385254146835, \quad c = 6\,399\,698.9017827 \text{ m}.$$

and:

$$S_m = 6\,367\,558.4968749794 \cdot B° \text{ arc } 1° - 16\,036.4802694138 \sin 2B$$
$$+ 16.8280668849 \sin 4B - 0.0219753092 \sin 6B \tag{9.3.12}$$
$$+ 0.311311 \times 10^{-4} \sin 8B - 0.46 \times 10^{-7} \sin 10B$$

Third, we form a linear equation system according to (9.1.1) and solve the system by means of Gaussian elimination of the principal element to determine the values of coefficients (a_i, b_i).

Fourth, we calculate the coordinates of points for the zone.

Selection of the auxiliary and common points   When the conformal polynomial (9.3.1) is set up by means of a numerical method, the auxiliary point 0 can be arbitrarily chosen.

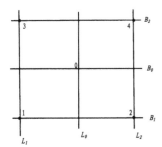

Figure 9.2 The distribution of auxiliary point and common points

Auxiliary point 0 should be chosen in accordance with the specific transformation area to achieve a minimum $(\Delta x, \Delta y)$ in the total area and furthermore to speed the convergence of the power series.

See Figure 9.2, assuming that the latitude range of the transformation area is B_1–B_2 and that its difference of longitude is l_1–l_2; then the auxiliary point 0 should be chosen at the center of the transformation area. That is,

$$B_0 = \frac{B_1 + B_2}{2}, \quad l_0 = \frac{l_1 + l_2}{2}$$

The number of common points should equal the degree of the power polynomial; in other words, a polynomial of degree n should have n common points.

Considering a uniform convergence of the power series in the inner convergence circle $|z - z_0| < R$, it would be better that the common points be well distributed around the area. As an example, the distribution of four common points 1, 2, 3, 4 for a polynomial of 4th degree is shown in Figure 9.2.

The scope of transformation area and analysis of its precision   To meet the requirements of surveying, the precision of a coordinate transformation should be within 0.001–0.002 m (or 1–2 mm).

It has been proven in practice that it is sufficient for the precision needs of zone transformation on the Gauss–Krüger projection to use a 4th-degree polynomial.

If the degree of the polynomial is determined, then the precision of coordinate transformation relates to the difference in longitude of the central meridians of the two co-ordinate systems and the size of the transformation area.

Now we consider transformation between 6° zones and transformation between 3° zones. We programmed with ALGOL using a numerical method for calculation and partitioned a 0°–60° range of latitude into 15 zones each with a 4° difference in latitude. For the choice of the auxiliary point and the common points for each zone, see Figures 9.3 and 9.4.

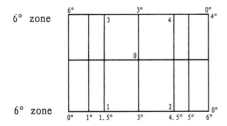

Figure 9.3 Distribution of auxiliary point and common points for two adjacent 6° zones

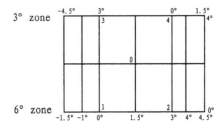

Figure 9.4 Distribution of auxiliary point and common points for two adjacent 3° zones

In order to check the precision of the zone transformation, we complete the zone transformation computation for each intersection point at every 0.5° difference in longitude and latitude, and obtain the deviations (δx_i, δy_i) between the computed values and the actual coordinates. i.e.,

$$\delta x_i = X_i - X_i', \quad \delta y_i = Y_i - Y_i' \tag{9.3.13}$$

where (X_i', Y_i') are the true coordinates of the points and (X_i, Y_i) are the zone transformation coordinates of the points.

For the 6° zone, there are $81 \times 15 = 1\,215$ points total within 0°–60° of latitude and 1°–5° of longitude, of which there are 1 203 points for which the absolute value of coordinate deviation is less than 1 mm and another 12 points which exceed 1 mm. There is only one point which has a maximum deviation of 1.5 mm. If the difference of longitude extends to 0.5°–5.5°, then a very few points have a maximum deviation of 2.4 mm.

For the 3° zone, there are $99 \times 15 = 1\,485$ points total within 0°–60° of latitude and from −1° to +5° of longitude. For almost all of the points, the absolute values of the coordinate deviation are less than 1 mm except for a single point that deviates up to 1.5 mm. If the difference of longitude extends from −1.5° to +4.5° then there are very few points which have a maximum deviation of 2 mm.

From the above results we conclude that the computing precision for transformation of 6° zones within a range of 2° from a border meridian is less 1 mm and the computing precision for the transformation of 3° zones within a range of 2.5° from a border meridian is within 1 mm. In the above zone transformation area, for the worst situation the precision is within 1.5 mm for all but a very few points.

9.3.3 Coefficient tables for the numerical method

Analysis of scheme

For a variable transformation area, for instance, in transformation between a local coordinate system and the national coordinate system, we can use the previous method and procedure to compute the zone transformation. For some fixed areas, however, for example coordinate transformation from a 6° zone to its adjacent 6° zone or from a 3° zone to its adjacent 3° zone, and vice versa, instead of using the previous method we may compute in advance (x_0, y_0), (X_0, Y_0), and coefficients (a_j, b_j) of a power polynomial (9.3.1) by partition, i.e., arrange a table of coefficients for zone transformation and then proceed with the zone transformation computation using (9.3.1).

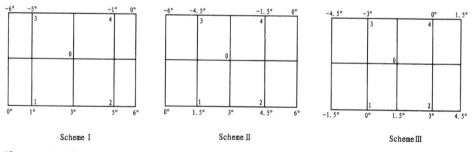

Figure 9.5   The arrangement of coefficient tables for 6° and 3° zone transformations

To complete the arrangement of coefficient tables for 6° and 3° zone transformations, we discuss results calculated for the following three schemes or approaches according to partitions using a 4° difference in latitude.

Scheme 1: Partition a latitude range of 0°–60° into 15 zones each with a 4° difference of latitude. In order to check the precision of zone transformation, we compute the deviations for intersecting points at every 0.5° difference of longitude and latitude. The result is as follows.

There are a total of $81 \times 15 = 1\,215$ points within a 1°–5° difference of longitude, of which there are 41 points for which the absolute value of coordinate deviation exceeds 1 mm, only one point of which has the maximum deviation of 1.5 mm. If the difference of longitude exceeds 0.5° (i.e., 0.5°–5.5°), then almost all of the coordinates of the points have deviations less than 2 mm except for a point that has a maximum deviation of 2.4 mm.

Scheme 2: This produces a total of $63 \times 15 = 945$ points within a 1.5°–4.5° difference of longitude. For only two points do the absolute values of coordinate deviation exceed 1 mm; one of them has a maximum value of 1.5 mm. If the difference of longitude exceeds 0.5° (i.e., 1°–5°), the coordinate deviations are still within 1.2 mm. Furthermore, if the difference of longitude exceeds 0.5° (i.e., 0.5°–5.5°), then the coordinate deviations are normally within 2 mm, and very few points have the maximum value of 2.4 mm.

Contrasting scheme 1 with scheme 2, we conclude that scheme 2 must be more compatible with a 6°-zone transformation, because it can retain 1 mm of precision in zone transformation for arbitrary points situated within a 2° difference of longitude from a border meridian, except for only a very few points that have the maximum difference value of 1.5 mm.

Scheme 3: This scheme is suitable for the transformation of a 3° zone. Within a 0°–3° difference of longitude the coordinate deviation is less than 1 mm. Within a −1.5° to +4.5° difference of longitude the deviations of the coordinate of only a few points reach 1.5 mm, and a very few points situated at the border exceed 2 mm.

In other words, we can compute by partition coefficients (a_i, b_i) according to scheme 2 or scheme 3 and compile coefficient tables for transformation of 6° and 3° zones.

Zone transformation precision from coefficient tables completely agrees with the current Gauss–Krüger coordinate conversion table for 6° (or 3°), but its transformation area is larger.

Coefficient tables for zone transformation

Coefficient tables for Gauss–Krüger zone transformation consist of Table A2.1 (Gauss–Krüger coordinate transformation coefficient table for 6°), and Table A2.2 (Gauss–Krüger coordinate transformation coefficient table for 3°), both in Appendix 2.

The value of each coefficient in those tables is computed with numerical zone transformation programs. They are suitable for computerized zone transformation of a large number of points. Commonly the precision is 1 mm. For the worst situation the precision of only a very few points exceeds 1.5 mm. For this reason the values can meet requirements for high precision in zone transformation.

Application examples

Example 1: Coordinate transformation between two adjacent 6° zones   In a transformation area ranging in B from 28° to 32°, given the coordinates for each point (x_i, y_i) in the western zone, let us find their coordinates (X_i, Y_i) in the eastern zone.

With $B_0 = 30°$ we can look up the zone transformation coefficients in Table A2.1 and carry out the computation using equations (9.3.1). If the transformation is made from coordinates (x_i, y_i) of the eastern zone to coordinates (X_i, Y_i) of the western zone, the same procedure as above is used, but (x_0, X_0) and b_i take opposite signs.

Example 2: Coordinate transformation between two adjacent 3° zones   In the above transformation area, using $B_0 = 30°$ we look up the zone transformation coefficients in Table A2.2. The computation procedure is the same as above. Table A2.2 is also applicable to transformation from a 6° zone to a 3° zone and vice versa.

In addition, the constant coefficients can be calculated according to a numerical method. An example will be given in Appendix 3, Section A3.6.

9.4   PLANE COORDINATE NETWORK TRANSFORMATION BETWEEN ADJACENT ZONES

A 6° zone is used for the Gauss–Krüger projection in Chinese topographic mapping. Since each zone has its own plane rectangular coordinate system, the plane coordinate networks of adjacent map sheets are independent of each other. To facilitate use of the adjacent map sheets, it has been specified in current Chinese standards that map sheets within a longitude difference of 30′ from the western border of each zone as well as within a longitude difference of 7.5′ (for the 1:25 000-scale series) and of 15° (for the 1:50 000-scale series) from the eastern border of each zone should contain an overprint of the grid of the adjacent zone.

Currently the conventional method of drawing a grid of the adjacent zone on the map is as follows. Check the coordinate values of points along the margin with reference to its adjacent zone using the margin coordinate table for the Gauss projection, and prepare the data. Then on a sheet plotted with points use dividers to determine the position of grid lines of the adjacent zone in the current zone.

To improve this manual operation and the accuracy as well as raising the working efficiency, a table of zone transformation for coordinates of the grid lines of the Gauss projection (Yang 1980a) has been developed by the first author. After straightforward use of the table to extract the data, we can directly use a coordinatograph to plot the grid of the adjacent zone while plotting marginal points and the grid of the current zone.

Because both the adjacent zone and the current zone have their own coordinate origin, and the map sheet has an appended grid of the adjacent zone, the grid lines of this zone intersect the grid lines of the adjacent zone at an angle. Hence we cannot directly use a coordinatograph to print a grid of the adjacent zone according to the coordinate values of the adjacent zone.

To print the grid of the adjacent zone using the coordinatograph we must first transform the coordinates of the adjacent zone to the coordinate system of the current zone.

When using the table of zone transformation for coordinates of grid lines on the Gauss projection, a computing equation using the linear interpolation method has been introduced. Here we present a more rigid method of numerical analysis to replace the previous linear interpolation. In addition, zone transformation computations can be accomplished using some BASIC programs from Yang (1987a) instead of overly laborious manual preparation of data.

Next we discuss a numerical analysis method for coordinate transformation of adjacent grid lines. From (9.3.1), the general equations for zone transformation of the Gauss–Krüger projection have the form:

$$x' = x'_0 + b_1\Delta y + a_1\Delta x + b_2(\Delta y^2 - \Delta x^2) + a_2 \cdot 2\Delta x\Delta y + a_3(3\Delta x\Delta y^2 - \Delta x^3)$$
$$+ b_3(\Delta x^3 - 3\Delta x^2\Delta y) + a_4(4\Delta x\Delta y^3 - 4\Delta x^3\Delta y) + b_4(\Delta y^4 - 6\Delta x^2\Delta y^2 + \Delta x^4)$$
$$y' = y'_0 + a_1\Delta y - b_1\Delta x + a_2(\Delta y^2 - \Delta x^2) - b_2 \cdot 2\Delta x\Delta y + a_3(\Delta y^3 - 3\Delta x^2\Delta y)$$
$$- b_3(3\Delta x\Delta y^2 - \Delta x^3) + a_4(\Delta y^4 - 6\Delta x^2\Delta y^2 + \Delta x^4) - b_4(4\Delta x\Delta y^3 - 4\Delta x^3\Delta y)$$

$$(9.4.1)$$

We look up the values of (x_0, y_0), (x'_0, y'_0), and (a_i, b_i) $(i = 1, 2, 3, 4)$ of equation (9.4.1) in Tables A2.1 and A2.2.

As shown in Figure 9.6, we are given the coordinates (x_i, y_i) $(i = 1, 2, 3, 4)$ of the margin of the sheet along the western border of the zone. Now the coordinate values of a series of points at the intersections of the vertical (or horizontal) grid lines and the abscissa line (or ordinate line) passing marginal points 2, 4 (or 1, 3) are desired, i.e., to obtain the transformation values for coordinates of the grid lines of the adjacent zone.

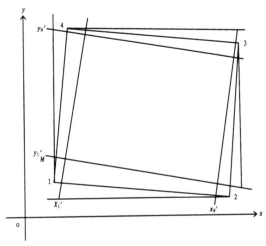

Figure 9.6 The coordinate systems of the margin of the sheet along the zone and adjacent zone

Assuming that the coordinate values of the points along the margin of the sheet for the adjacent zone are (x_i, y_i) $(i = 1, 2, 3, 4)$, then the desired coordinate values of the initial and final vertical grid lines result from (x'_4, x'_2), and the desired coordinate values of the initial and final horizontal grid lines are found from (y_1, y'_3).

For this reason the problem of zone transformation of coordinates can be treated as the third type of coordinate transformation in finding x from (x', y) or finding y from (x, y').

9.4.1 Computing x from (x', y)

Introducing the following symbols,

$$Dx = x' - x'_0 - (b_1\Delta y + b_2\Delta y^2 + b_3\Delta y^3 + b_4\Delta y^4)$$
$$c_1 = a_1 + 2a_2\Delta y + 3a_3\Delta y^2 + 4a_4\Delta y^3$$
$$c_2 = -b_2 - 3b_3\Delta y - 6b_4\Delta y^2$$
$$c_3 = -a_3 - 4a_4\Delta y$$
$$c_4 = b_4$$

$$(9.4.2)$$

we rewrite the second expression of (9.4.1) as:

$$Dx = c_1\Delta x + c_2\Delta x^2 + c_3\Delta x^3 + c_4\Delta x^4$$

(9.4.3)

The inverse equation of series (A1.6) in Appendix 1 gives:

$$\Delta x = D_1 Dx + D_2 Dx^2 + D_3 Dx^3 + D_4 Dx^4$$

(9.4.4)

where:

$$D_1 = \frac{1}{c_1}, \quad D_2 = -\frac{c_2}{c_1^3}, \quad D_3 = \frac{2c_2^2}{c_1^5} - \frac{c_3}{c_1^4}, \quad D_4 = -\frac{5c_2^3}{c_1^7} + \frac{5c_2 c_3}{c_1^6} - \frac{c_4}{c_1^5}$$

Then:

$$x = \Delta x + x_0$$

(9.4.5)

9.4.2 Computing y from (x, y')

Considering the first expression of (9.4.1), we have:

$$\left.\begin{aligned}
Dy &= y' - y_0' - (b_1\Delta x + a_2\Delta x^2 - b_3\Delta x^3 - a_4\Delta x^4) \\
c_1 &= a_1 - 2b_2\Delta x - 3a_3\Delta x^2 + 4b_4\Delta x^3 \\
c_2 &= a_2 - 3b_3\Delta x - 6a_4\Delta x^2 \\
c_3 &= a_3 - 4b_4\Delta x \\
c_4 &= a_4
\end{aligned}\right\}$$

(9.4.6)

Hence we have:

$$Dy = c_1\Delta y + c_2\Delta y^2 + c_3\Delta y^3 + c_4\Delta y^4$$

(9.4.7)

The inverse equation of series (9.4.4) yields:

$$\Delta y = D_1 Dy + D_2 Dy^2 + D_3 Dy^3 + D_4 Dy^4$$

(9.4.8)

Then:

$$y = \Delta y + y_0$$

(9.4.9)

With the above equations zone transformation for adjacent grid lines can be realized according to the following procedure:

- First, we separately compute the Gauss–Krüger coordinates of marginal points in the adjacent zone and the current zone by using the direct transformation equations for the Gauss–Krüger projection with constant coefficients.

- Second, we look up (x_0, y_0), (x_0', y_0'), and (a_i, b_i) in the coefficient table for 6°-zone transformation of Gauss–Krüger coordinates.

- Third, we compute coordinate values of the initial and final grid lines and the number of grid lines needed to be overprinted.

- Fourth, we individually compute coordinate values of grid lines according to equations (9.4.2) to (9.4.9). For examples of zone transformation of adjacent grid lines see Appendix 3, Section A3.7.

9.5 APPLICATION OF DOUBLE PROJECTION

9.5.1 Mercator projection

The normal cylindrical conformal projection is referred to as the Mercator projection, and the transverse cylindrical conformal projection is referred to as the transverse Mercator projection. For the sphere, the equations for the Mercator projection coordinates are:

$$x_M = R\lambda, \quad y_M = R\ln\tan\left(\frac{\pi}{4} + \frac{\varphi}{2}\right) \tag{9.5.1}$$

Introducing spherical rectangular coordinates (X, Y), the equations for the spherical transverse Mercator projection can be written:

$$x_T = R\ln\tan\left(\frac{\pi}{4} + \frac{X}{2}\right), \quad y_T = RY \tag{9.5.2}$$

Using hyperbolic functions, we have:

$$\left.\begin{aligned} \sinh y &= \frac{e^y - e^{-y}}{2}, \quad \cosh y = \frac{e^y + e^{-y}}{2} \\[2mm] \tanh y &= \frac{\sinh y}{\cosh y}, \quad \tanh y = \sin\theta \\[2mm] y &= \ln\tan\left(\frac{\pi}{4} + \frac{\theta}{2}\right) \end{aligned}\right\} \tag{9.5.3}$$

Hence equation (9.5.2) may be rewritten:

$$\tan\frac{x_T}{R} = \sin X, \quad \tan\frac{y_T}{R} = \tan Y \tag{9.5.4}$$

The relation between spherical rectangular coordinates and geographic coordinates is given by equation (3.4.37):

$$\sin X = \cos\varphi\sin\lambda, \quad \tan Y = \tan\varphi\sec\lambda \tag{9.5.5}$$

Then we obtain the following equations for the transverse Mercator projection coordinates:

$$\tanh\frac{x_T}{R} = \cos\varphi\sin\lambda, \quad \tan\frac{y_T}{R} = \tan\varphi\sec\lambda \tag{9.5.6}$$

From (9.5.1) we have:

$$\tan\frac{x_M}{R} = \tan\lambda, \quad \tanh\frac{y_M}{R} = \sin\varphi \tag{9.5.7}$$

From (3.4.38):

$$\sin\varphi = \cos X\sin Y, \quad \tan\lambda = \tan X\sec Y \tag{9.5.8}$$

So we obtain:

$$\tan\frac{x_M}{R} = \tan X\sec Y, \quad \tanh\frac{y_M}{R} = \cos X\sin Y \tag{9.5.9}$$

Noting that:

$$Y = \frac{y_T}{R}$$

and:

$$\tan X = \sinh \frac{x_T}{R}, \quad \cos X = \text{sech} \frac{x_T}{R}$$

we obtain the relationship for transformation from transverse Mercator coordinates into Mercator coordinates:

$$\tan \frac{x_M}{R} = \sinh \frac{x_T}{R} \sec \frac{y_T}{R}, \quad \tanh \frac{y_M}{R} = \text{sech} \frac{x_T}{R} \sin \frac{y_T}{R} \tag{9.5.10}$$

Considering equation (9.5.6) we have:

$$\tan \varphi = \sinh \frac{y_M}{R}, \quad \cos \varphi = \text{sech} \frac{y_M}{R} \tag{9.5.11}$$

Then we can obtain the relationship for transformation from Mercator coordinates into transverse Mercator coordinates:

$$\tan \frac{x_T}{R} = \sin \frac{x_M}{R} \sec \frac{y_M}{R}, \quad \tanh \frac{y_T}{R} = \sec \frac{x_M}{R} \sinh \frac{y_M}{R} \tag{9.5.12}$$

9.5.2 Double transverse Mercator projection and the Gauss–Krüger projection

The spherical Gauss projection is considered the same as the spherical transverse Mercator projection. But the spherical transverse Mercator projection is not of the same type as the Gauss–Krüger projection. In order to replace approximately Gauss–Krüger coordinates with transverse Mercator coordinates, the double transverse Mercator projection is needed, i.e., we conformally project the surface of the ellipsoid onto the surface of the sphere according to the second projection method of Gauss, and then project the surface of the sphere onto a plane according to the transverse Mercator projection (Hua 1983b). This is the so-called double-transverse Mercator projection.

The second projection equations of Gauss used for conformal projection of the surface of the ellipsoid onto the surface of the sphere are given as (2.10.1) and (2.10.2):

$$\tan \left(\frac{\pi}{4} + \frac{\varphi}{2} \right) = \beta U^{\alpha}, \quad \lambda = \alpha l \tag{9.5.13}$$

where:

$$\tan \varphi_0 = \frac{R}{N_0} \tan B_0, \quad \alpha = \frac{\sin B_0}{\sin \varphi_0}, \quad \beta = \tan \left(\frac{\pi}{4} + \frac{\varphi_0}{2} \right) U_0^{-\alpha},$$

$$R = \sqrt{M_0 N_0}, \quad a = \sqrt{1 + e'^2 \cos^4 B_0}$$

B_0 is referred to as the standard parallel. If the point to be transformed is near the standard parallel, when the Gauss–Krüger coordinates are replaced with double Mercator coordinates the precision is 0.1 mm, and their abscissas are identical, i.e., $x_G = x_T$, but their ordinates have the relation:

$$y_G = y_T - \Delta y$$

where

$$\Delta y = R\varphi_0 - S_0$$

and S_0 is the arc length of the meridian from the equator to latitude B_0. Thus we obtain the transformation for double transverse Mercator coordinates into Gauss–Krüger coordinates:

$$x_G = x_T, \quad y_G = y_T - \Delta y \qquad (9.5.14)$$

where:

$$\Delta y = R\varphi_0 - S_0, \quad R = \sqrt{M_0 N_0}$$

$$\sin\varphi_0 = \frac{\sin B_0}{\alpha}, \quad \alpha = \sqrt{1 + e'^2 \cos^4 B_0}$$

Substituting (9.5.6) into the above gives approximate equations for the Gauss–Krüger projection coordinates:

$$x_G = R\tanh^{-1}(\cos\varphi\sin\lambda), \quad y_G = R\arctan(\tan\varphi\sec\lambda) - \Delta y \qquad (9.5.15)$$

From (9.5.14) we have:

$$x_T = x_G, \quad y_T = y_G + \Delta y \qquad (9.5.16)$$

From (9.5.7) and (9.5.10),

$$\sin\varphi = \operatorname{sech}\frac{x_T}{R}\sin\frac{y_T}{R}, \quad \tan\lambda = \sinh\frac{x_T}{R}\sec\frac{y_T}{R} \qquad (9.5.17)$$

Substituting (9.5.16) into the above we obtain:

$$\left.\begin{aligned}
\varphi &= \arcsin\left(\operatorname{sech}\frac{x_G}{R}\sin\frac{y_G + \Delta y}{R}\right) \\
\lambda &= \arctan\left(\sinh\frac{x_G}{R}\sec\frac{y_G + \Delta y}{R}\right)
\end{aligned}\right\} \qquad (9.5.18)$$

(B, l) can be computed from (5.5.4) and (3.7.18), i.e., inversely solving the geographic coordinates of Gauss–Krüger coordinates.

9.5.3 Direct zone transformation of Gauss–Krüger coordinates

Assume that (x_G, y_G) are the coordinates of the Gauss–Krüger projection in the current zone and (x_G', y_G') are the coordinates in adjacent zone, (x_T, y_T) are the coordinates of double transverse Mercator projection in the current zone, and (x_T', y_T') are the coordinates of the double transverse Mercator projection in an adjacent zone. Now we derive the zone transformation equations for the Gauss–Krüger projection. It is completed with the following steps:

1. Transform the Gauss–Krüger coordinates into double transverse Mercator coordinates:

$$x_T = x_G, \quad y_T = y_G + \Delta y \tag{9.5.19}$$

2. Inverse transformation from (9.5.18) yields:

$$\left.\begin{aligned}
\sin\varphi &= \operatorname{sech}\frac{x_G}{R}\sin\frac{y_G + \Delta y}{R} \\[2mm]
\tan\lambda &= \sinh\frac{x_G}{R}\sec\frac{y_G + \Delta y}{R}
\end{aligned}\right\} \tag{9.5.20}$$

Suppose that Δl is the difference between the difference of longitude of a point in the new (adjacent) zone and in the old (current) zone. Δl takes the plus sign for transformation from east to west and the minus sign from west to east.

3. Once again we compute the coordinates of the points of the double transverse Mercator in the new (adjacent) zone.

   Since the spherical difference of longitude for the point in the new zone is given by:

$$\lambda' = \lambda + \Delta\lambda = \lambda + \alpha\Delta l \tag{9.5.21}$$

substituting λ' and spherical latitude φ into the equations for direct solution of the double transverse Mercator coordinate yields:

$$\left.\begin{aligned}
\tanh\frac{x_T'}{R} &= \cos\varphi\sin(\lambda + \Delta\lambda) \\[2mm]
\tan\frac{y_T'}{R} &= \tan\varphi\sec(\lambda + \Delta\lambda)
\end{aligned}\right\} \tag{9.5.22}$$

4. By substituting (9.5.20) into (9.5.22), i.e., eliminating spherical latitude φ and spherical difference of longitude λ, we can obtain equations for calculating the coordinates of the new Mercator zone when presented with coordinates of the old Gauss–Krüger zone.

Rewriting (9.5.22) as:

$$\left.\begin{aligned}
\tanh\frac{x_T'}{R} &= \cos\varphi(\sin\lambda\cos\Delta\lambda + \cos\lambda\sin\Delta\lambda) \\[2mm]
\cot\frac{y_T'}{R} &= \cot\varphi(\cos\lambda\cos\Delta\lambda - \sin\lambda\sin\Delta\lambda)
\end{aligned}\right\} \tag{9.5.23}$$

from (9.5.20) we have:

$$\begin{aligned}
\cot\varphi &= \frac{\cos\varphi}{\sin\varphi} = \frac{\sqrt{1 - \sin^2\varphi}}{\sin\varphi} \\[2mm]
&= \csc\frac{y_G + \Delta y}{R}\sqrt{\cosh^2\frac{x_G}{R} - \sin^2\frac{y_G + \Delta y}{R}}
\end{aligned} \tag{9.5.24}$$

$$\cos\varphi = \sqrt{1 - \sin^2\varphi} = \operatorname{sech}\frac{x_G}{R}\sqrt{\cosh^2\frac{x_G}{R} - \sin^2\frac{y_G + \Delta y}{R}} \tag{9.5.25}$$

$$\sin\lambda = \frac{\tan\lambda}{\sqrt{1 + \tan^2\lambda}} = \frac{\sinh\dfrac{x_G}{R}}{\sqrt{\cosh^2\dfrac{x_G}{R} - \sin^2\dfrac{y_G + \Delta y}{R}}} \tag{9.5.26}$$

$$\cos\lambda = \frac{1}{\sqrt{1 + \tan^2\lambda}} = \frac{\cos\dfrac{y_G + \Delta y}{R}}{\sqrt{\cosh^2\dfrac{x_G}{R} - \sin^2\dfrac{y_G + \Delta y}{R}}} \tag{9.5.27}$$

Substituting equations (9.5.24) to (9.5.27) into (9.5.23) and rearranging yields:

$$\left.\begin{aligned}
\tanh\frac{x_T'}{R} &= \tanh\frac{x_G}{R}\cos\Delta\lambda + \cos\frac{y_G + \Delta y}{R}\operatorname{sech}\frac{x_G}{R}\sin\Delta\lambda \\
\cot\frac{y_T'}{R} &= \cot\frac{y_G + \Delta y}{R}\cos\Delta\lambda - \csc\frac{y_G + \Delta y}{R}\operatorname{sech}\frac{x_G}{R}\sin\Delta\lambda
\end{aligned}\right\} \tag{9.5.28}$$

Considering

$$x_T' = x_G', \quad y_T' = y_G' + \Delta y \tag{9.5.29}$$

and substituting (9.5.29) into (9.5.28), we obtain the equations for direct calculation of coordinates of the new Gauss–Krüger zone when presented with coordinates of the old Gauss–Krüger zone:

$$\left.\begin{aligned}
x_G' &= R\tanh^{-1}\left(\tanh\frac{x_G}{R}\cos(\alpha\Delta l) + \cos\frac{y_G + \Delta y}{R}\operatorname{sech}\frac{x_G}{R}\sin(\alpha\Delta l)\right) \\
y_G' &= R\cot^{-1}\left(\cot\frac{y_G + \Delta y}{R}\cos(\alpha\Delta l) - \csc\frac{y_G + \Delta y}{R}\sinh\frac{x_G}{R}\sin(\alpha\Delta l)\right) - \Delta y
\end{aligned}\right\} \tag{9.5.30}$$

Note the sign of Δl during computation. For example, transformation from zone 20 to 21 belongs to the 6° zone transformation from west to east. As a result, $\Delta l = -6°$. In addition, there exist the following.

Example 1: Direct transformation of Gauss–Krüger coordinates

Considering a map sheet area with a range from 41°15′ N to 44°10′ N and from 122°00′ E to 126°30′ E, as well as a central meridian at 123° E and a scale of 1:500 000, we compute coordinates of points along the margin for the transverse Mercator projection.

The final results are obtained as follows:

$B_0 = 43°, \quad L_0 = 123°, \quad \varphi_0 = 42°56′54″.9382, \quad \alpha = 1.000963464,$

$\beta = 1.002548622, \quad R = 6\ 376\ 715.366\ \text{m}$

In addition, when $B = 41°15′, L = 122°$, then $x_G = -83\ 818.45136$ m and $y_G = 4\ 568\ 900.419$ m are obtained.

Example 2: Inverse transformation of the Gauss–Krüger projection

Inversely computing the coordinates of above example, we obtain the results

$B = 41°14′59.9994″, \quad L = 122°00′00.002″.$

Example 3: Zone transformation for Gauss–Krüger coordinates

Given the coordinates of a point in zone 20 as $x_G = 100\ 640.3$ m, $y_G = 6\ 020\ 222.8$ m, compute its coordinates in zone 21, and then inversely transform them from zone 21 to zone 20.

The result is that with the central meridian of the old zone at $L_0 = 117°$ and the central meridian of the new zone at $L_{0'} = 123°$; when $B_0 = y_G/6\ 378\ 245$, then $x_T = -289\ 914.3709$ m, $y_T = 6\ 028\ 279.13$ m.

The inverse result is $x_T = 100\ 640.2722$ m, $y_T = 6\ 020\ 222.767$ m.

CHAPTER TEN

New Map Projections

10.1 AFFINE TRANSFORMATION OF EQUAL-AREA PROJECTIONS AND SEEKING OF MODIFIED EQUAL-AREA PROJECTIONS (YANG 1992)

The general equations for map projections, including equal-area, have the form:

$$x = f_1(\varphi, \lambda), \quad y = f_2(\varphi, \lambda) \tag{10.1.1}$$

Providing conditions for equal-area projections, the formula takes the following form:

$$H = x_\lambda y_\varphi - x_\varphi y_\lambda = R^2 \cos\varphi \tag{10.1.2}$$

After affine transformation, we obtain the following form:

$$\left. \begin{array}{l} x' = a_1 x + a_2 y + a_3 \\ y' = b_1 x + b_2 y + b_3 \end{array} \right\} \tag{10.1.3}$$

Partially differentiating the above equations, they take the following form:

$$\left. \begin{array}{l} x'_\varphi = a_1 x_\varphi + a_2 x_\varphi, \quad x'_\lambda = a_1 x_\lambda + a_2 y_\lambda \\ y'_\varphi = b_1 x_\varphi + b_2 y_\varphi, \quad y'_\lambda = b_1 x_\lambda + b_2 y_\lambda \end{array} \right\} \tag{10.1.4}$$

Hence:

$$x'_\varphi y'_\lambda = a_1 b_1 x_\varphi x_\lambda + a_1 b_2 x_\varphi y_\lambda + a_2 b_1 y_\varphi x_\lambda + a_2 b_2 y_\varphi y_\lambda$$

$$y'_\varphi x'_\lambda = a_1 b_1 x_\varphi x_\lambda + a_2 b_1 x_\varphi y_\lambda + a_1 b_2 y_\varphi x_\lambda + a_2 b_2 y_\varphi y_\lambda$$

Subtracting the two equations, the formula becomes:

$$H' = x'_\varphi y'_\lambda - y'_\varphi x'_\lambda = (x_\varphi y_\lambda - y_\varphi x_\lambda)(a_1 b_2 - a_2 b_1)$$

or:

$$H' = H(a_1 b_2 - a_2 b_1) \tag{10.1.5}$$

If the property of equal-area is maintained after an affine transformation, the following condition must be met:

$$a_1 b_2 - a_2 b_1 = 1 \tag{10.1.6}$$

Accordingly, there are some exceptional cases as follows,

$$a_1 b_2 = 1 \quad \text{and} \quad \begin{cases} b_1 = 0, a_2 = 0 \\ b_1 = 0, a_2 \neq 0 \\ b_1 \neq 0, a_2 = 0 \end{cases} \tag{10.1.7}$$

In particular,

$$\left. \begin{array}{l} a_1 = b_2 = \sin\alpha \\ a_2 = -\cos\alpha, \quad b_1 = \cos\alpha \end{array} \right\} \tag{10.1.8}$$

The above formula rotates the orientation.
   Noting that

$$\frac{h}{\sqrt{hk}} \frac{k}{\sqrt{hk}} = 1$$

and letting

$$a_1 = \frac{h}{\sqrt{hk}}, b_2 = \frac{k}{\sqrt{hk}}$$

we obtain several forms of affine transformation that provide equations satisfying conditions for equal-area projections:

$$\left. \begin{array}{l} x' = \dfrac{k}{\sqrt{hk}} x + b_3 \\[3mm] y' = \dfrac{h}{\sqrt{hk}} y + a_3 \end{array} \right\} \tag{10.1.9}$$

$$\left. \begin{array}{l} x' = \dfrac{k}{\sqrt{hk}} x + b_3 \\[3mm] y' = \dfrac{h}{\sqrt{hk}} y + a_2 x + a_3 \end{array} \right\} \tag{10.1.10}$$

or:

$$\left. \begin{array}{l} x' = b_1 y + \dfrac{k}{\sqrt{hk}} x + b_3 \\[3mm] y' = \dfrac{h}{\sqrt{hk}} y + a_3 \end{array} \right\} \tag{10.1.11}$$

To rotate the orientation:

$$\left. \begin{array}{l} x' = y\cos\alpha + x\sin\alpha + b_3 \\ y' = y\sin\alpha - x\cos\alpha + a_3 \end{array} \right\} \tag{10.1.12}$$

The general formulae for two-dimensional affine transformation are:

$$\left. \begin{array}{l} x' = b_1 y + b_2 x + b_3 \\ y' = a_1 y + a_2 x + a_3 \end{array} \right\} \quad \text{and} \quad \begin{bmatrix} a_1 & a_2 \\ b_1 & b_2 \end{bmatrix} = 1 \tag{10.1.13}$$

Equations (10.1.9) to (10.1.13) are all coordinate transformations for modified equal-area projections after an affine transformation (except for the rotational transformation, which preserves area but does not alter the map shape).

Taking the azimuthal equal-area projection as an example, the coordinate formulae have the form:

$$x = 2R\sin\frac{Z}{2}\sin\alpha, \quad y = 2R\sin\frac{Z}{2}\cos\alpha \qquad (10.1.14)$$

In equation (10.1.9), letting

$$\frac{h}{\sqrt{hk}} = p, \quad \frac{k}{\sqrt{hk}} = q, \quad a_3 = b_3 = 0$$

then the modified formulae for an azimuthal equal-area projection with affine transformation are obtained:

$$\left.\begin{array}{l} x' = qx = 2qR\sin\dfrac{Z}{2}\sin a \\[3mm] y' = py = 2pR\sin\dfrac{Z}{2}\cos a \end{array}\right\} \qquad (10.1.15)$$

where $pq = 1$.

With partial differentiation of the above equations, respectively, they take the following forms:

$$x'_z = qR\cos\frac{Z}{2}\sin a, \quad x'_a = 2qR\sin\frac{Z}{2}\cos a$$

$$y'_z = pR\cos\frac{Z}{2}\cos a, \quad y'_a = -2pR\sin\frac{Z}{2}\sin a$$

so the transformation formulae are obtained as:

$$\left.\begin{array}{l} \mu_1 = \dfrac{\sqrt{E}}{R} = \cos\dfrac{Z}{2}(p^2\cos^2 a + q^2\sin^2 a)^{1/2} \\[3mm] \mu_2 = \dfrac{\sqrt{G}}{R\sin Z} = \sec\dfrac{Z}{2}(p^2\sin^2 a + q^2\cos^2 a)^{1/2} \\[3mm] P = \dfrac{H}{R^2\sin Z} = \dfrac{x'_z y'_a - y'_z x'_a}{R^2\sin Z} = \dfrac{R^2\sin Z}{R^2\sin Z} = 1 \\[3mm] \tan\dfrac{\omega}{2} = \dfrac{1}{2}\sqrt{\mu_1^2 + \mu_2^2 - 2} \end{array}\right\} \qquad (10.1.16)$$

Now we will discuss how to determine the parameters p, q of equation (10.1.15).

From equation (10.1.16), we know that when $a = 0°$,

$$\mu_1 = p\cos\frac{Z}{2}, \quad \mu_2 = q\sec\frac{Z}{2}$$

Then:

$$\tan\left(\frac{\pi}{4} + \frac{\omega}{4}\right) = \frac{\sqrt{\mu_2}}{\sqrt{\mu_1}} = \sqrt{\frac{q}{p}}\sec\frac{Z}{2} = q\sec\frac{Z}{2}$$

When $a = 90°$

$$\mu_1 = q\cos\frac{Z}{2}, \quad \mu_2 = p\sec\frac{Z}{2}, \quad \text{then } \tan\left(\frac{\pi}{4} + \frac{\omega}{4}\right) = \sqrt{\frac{p}{q}}\sec\frac{Z}{2} = p\sec\frac{Z}{2}$$

If we let

$$q\sec\frac{Z_{max}}{2} = p\sec\frac{Z_{min}}{2}$$

then we obtain:

$$p = \sqrt{\sec\frac{Z_{max}}{2}\cos\frac{Z_{min}}{2}}, \quad q = \frac{1}{p} \qquad (10.1.17)$$

Distortion isograms of maximum angular distortion ω can be determined using the following method.

From equation (10.1.16), we know:

$$\tan\frac{\omega}{2} = \frac{1}{2}\sqrt{\mu_1^2 + \mu_2^2 - 2}$$

that is,

$$2 + 4\tan^2\frac{\omega}{2} = \mu_1^2 + \mu_2^2 \qquad (10.1.18)$$

Introducing the following symbols,

$$C = 2 + 4\tan^2\frac{\omega}{2}; \quad t_1 = p^2\cos^2 a + q^2\sin^2 a; \quad t_2 = p^2\sin^2 a + q^2\cos^2 a$$

then equation (10.1.18) can be rewritten:

$$C = t_1\cos^2\frac{Z}{2} + \frac{t_2}{\cos^2\frac{Z}{2}}$$

Letting

$$\cos^2\frac{Z}{2} = y$$

we have:

$$t_1 y^2 - Cy + t_2 = 0 \qquad (10.1.19)$$

The solution is:

$$y = \frac{C \pm \sqrt{C^2 - 4t_1 t_2}}{2t_1} \quad \text{(let } 0 \le y \le 1\text{)}$$

We obtain:

$$Z = 2\arccos\sqrt{y} \qquad (10.1.20)$$

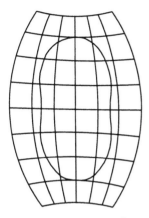

Figure 10.1 Geographic graticule for the modified equal-area projection with a distortion isogram for $\omega = 6°$

Since a_i is given, we can find Z_i. Then based on a_i, Z_i, we can find x'_i, y'_i, so that we can draw distortion isograms of the maximum angular distortion ω.

Now taking the Pacific region as an example, its mapping area is ϕ: 60°S–60°N, λ: 75°W–15°E. The centerpoint of the projection: $\phi_0 = 0°$, $\lambda_0 = 30°W$. The radius of the sphere is taken as $R = 6\ 371\ 116$ m.

Using $Z_{max} = 44°.5$, $Z_{min} = 21°.5$, parameter p is determined. From equation (10.1.17), $p = 1.0303$ and $q = 1/p$; then $\omega = 5.4458°$. If instead we choose the azimuthal equal-area projection, then $\omega = 8.858°$.

Figure 10.1 shows the geographic graticule for the modified equal-area projection, with a distortion isogram for $\omega = 6°$. We can see that the distortion isogram is oval-like. This projection can be called a modified azimuthal equal-area projection.

10.2   LINEAR TRANSFORMATION OF THE GNOMONIC PROJECTION AND SEEKING THE DOUBLE-AZIMUTHAL PROJECTION (YANG 1990b)

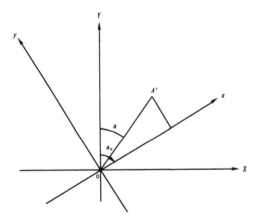

Figure 10.2   Coordinate system for modified gnomonic projection

As shown in Figure 10.2, in the coordinate system xOy for which the vertical circle projection of azimuthal angle a_0 is regarded as the axis for the abscissa and the new pole projection O as origin, the coordinate formulae for the gnomonic projection have the form:

$$x = R \tan Z \cos (a_0 - a)\ y$$
$$= R \tan Z \sin (a_0 - a) \qquad (10.2.1)$$

After linear transformation, we obtain a modified gnomonic projection:

$$\left.\begin{array}{l} x' = hy = kR \tan Z \cos(a_0 - a) \\ y' = kx = kR \tan Z \sin(a_0 - a) \end{array}\right\} \qquad (10.2.2)$$

where $h \neq k$.

The transformation equation can be written in following form:

$$\mu_1 = \frac{\sqrt{E}}{R} = \sec^2 Z [h^2 \sin^2 (a_0 - a) + k^2 \cos^2 (a_0 - a)]^{1/2}$$

$$\mu_2 = \frac{\sqrt{G}}{R \sin Z} = \sec Z [h^2 \cos^2 (a_0 - a) + k^2 \sin^2 (a_0 - a)]^{1/2}$$

$$\left.\begin{array}{l} \tan\varepsilon = -\dfrac{F}{H} = \dfrac{k^2 - h^2}{hk}\cos(a_0 - a)\sin(a_0 - a) \\[4mm] P = \dfrac{H}{R^2 \sin Z} = hk\sec^3 Z \\[4mm] \tan\dfrac{\omega}{2} = \dfrac{1}{2}\sqrt{\dfrac{\mu_1^2 + \mu_2^2}{p} - 2} \end{array}\right\} \qquad (10.2.3)$$

From equation (10.2.2), we have:

$$\frac{y'}{x'} = \frac{h}{k}\tan(a_0 - a) \qquad (10.2.4)$$

Note that:

$$\tan a' = \frac{y'}{x'}$$

then:

$$\tan a' = \frac{h}{k}\tan(a_0 - a) \qquad (10.2.5)$$

It is thus clear that usually $a' \neq a_0 - a$, i.e. there is angular distortion at the centerpoint of the projection, except at the angle along the coordinate axis.

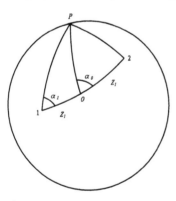

Figure 10.3 Spherical elements for calculating the double-azimuthal projection

From equation (10.2.3), when $a_0 - a = 0°, 180°$, $\mu_1 = k\sec^2 Z_i$, $\mu_2 = h\sec Z_i$, $\varepsilon = 0$. The conditions for no angular distortion are: $\varepsilon = 0$, $k\sec^2 Z_i = h\sec Z_i$, that is,

$$k = h\cos Z_i \qquad (10.2.6)$$

At this moment, the two points (Z_i, a_0) and $(Z_i, a_0 - 180°)$ do not have angular distortion. This modified gnomonic projection which maintains azimuths without distortion from two points is called the two-point azimuthal or double-azimuthal projection.

Here we introduce the method for calculating the double-azimuthal projection.

As shown in Figure 10.3, the values of point 1 (φ_1, λ_1), point 2 (φ_2, λ_2), and the distance $2Z_i$ between points 1 and 2 are given. Finding the middle point O (φ_0, λ_0) of distance $2Z_i$ and the value of azimuthal angle a_0 of $\overline{O2}$ can used with the following formulae:

$$\left.\begin{array}{l} \sin\varphi_0 = \cos Z_1 \sin\varphi_1 + \sin Z_1 \cos\varphi_1 \cos a_1 \\[3mm] \tan(\lambda_0 - \lambda_1) = \dfrac{\sin Z_1 \sin a_1}{\cos\varphi_1 \cos Z_1 - \sin\varphi_1 \sin Z_1 \cos a_1} \\[4mm] \tan a_0 = \dfrac{\cos\varphi_2 \sin(\lambda_2 - \lambda_0)}{\cos\varphi_0 \sin\varphi_2 - \sin\varphi_0 \cos\varphi_2 \cos(\lambda_2 - \lambda_0)} \end{array}\right\} \qquad (10.2.7)$$

where:

$$\cos 2Z_1 = \sin \varphi_1 \sin \varphi_2 + \cos\varphi_1 \cos \varphi_2 \cos(\lambda_2 - \lambda_1)$$

$$\tan a_1 = \frac{\cos\varphi_2 \sin(\lambda_2 - \lambda_1)}{\cos\varphi_1 \sin\varphi_2 - \sin\varphi_1 \cos\varphi_2 \cos(\lambda_2 - \lambda_1)}$$

In accordance with equation (10.2.6), we select h and k. If we let $h = 1$, then $k = \cos Z_1$.

Hence the coordinate transformation formulae for the double-azimuthal projection which maintains undistorted angles from points 1 and 2 take the form:

$$x' = hR \tan Z \cos(a_0 - a), \quad y' = kR \tan Z \sin(a_0 - a) \qquad (10.2.8)$$

where $h = 1$, $k = \cos Z_1$.

Spherical coordinate transformation formulae have the form:

$$\left. \begin{array}{l} \cos Z = \sin\varphi_0 \sin\varphi + \cos\varphi_0 \cos\varphi \cos(\lambda - \lambda_0) \\[2mm] \tan a = \dfrac{\cos\varphi \sin(\lambda - \lambda_0)}{\cos\varphi_0 \sin\varphi - \sin\varphi_0 \cos\varphi \cos(\lambda - \lambda_0)} \end{array} \right\} \qquad (10.2.9)$$

CHAPTER ELEVEN

Variable-scale Map Projections

Usually we try to choose the map projection that has the smallest deformation. But some urban maps for travel need to show detailed streets in the business center, in scenic spots and at historical sites, and for other service facilities such as restaurants and hotels. The key elements of such detail are nearly always focused in the center of a city or region. So this area needs to be expressed at a larger scale. Applying the deformation information for a map projection, we can edit a map to have arbitrary different scales. On the map we can make the scale of every region vary, one at a time, two at a time or not the same everywhere. In addition, for a large regional map if we need to stress an important area we can use the surrounding area as a background, or we can make the different areas have a different kind of areal deformation to satisfy the requirement of various thematic map uses. For example, in epidemiology rates can be shown on a base stretched to equal the number of susceptibles, or, for political purposes on a base of voters. This kind of map is usually called a variable-scale map. In Russia they are known as varivalent projections, in France as anamorphoses, in the Anglo-American literature as 'area cartograms' (Dorling 1996; Tobler 1963, 1986a). It can be shown that all equal-area map projections are the special case of 'constant importance'. We can obtain variable-scale map projections in many ways. These will be introduced as follows.

11.1 VARIABLE-SCALE PROJECTIONS

11.1.1 Double projection method

Project, in contrast to an ordinary map, to an interim spherical surface, then from the interim spherical surface to a plane; this may result in a variable-scale map.

Assume that the rectangular coordinates of an arbitrary point in an ordinary urban plane map are x, y, and its spherical coordinates on an interim spherical surface are Z, a; after projecting to a plane, its rectangular coordinates are x', y'. Then the general relation formulae are:

$$Z = \phi_1(x, y), \quad a = \phi_2(x, y) \tag{11.1.1}$$

$$x' = f_3(Z, a), \quad y' = f_4(Z, a) \tag{11.1.2}$$

Substituting (11.1.1) into (11.1.2) we have:

$$x' = f_1(x, y), \quad y' = f_2(x, y)$$

(11.1.3)

By its nature it is still a projection transformation between two planes.

Now we give some examples to illustrate this method of coordinate transformation.

Example 1: Coordinate transformation of an inverse azimuthal equidistant projection – using the orthographic projection

Given the coordinate formulae of an azimuthal equidistant projection:

$$x = RZ \sin a, \quad y = RZ \cos a$$

(11.1.4)

the coordinate formulae of an inverse azimuthal equidistant projection are:

$$Z = \frac{\sqrt{x^2 + y^2}}{R}$$

(11.1.5)

and:

$$\sin a = \frac{x}{\sqrt{x^2 + y^2}}, \quad \cos a = \frac{y}{\sqrt{x^2 + y^2}}$$

(11.1.6)

Then project this orthographically to an interim spherical surface then onto a plane. We have:

$$x' = R \sin Z \sin a, \quad y' = R \sin Z \cos a$$

(11.1.7)

or rewritten as:

$$y' = R \sin\left(\frac{\sqrt{x^2 + y^2}}{R}\right) \frac{y}{\sqrt{x^2 + y^2}}, \quad x' = R \sin\left(\frac{\sqrt{x^2 + y^2}}{R}\right) \frac{x}{\sqrt{x^2 + y^2}}$$

(11.1.8)

The above is an example of a variable-scale projection.

Example 2: Coordinate transformation of an inverse azimuthal equidistant projection – onto a transverse cylindrical conformal projection

Applying the coordinate formulae of a transverse cylindrical conformal projection, to project from the interim spherical surface to a variable-scale map plane as:

$$x' = \frac{1}{2} R \ln\left(\frac{1 + \sin z \sin a}{1 - \sin z \sin a}\right), \quad y' = R \arctan(\cot Z \sec a)$$

(11.1.9)

Substituting (11.1.5), (11.1.6) into the above equation we have:

$$\left.
\begin{aligned}
y' &= R \arctan\left(\frac{\sqrt{x^2 + y^2}}{y} \cot \frac{\sqrt{x^2 + y^2}}{R}\right) \\
x' &= \frac{1}{2} R \left[\ln\left(1 + \frac{x}{\sqrt{x^2 + y^2}} \sin \frac{\sqrt{x^2 + y^2}}{R}\right) - \ln\left(1 - \frac{x}{\sqrt{x^2 + y^2}} \sin \frac{\sqrt{x^2 + y^2}}{R}\right)\right]
\end{aligned}
\right\}$$

(11.1.10)

This is also a type of variable-scale projection.

Example 3: Coordinate transformation of an inverse tangent cylindrical equidistant projection – onto a general polyconic projection

Given the coordinate formulae of a tangent cylindrical equidistant projection:

$$x = Ra, \quad y = R\left(\frac{\pi}{2} - Z\right)$$ (11.1.11)

the formulae for an inverse tangent cylindrical equidistant projection are:

$$Z = \frac{\pi}{2} - \frac{y}{R}, \quad a = \frac{x}{R}$$ (11.1.12)

The coordinate formulae to go from the inverse tangent cylindrical equidistant projection to a general polyconic projection are:

$$x' = \rho \sin \delta, \quad y' = S + \rho(1 - \cos \delta)$$ (11.1.13)

$$\rho = R \cot\left(\frac{\pi}{2} - Z\right), \quad \delta = a \sin\left(\frac{\pi}{2} - Z\right)$$ (11.1.14)

Where:

$$S = R\left(\frac{\pi}{2} - Z\right)$$

On substitution of equation (11.1.12) into the above we get:

$$y' = R\left(\frac{\pi}{2} - \frac{y}{R}\right) + R \cot\frac{y}{R}\left[1 - \cos\left(\frac{x}{R}\sin\frac{y}{R}\right)\right]$$

$$x' = R \cot\frac{y}{R}\sin\left(\frac{x}{R}\sin\frac{y}{R}\right)$$ (11.1.15)

These are the formulae to go from an inverse tangent cylindrical equidistant projection to a general polyconic projection. It is also a variable-scale projection.

11.1.2   The coordinate transformation of plane figures

We know from Section 3.3 that the term plane figure has a broad meaning. A plane figure transformation lies in building a function corresponding to a one to one relation between two planes, i.e.,

$$x' = f_1(x, y), \quad y' = f_2(x, y)$$ (11.1.16)

To realize the above transformation of a plane figure we can introduce an intermediate variable and use many kinds of map projection.

Rectangular or polar coordinates can describe any plane figure. The plane rectangular coordinate grid or plane polar coordinate grid can be looked at as the control net of the plane figure. For this reason, the study of a coordinate transformation of a plane figure can be identified with the study of a transformation, of the rectangular grid or the polar grid, going from the first plane to the second, or its inverse.

The coordinate transformation of the rectangular coordinate grid net

The rectangular coordinates may be taken as the mapping of meridians and parallels of the normal cylindrical projection, i.e.,

$$x = C\lambda, \quad y = f(\varphi) \tag{11.1.17}$$

It is clear that the mapping of a rectangular coordinate grid from the first plane to the second may be taken to be the same as the transformation from a cylindrical projection to other projections; that is, the shape of meridians and parallels on any other projection may be taken as a mapping of the rectangular coordinate grid onto the second plane, i.e.,

$$x' = f_1(\varphi, \lambda), \quad y' = f_2(\varphi, \lambda) \tag{11.1.18}$$

A way of realizing the above transformation is:

$$x, y \xrightarrow{\text{contrary cylindrical projection}} \varphi, \lambda \xrightarrow{\text{new projection}} x', y'$$

and its inverse transformation is:

$$x', y' \xrightarrow{\text{contrary new projection}} \varphi, \lambda \xrightarrow{\text{cylindrical projection}} x, y$$

It must be pointed out that here the spherical surface is only an interim supplementary surface and does not represent the earth.

The coordinate transformation of a polar coordinate grid net

The polar coordinate grid may be taken as the mapping of meridians and parallels of a normal azimuthal projection, i.e.,

$$\rho = f(\varphi), \quad \delta = \lambda \tag{11.1.19}$$

From this we can see that the mapping of a polar coordinate net from the first plane to the second may be taken as the transformation from a normal azimuthal projection to other projections; that is, the shape of meridians and parallels of any other projection may be taken as the mapping of a polar coordinate grid, i.e.,

$$x' = f_1(\varphi, \lambda), \quad y' = f_2(\varphi, \lambda)$$

A way of realizing above transformation is:

$$\rho, \delta \xrightarrow{\text{contrary azimuthal projection}} \varphi, \lambda \xrightarrow{\text{new projection}} x', y'$$

By way of its inverse transformation we get:

$$x', y' \xrightarrow{\text{contrary new projection}} \varphi, \lambda \xrightarrow{\text{azimuthal projection}} \rho, \delta$$

Examples of the plane figure transformations are shown in Section 3.3.

The method of variable-scale projections may be applied to other subjects such as improving the measurement precision of a nomograph. For the application examples see Section 3.3.

11.1.3   Polyfocal projections

Polyfocal projections (Kadmon 1978) are also a kind of variable-scale map projection involving simple and flexible calculations.

For the sake of convenience we begin with a single focal projection.

Assuming the scale S_0 of an original map. R is the distance from an arbitrary point to the center of focus O_1, $f(R)$ is a function of distance.

Assume that the scale of the variable-scale map is S and $S = S_0 + S_0 f(R)$. We want to reduce the scale S with increasing R, and there is to be a maximum scale value at the focus. So we let:

$$f(R) = \frac{A}{1 + CR^2} \tag{11.1.20}$$

Then we get:

$$S = S_0 + \frac{AS_0}{1 + CR^2} \tag{11.1.21}$$

As seen from above formula, $S = S_0(1 + A)$ when $R = 0$, that is, the scale of center is enlarged by $(1 + A)$ times. So we can take A as the magnification coefficient of the scale. The magnitude of S is in inverse ratio with C when A is a fixed value. So controlling the magnitude of area size is the action of C.

The distance R on the original map corresponds to the radial distance r of the variable-scale map. Let:

$$r = SR \tag{11.1.22}$$

Letting $S_0 = 1$ and introducing (11.1.21) into above we have:

$$r = R + \frac{AR}{1 + CR^2} \tag{11.1.23}$$

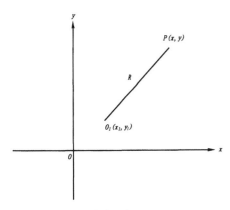

Figure 11.1 Polyfocal projection coordinate system

As shown in Figure 11.1, the coordinates of the focus O_1 of the original map are (x_1, y_1); the distance from any point $P(x, y)$ to focus is:

$$R = \sqrt{(x - x_1)^2 + (y - y_1)^2}$$

Then we have:

$$x = x_1 + R \sin \alpha, \tag{11.1.24}$$
$$y = y_1 + R \cos \alpha$$

Assuming that the original map takes O_1 as the projection center and is transformed to a variable-scale map according the normal azimuthal projection, then the coordinates of a point P' are:

$$x' = x_1 + r \sin \alpha, \tag{11.1.25}$$
$$y' = y_1 + r \cos \alpha$$

Introducing equation (11.1.23) into the above, we have:

$$\left.\begin{array}{l} y' = y_1 + \left(R + \dfrac{AR}{1 + CR^2}\right)\cos\alpha = y_1 + R\cos\alpha + \dfrac{AR\cos\alpha}{1 + CR^2} \\[4mm] x' = x_1 + \left(R + \dfrac{AR}{1 + CR^2}\right)\sin\alpha = x_1 + R\sin\alpha + \dfrac{AR\sin\alpha}{1 + CR^2} \end{array}\right\} \tag{11.1.26}$$

From equation (11.1.24) we have:

$$R\cos\alpha = y - y_1, \quad R\sin\alpha = x - x_1$$

(11.1.27)

Introducing (11.1.24) and (11.1.27) into (11.1.26) we get:

$$\left.\begin{array}{l} x' = x + \dfrac{A(x - x_1)}{1 + CR^2} = x + f(R)(x - x_1) = x + \Delta x \\[3mm] y' = y + \dfrac{A(y - y_1)}{1 + CR^2} = y + f(R)(y - y_1) = y + \Delta y \end{array}\right\}$$

(11.1.28)

From equation (11.1.28) we see that the coordinates of point P' are made up of the coordinates of the original P, the increment of the radial function, and some parameters.

Considering the possibility of n foci $O_1, O_2, \ldots, O_n$, the coordinate formulae of these polyfocal projections can be written as:

$$\left.\begin{array}{l} x' = x + \Delta x_1 + \Delta x_2 + \ldots + \Delta x_n = x + \displaystyle\sum_{i=1}^{n} \Delta x_i \\[4mm] y' = y + \Delta y_1 + \Delta y_2 + \ldots + \Delta y_n = y + \displaystyle\sum_{i=1}^{n} \Delta y_i \end{array}\right\}$$

(11.1.29)

Assuming the coordinates of the n foci are (x_i, y_i) $(i = 1, 2, \ldots, n)$, and that their parameters are A_i, C_i, then the coordinate formula of polyfocal projections is:

$$x' = x + \sum_{i=1}^{n} \frac{A_i(x - x_i)}{1 + C_i R_i^2}, \quad y' = y + \sum_{i=1}^{n} \frac{A_i(y - y_i)}{1 + C_i R_i^2}$$

(11.1.30)

where:

$$R_i = \sqrt{(x - x_i)^2 + (y - y_i)^2}$$

Examples

Assume the square grid of an original map can be treated as the cylindrical equidistant projection grid, the radius of a corresponding supplementary sphere is 0.3 cm, the mapping area range is φ: $-60°$ to $+60°$, λ: $-80°$ to $+80°$.

If the coordinates of two foci are respectively at O_1 $(-20°, -40°)$, O_2 $(30°, 40°)$, and the parameters are:

$$A_1 = 2.0, \quad C_1 = \frac{1}{100}, \quad A_2 = 3.0, \quad C_2 = \frac{1}{300}$$

the calculation formulae for the coordinates of the original map are:

$$x = R\lambda, \quad y = R\varphi$$

(11.1.31)

The projection coordinate formulae of double foci are:

$$x' = x + \frac{A_1(x - x_1)}{1 + C_1 R_1^2} + \frac{A_2(x - x_2)}{1 + C_2 R_2^2}, \quad y' = y + \frac{A_1(y - y_1)}{1 + C_1 R_1^2} + \frac{A_2(y - y_2)}{1 + C_2 R_2^2}$$

(11.1.32)

where:

$$R_1 = \sqrt{(x - x_1)^2 + (y - y_1)^2}, \quad R_2 = \sqrt{(x - x_2)^2 + (y - y_2)^2}$$

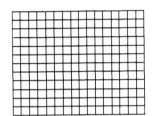

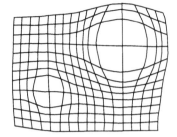

Figure 11.2   Equidistant cylindrical projection grid

Figure 11.3   Double foci projection grid

Figure 11.2 shows the case in which the original map is a cylindrical equidistant projection grid. Figure 11.3 is the transformed double foci projection grid. From that we can see that the grid at the foci is enlarged. This transformation method has the virtues of simplicity and flexible calculation; the value of the parameters A, C can be adjusted according to requirements.

11.1.4   Modified polyfocal projection

In the single focal projection (11.1.28) of the last section, if we make $f(R)$ another function of R, such as:

$$f(R) = \alpha(1 + R)^\beta, \quad f(R) = \frac{2\alpha}{\sqrt{(2\alpha)^2 + R^2}}, \quad f(R) = \alpha e^{-\beta R} \quad \text{etc.}$$

and let $R = \varphi(x, y)$, then (11.1.3) can be rewritten as:

$$\left.\begin{array}{l} x' = x_1 + f(\varphi(x, y))(x - x_1) \\ y' = y_1 + f(\varphi(x, y))(y - y_1) \end{array}\right\} \tag{11.1.33}$$

To control the change of the scale, $f(\varphi(x, y))$ can also be written in the form of an auxiliary function as follows

$$f(\varphi(x, y)) = \begin{cases} f_1(\varphi(x, y)), & 0 < \varphi(x, y) \le a_k \\ f_1(\varphi(x_k, y_k)), & \varphi(x, y) > a_k \end{cases} \tag{11.1.34}$$

where the meaning of f_1 is the same as f, it shows the change of the scale. a_k is the critical value; it is the controlling parameter of the range of the scale change. And:

$$\varphi(x_k, y_k) = a_k$$

We call equations (11.1.33) and (11.1.34) the modified single focal projection. It gives a map projection a 'magnification' effect. In the magnification zone (i.e., $0 < \varphi(x, y) \le a_k$), the scale changes by $f_1(\varphi(x, y))$. Outside the magnified zone (i.e., $(x, y) > a_k$), the scale is the constant given by $f_1(\varphi(x_k, y_k))$ and does not change. One major virtue of this kind of modified single focus projection is its adjustability, that is, we can control the magnification and the range of magnification by designing f_1, and can control the shape of the magnified zone by designing φ, to control the scope of magnifier by choosing a_k. For example, if the original map is the equidistant projection, we may use:

$$\varphi(x, y) = R = \sqrt{(x - x_1)^2 + (y - y_1)^2}$$

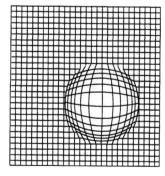

Figure 11.4 Circle magnifier grid

to design a circular magnification zone. The design of the magnification factor is as follows: if we need the scale to change faster in the smaller portion near the focal point, then f in (11.1.33) can be $f(R) = \alpha(1 + R)^\beta$. Here α, β are the control parameters of the extent of enlargement. Figure 11.4 illustrates one kind of circle 'magnifier'.

The general formula of the modified polyfocal projection is as follows:

$$\left.\begin{array}{l} x' = x + \sum_{i=1}^{n} F_i(\varphi_i(x, y))(x - x_i) \\[3mm] y' = y + \sum_{i=1}^{n} F_i(\varphi_i(x, y))(y - y_i) \end{array}\right\} \quad (11.1.35)$$

$$F_i(\varphi_i(x, y)) = \begin{cases} G_i(\varphi_i(x, y)), & 0 < \varphi_i(x, y) \le a_i \\ G_i(\varphi_i(x_i, y_i)), & \varphi_i(x, y) > a_i \end{cases}$$

where F_i presents the change of scale. $R_i = \varphi_i(x, y)$ is the distance from an arbitrary point to the focus. The meaning of G_i is the same as F_i; a_i is the critical value corresponding to the polyfocus $O_i(x_i, y_i)$, $\varphi_i(x_i, y_i) = a_i$, $i = 1, 2, \ldots, n$.

In Wang (1993a) this modified polyfocal projection is referred to as one kind of adjustable map projection with a magnifying glass effect.

11.2   COMPOSITE PROJECTIONS

11.2.1   Composite azimuthal projection

According to Yang (1988), the fundamentals of a composite azimuthal projection can be described as follows.

Assume the projection area is $0° - Z_k$, if we demand that we use an azimuthal projection having a property in the zone $0° - Z_i$, i.e., $\rho = f_1(Z)$, and then we use an azimuthal projection having different nature in the interval $Z_i - Z_k$, i.e., $\rho = f_2(Z)$, we can obtain a general formula of a composite azimuthal projection:

$$\rho = \begin{cases} f_1(Z), & (0 \le Z \le Z_i) \\ f_2(Z), & (Z_i < Z \le Z_k) \end{cases}, \quad \delta = al \quad (11.2.1)$$

For example, the polar coordinate formulae of a gnomonic–stereographic composite azimuthal projection are:

$$\rho = \begin{cases} R \tan Z, & (0 \le Z \le Z_i) \\ 2R \tan \dfrac{Z}{2}, & (Z_i < Z \le Z_k) \end{cases}, \quad \delta = al \quad (11.2.2)$$

The polar coordinate formulae of a conformal–equivalent composite azimuthal projection are:

$$\rho = \begin{cases} 2R\tan\dfrac{Z}{2}, & (0 \le Z \le Z_i) \\ \\ 2R\sin\dfrac{Z}{2}, & (Z_i < Z \le Z_k) \end{cases}, \quad \delta = al \qquad (11.2.3)$$

If we want to build the above composite azimuthal projection we need to give boundary conditions so that the two functions meet properly at Z_i and this must be done in a common coordinate system. Now we discuss two separate methods for this as follows.

The equal boundary function value method

Because $f_1(Z_i) \ne f_2(Z_i)$, letting $f_1(Z_i) = f_2(Z_i')$, then the required polar distance of the equal value on the boundary becomes:

$$Z_i' = f_2^{-1}(f_1(Z_i)) \qquad (11.2.4)$$

To equation (11.2.2), we have:

$$Z_i' = 2\arctan\left(\frac{1}{2}\tan Z_i\right) \qquad (11.2.5)$$

Thus we can get the calculation formulae for the gnomonic–stereographic composite azimuthal projection:

$$\rho = \begin{cases} R\tan Z', & (0 \le Z \le Z_i) \\ \\ 2R\tan\dfrac{Z_i' - Z_i + Z}{2}, & (Z_i < Z \le Z_k) \end{cases}, \quad \delta = al \qquad (11.2.6)$$

where Z_i' is obtained from equation (11.2.5).
    To formula (11.2.3), we have:

$$Z_i' = 2\arcsin\left(\tan\frac{Z_i}{2}\right) \qquad (11.2.7)$$

So we can get the calculation formula of a conformal–equivalent composite azimuthal projection:

$$\rho = \begin{cases} 2R\tan\dfrac{Z}{2}, & (0 \le Z \le Z_i) \\ \\ 2R\sin\dfrac{Z_i' - Z_i + Z}{2}, & (Z_i < Z \le Z_k) \end{cases}, \quad \delta = al \qquad (11.2.8)$$

where Z_i is obtained from equation (11.2.7).

The translation method

Using the translation method, the boundary values will be forced to be the same, then the formulae for composite azimuthal projections in a common polar coordinate system take the form:

$$\rho = \begin{cases} f_1(Z), & (Z \le Z_i) \\ f_1(Z_i) + K \cdot [f_2(Z) - f_2(Z_i)], & (Z > Z_i) \end{cases} \qquad (11.2.9)$$

where K is the scale parameter. When $K = 1$, letting $D = f_1(Z_i) - f_2(Z_i)$, then polar coordinate formulae for the gnomonic–equidistant composite projection have the form:

$$\rho = \begin{cases} RZ, & (Z \leq Z_i) \\ D + RZ, & (Z > Z_i) \end{cases}, \quad \delta = al \tag{11.2.10}$$

where the translation constant $D = R\tan Z_i - RZ_i$; R is the earth radius.

In composite azimuthal projections, if we use the same projection but with changed scales, the scale adjusting coefficient is K; between the inner circle, from $Z \leq Z_i$ and through the outside circle where $Z > Z_i$, then the formula for this kind of composite azimuthal projection with radius ρ can be written as:

$$\rho = \begin{cases} f_1(Z), & (Z \leq Z_i) \\ f_1(Z_i) + K \cdot [f_1(Z) - f_1(Z_i)], & (Z > Z_i) \end{cases} \tag{11.2.11}$$

For example, the formula of equidistant composite azimuthal projection ρ is:

$$\rho = \begin{cases} RZ, & (Z \leq Z_i) \\ R[Z_i + K \cdot (Z - Z_i)], & (Z > Z_i) \end{cases} \tag{11.2.12}$$

And the formula of the equivalent composite azimuthal projection ρ is:

$$\rho = \begin{cases} 2R\sin\dfrac{Z}{2}, & (Z \leq Z_i) \\ 2R\sin\dfrac{Z_i}{2} + 2R\left(\sin\dfrac{Z}{2} - \sin\dfrac{Z_i}{2}\right)K, & (Z > Z_i) \end{cases} \tag{11.2.13}$$

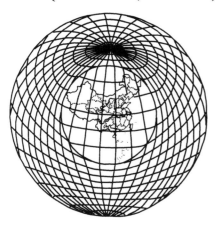

Figure 11.5 The grid for equidistant composite azimuthal projection

In an equidistant composite azimuthal projection used in North America, with projection center at: $\varphi_0 = 40°N$, $\lambda_0 = 90°W$, and the radius of the inner circle set at 30°, and the outer circle extending to 20°, then the adjusted coefficient of scale is $K = 0.35$. Figure 11.5 shows the grid for equidistant composite azimuthal projection.

This kind of composite azimuthal projection stresses the important area and uses the surrounding area as a foil. In Snyder (1987b) it is called a 'magnified' azimuthal projection.

Distortion analysis

In 'the equal boundary values method', above, it can be proven that in $[0, Z_i]$ the projection keeps the original properties, but in $[Z_i, Z_k]$ it does not keep the original projection properties and instead has an arbitrary quality. In 'the method of translation', the region $Z > Z_i$ does not retain the distortion characteristics of the original projection.

Because

$$\rho = D + f_2(Z), \quad \mu_1 = \frac{d(D + f_2(Z))}{RdZ} = \frac{df_2(Z)}{RdZ}$$

the length ratio of the original projection is maintained. But

$$\mu_2 = \frac{\rho}{R\sin Z} = \frac{D}{R\sin Z} + \frac{f_2(Z)}{R\sin Z}$$

the length ratio of almucantars of the original projection is not kept.

11.2.2 Composite cylindrical projections

The general formulae of cylindrical projections are:

$$x = r_0(\lambda - \lambda_0), \quad y = f(\varphi) \tag{11.2.14}$$

where $r_0 = R\cos\varphi_0$, φ_0 is the latitude of the standard parallels.

From the above we can see that the boundary condition of composite cylindrical projections needs to keep the standard parallels the same. Using the 'translation method', we can write out the general formulae of composite cylindrical projections having different properties in a rectangular coordinate system:

$$y = \begin{cases} f_1(\varphi), & (\varphi \le \varphi_i) \\ D + f_2(\varphi), & (\varphi > \varphi_i) \end{cases}, \quad x = r_0(\lambda - \lambda_0) \tag{11.2.15}$$

where $D = f_1(\varphi_i) - f_2(\varphi_i)$; latitude φ_i is the boundary value.

The general formula of composite cylindrical projections having same properties is:

$$y = \begin{cases} f_1(\varphi), & (\varphi \le \varphi_i) \\ f_1(\varphi_i) + K[f_1(\varphi) - f_1(\varphi_i)], & (\varphi > \varphi_i) \end{cases}, \quad x = r_0(\lambda - \lambda_0) \tag{11.2.16}$$

where K is the scale coefficient.

For example, the calculation formulae of a conformal–equivalent composite cylindrical projection are:

$$y = \begin{cases} r_0\ln\tan\left(\dfrac{\pi}{4} + \dfrac{\varphi}{2}\right), & (\varphi \le \varphi_i) \\ D + \dfrac{R^2}{r_0}\sin\varphi, & (\varphi > \varphi_i) \end{cases}, \quad x = r_0(\lambda - \lambda_0) \tag{11.2.17}$$

where:

$$D = r_0\ln\tan\left(\frac{\pi}{4} + \frac{\varphi_i}{2}\right) - \frac{R^2}{r_0}\sin\varphi_i$$

Figure 11.6 shows the grid for conformal–equivalent composite cylindrical projection, with standard parallel $\varphi_0 = \pm50°$, and is at the boundary.

In the same way, the calculation formulae of a conformal–equidistant composite cylindrical projection are:

$$y = \begin{cases} r_0\ln\tan\left(\dfrac{\pi}{4} + \dfrac{\varphi}{2}\right), & (\varphi \le \varphi_i) \\ D + R\varphi, & (\varphi > \varphi_i) \end{cases}, \quad x = r_0(\lambda - \lambda_0) \tag{11.2.18}$$

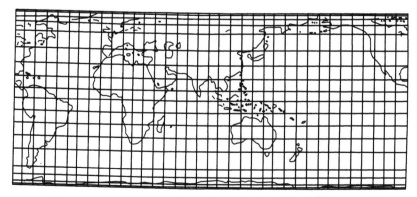

Figure 11.6 The grid for conformal–equivalent composite cylindrical projection

Furthermore, the calculation formulae of an equidistant composite cylindrical projection become:

$$y = \begin{cases} R\varphi, & (\varphi \leq \varphi_i) \\ R[\varphi_i + K \cdot (\varphi - \varphi_i)], & (\varphi > \varphi_i) \end{cases}, \quad x = r_0(\lambda - \lambda_0) \tag{11.2.19}$$

where:

$$D = r_0 \ln\tan\left(\frac{\pi}{4} + \frac{\varphi_i}{2}\right) - R\varphi_i$$

By the same method we can get the composite oblique (or transverse) cylindrical projection.

Distortion analysis

Using the 'translation' method for cylindrical projections, within the region of $\varphi > \varphi_i$, the original projection's properties are kept. For example,

$$y = D + f_2(\varphi), \quad x = r_0(\lambda - \lambda_0)$$

then the length ratio of the meridian is:

$$m = \frac{dy}{Rd\varphi} = \frac{df_2(\varphi)}{Rd\varphi}$$

and the length ratio of the parallel is:

$$n = \frac{dx}{rd\lambda} = \frac{r_0}{r}$$

11.2.3 Composite cylindrical–pseudocylindrical projections

From the general formulae (11.2.14) of cylindrical projections we know that in cylindrical–pseudocylindrical projections the necessary condition of a pseudocylindrical projection is the pseudocylindrical projection with equally divided parallels, i.e.,

$$y = f_2(\varphi) \cdot (\lambda - \lambda_0) \tag{11.2.20}$$

Using the 'translation method', we can write out the general formulae of composite cylindrical–pseudocylindrical projections:

$$y = \begin{cases} f(\varphi), & (\varphi \le \varphi_i) \\ D + f_1(\varphi), & (\varphi > \varphi_i) \end{cases}, \quad x = \begin{cases} R\cos\varphi_0 \cdot (\lambda - \lambda_0), & (\varphi \le \varphi_i) \\ f_2(\varphi) \cdot (\lambda - \lambda_0), & (\varphi > \varphi_i) \end{cases}$$ (11.2.21)

where $D = f(\varphi_i) - f_1(\varphi_i)$.
The boundary condition is:

$$R\cos\varphi_0 = f_2(\varphi_i)$$ (11.2.22)

Fixing φ_i we have:

$$\varphi_0 = \arccos\left(\frac{1}{R}f_2(\varphi_i)\right)$$ (11.2.23)

Fixing φ_0 we have:

$$\varphi_i = f_2^{-}(R\cos\varphi_0)$$ (11.2.24)

Hence the coordinate formulae of the Sanson pseudocylindrical projection can be given as:

$$y = R\varphi, \quad x = R(\lambda - \lambda_0)\cos\varphi$$ (11.2.25)

The boundary condition is:

$$\varphi_0 = \varphi_i$$ (11.2.26)

Using the 'translation method', we get the coordinate formulae of the composite cylindrical–Sanson pseudocylindrical projection:

$$y = \begin{cases} f(\varphi), & (\varphi \le \varphi_i) \\ D + R\varphi, & (\varphi > \varphi_i) \end{cases}, \quad x = \begin{cases} R\cos\varphi_i \cdot (\lambda - \lambda_0), & (\varphi \le \varphi_i) \\ R\cos\varphi \cdot (\lambda - \lambda_0), & (\varphi > \varphi_i) \end{cases}$$ (11.2.27)

where f is an arbitrary cylindrical projection, $D = f(\varphi_i) - R\varphi_i$.
As another example, the coordinate formulae of the Eckert pseudocylindrical projection are:

$$\left. \begin{array}{l} y = \dfrac{2R}{\sqrt{\pi + 2}}\psi, \quad x = \dfrac{2R}{\sqrt{\pi + 2}}(\lambda - \lambda_0)\cos^2\dfrac{\psi}{2} \\[2mm] \sin\psi + \psi = \dfrac{\pi + 2}{2}\sin\varphi \end{array} \right\}$$ (11.2.28)

Fixing φ_i, we get the boundary conditions as:

$$\left. \begin{array}{l} \varphi_0 = \arccos\left(\dfrac{2}{\sqrt{\pi + 2}}\cos^2\dfrac{\psi_i}{2}\right) \\[2mm] \sin\psi_i + \psi_i = \dfrac{\pi + 2}{2}\sin\varphi_i \end{array} \right\}$$ (11.2.29)

And if fixing φ_0 we have the boundary conditions:

$$\left. \begin{array}{l} \psi_i = 2\arccos\left(\dfrac{\sqrt{\pi + 2}}{2}\cos\varphi_0\right)^{1/2} \\[2mm] \dfrac{\pi + 2}{2}\sin\varphi_i = \sin\psi_i + \psi_i \end{array} \right\}$$ (11.2.30)

Using the 'translation method', we get the coordinate formulae of the composite cylindrical–Eckert pseudocylindrical projection:

$$y = \begin{cases} f(\varphi), & (\varphi \le \varphi_i) \\ D + \dfrac{2R}{\sqrt{\pi + 2}}\psi, & (\varphi > \varphi_i) \end{cases} \quad x = \begin{cases} R\cos\varphi_0 \cdot (\lambda - \lambda_0), & (\varphi \le \varphi_i) \\ \dfrac{2R}{\sqrt{\pi + 2}}(\lambda - \lambda_0)\cos^2\dfrac{\psi}{2}, & (\varphi > \varphi_i) \end{cases} \quad (11.2.31)$$

$$\sin\psi + \psi = \frac{\pi + 2}{2}\sin\varphi$$

where φ_0 or φ_i can be determined by equation (11.2.29) or (11.2.30).

$$D = f(\varphi_i) - \frac{2R}{\sqrt{\pi + 2}}\psi_i$$

Distortion analysis

According to the distortion formula (Li *et al.* 1993) of pseudocylindrical projections, we can prove that using translation in pseudocylindrical projections, within the region $\varphi > \varphi_i$, the original projection's nature is kept.

11.2.4   Composite conic projections

Similarly to composite cylindrical projections, conic projections with different qualities or the same quality can also be combined to: a conformal–equivalent composite conic projection, an equidistant composite conic projection, etc.

The boundary condition of composite conic projections needs to meet the conditions of a tangent conic projection, that is to maintain the uniformity of B_0, and $n_0 = 1$.

Using the translation method, the general formulae for composite tangent conic projections with different qualities in a polar coordinate system have the form:

$$\rho = \begin{cases} f_1(B), & (B \le B_i) \\ D + f_2(B), & (B > B_i) \end{cases}, \quad \delta = \alpha l \qquad (11.2.32)$$

where $D = f_1(B_i) - f_2(B_i)$, $\alpha = \sin B_0$.

For example, the polar coordinate formulae for the conformal–equivalent composite conic projection have the form:

$$\rho = \begin{cases} \dfrac{C_1}{U^\alpha}, & (B \le B_i) \\ D + \sqrt{\dfrac{2}{\alpha}(C_2 - F)}, & (B > B_i) \end{cases}, \quad \delta = \alpha l \qquad (11.2.33)$$

where:

$$D = \frac{C_1}{U_1^\alpha} - \sqrt{\frac{2}{\alpha}(C_2 - F_i)}, \quad \alpha = \sin B_0, \quad C_1 = \rho_0 U_0^\alpha,$$

$$\rho_0 = N_0\cot B_0, \quad C_2 = F_0 + \frac{1}{2}\rho_0 r_0$$

When the boundary is $B = B_0$ in the above equation, $D = 0$.

The general polar coordinate formulae for composite conic projections with the same qualities have the form:

$$\rho = \begin{cases} f_1(B), & (B \le B_i) \\ f_1(B_i) + K[f_1(B) - f_1(B_i)], & (B > B_i) \end{cases}, \quad \delta = \alpha l \tag{11.2.34}$$

where K is the scale parameter.

For example, the standard parallel for the equidistant conic projection is B_1, B_2, the polar coordinate formulae for the equidistant composite conic projection have the form:

$$\rho = \begin{cases} C_3 - S, & (B \le B_i) \\ C_3 - S_i + K(S_i - S), & (B > B_i) \end{cases}, \quad \delta = \alpha l \tag{11.2.35}$$

where:

$$\alpha = \frac{r_1 - r_2}{S_2 - S_1}, \quad C_3 = \frac{r_2 S_2 - r_2 S_1}{r_1 - r_2}$$

For the tangent conic projection, the projection constant in the above equation is

$$\alpha = \sin B_0, \quad C_3 = S_0 + \rho_0$$

Distortion analysis

In the composite tangent conic projection, the projection maintains the distortion properties of the original projection within the region $B > B_i$. Because

$$\rho = D + f_2(B), \quad m = -\frac{d\rho}{MdB} = -\frac{df_2(B)}{MdB}$$

i.e., length ratio of the meridian of the original projection is maintained. But:

$$n = \frac{\alpha\rho}{r} = \frac{\alpha D}{r} + \frac{\alpha f_2(B)}{r}$$

i.e., length ratio of the meridian of the original projection cannot be maintained. In particular, taking the tangent parallel B_0 as boundary, the distortion nature of the original projection is maintained.

11.2.5 Composite conic–Bonne equivalent pseudoconic projection

Boundary conditions of the composite projection are: take the tangent parallel B_0 of the tangent conic projection as the boundary parallel. Its coordinate formulae are written as:

$$y = \rho_S - \rho\cos\delta, \quad x = \rho\sin\delta$$

$$\rho = \begin{cases} f(B), & (B \le B_0) \\ C - S, & (B > B_0) \end{cases}, \quad \delta = \begin{cases} \alpha l, & (B \le B_0) \\ \dfrac{r}{\rho}l, & (B > B_0) \end{cases} \tag{11.2.36}$$

where $\alpha = \sin B_0$, $C = \rho_0 + S_0$, $\rho_0 = N_0 \cot B_0$.

$f(B)$ are conic projections of different natures, for example with the conformal conic projection, we have:

$$f(B) = \frac{C_1}{U^\alpha}, \quad C_1 = \rho_0 U_0^\alpha$$

Distortion analysis

On the composite projection the function values on the bounding tangent parallel B_0 are equal, so within the region $B > B_0$, the original projection properties are kept.

11.2.6 Composite conic–ordinary polyconic projection

Boundary conditions of the composite projection are: take the tangent parallel B_0 of the tangent conic projection as the boundary parallel. The coordinate formulae are written as:

$$y = q - \rho\cos\delta, \quad x = \rho\sin\delta$$

$$\rho = \begin{cases} f(B), & (B \le B_0) \\ N\cot B, & (B > B_0) \end{cases}, \quad \delta = \begin{cases} \alpha l, & (B \le B_0) \\ l\sin B, & (B > B_0) \end{cases} \tag{11.2.37}$$

where $\alpha = \sin B_0$, $q = N_0\cot B_0$ $(B \le B_0)$, $q = S - S_0 + N\cot B$ $(B > B_0)$.

$f(B)$ are conic projections of different types, for example using the equivalent conic projection, we have:

$$f(B) = \sqrt{\frac{2}{\alpha}(C_2 - F)}, \quad C_2 = \frac{\alpha\rho_0^2}{2} + F_0$$

Distortion analysis

In the composite projection, the function values on the bounding tangent parallel B_0 are equal, so within the region $B > B_0$, the original projection properties are kept.

CHAPTER TWELVE

Position Lines

▌

12.1 TYPES OF POSITION LINES AND THEIR PROJECTION

12.1.1 The small circle and great circle on the projection plane

Since the distances from points of a small circle to the center of the same small circle are all equivalent, the small circle can be also called an equidistant line.

The shortest distance between two points on the sphere is along a great circle of the sphere, so the ground locus of most flying is along a great circle path. Since the locus of an azimuth that has a fixed value is also a great circle, so the great circle is also called an equiazimuth line.

A position line is defined as an auxiliary line that is used to be certain of the position of a ship (or plane) at a particular time. That is, any point on this line is possibly the position of the ship (or plane), so the line is called a position line. The great circles and small circles are all position lines (Yang 1991).

For example, if the azimuths of a target point measured synchronously from two fixed stations are (α, β), then the target point we want to locate is the point of intersection of the equiazimuth lines (great circles) so that the azimuths are α and β.

The small circle on the projection plane

Assume that on the sphere with point $P(\varphi, \lambda)$ along a small circle, the centerpoint of the small circle is $A(\varphi_A, \lambda_A)$ and the radius is k. Then from the spherical triangle we can get the small-circle equation:

$$\cos k = \sin \varphi_A \sin \varphi + \cos \varphi_A \cos \varphi \cos (\lambda - \lambda_A) \tag{12.1.1}$$

It can be written as follows after rearrangement:

$$\cos (\lambda - \lambda_A) = \cos k \sec \varphi_A \sec \varphi - \tan \varphi_A \tan \varphi \tag{12.1.2}$$

where the radius of the small circle is:

$$k = \frac{S}{R} \cdot \frac{180°}{\pi}$$

the unit of S is km, and R is the radius of the earth in km.

From equation (12.1.2), for a small circle having a fixed radius k and given a series of latitudes φ_i, we can get the differences of longitude $\Delta\lambda_i = (\lambda - \lambda_A)$.

The changing interval of φ_i is $[\varphi_A - k, \varphi_A + k]$. Given the general equation for map projections:

$$x = f_1(\varphi, \lambda) \quad y = f_2(\varphi, \lambda) \tag{12.1.3}$$

Introducing (φ_i, λ_i) into the above equations we can get (x_i, y_i). Then on the projection plane the image of the small circle, for which the radius is k, can be obtained.

The great circle on the projection plane

Now assume that on the sphere the movable point $P(\varphi, \lambda)$ is along a great circle which passes through point $A(\varphi_A, \lambda_A)$ at an azimuth of a. Then from the spherical triangle we can get the great-circle equation:

$$\tan a = \frac{\cos\varphi \sin(\lambda - \lambda_A)}{\cos\varphi_A \sin\varphi - \sin\varphi_A \cos\varphi \cos(\lambda - \lambda_A)} \tag{12.1.4}$$

After rearrangement it becomes:

$$\tan\varphi = \tan\varphi_A \cos(\lambda - \lambda_A) + \sec\varphi_A \cot a \sin(\lambda - \lambda_A)$$

and then:

$$\tan\varphi_i = \tan\varphi_A \cos(\lambda_i - \lambda_A) + \sec\varphi_A \cot a \sin(\lambda_i - \lambda_A) \tag{12.1.5}$$

From equation (12.1.5), for a great circle having a fixed azimuth a and given a series of longitudes λ_i, we can obtain latitudes φ_i, using the full range of the longitude difference $(\lambda_i - \lambda_A) = -180°$ to $+180°$.

Substituting (φ_i, λ_i) into equation (12.1.3) for map projections, we can obtain the image on the projection plane of the great circle for which the azimuth is a.

A convenient mathematical model for projecting great and small circles

As shown in Figure 12.1, if fixed point A is the pole of a pseudo-geographical coordinate system, then the small circles and great circles with point A as the pole are called pseudo-parallels and pseudo-meridians.

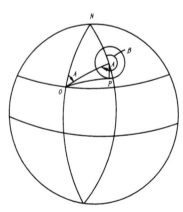

Figure 12.1 Pseudo-spherical coordinate system

The center of projection O is the new pole of a pseudo-spherical coordinate system. Applying this and the formula for oblique projections we can project a pseudo-meridian and pseudo-parallel graticule (i.e., great and small circle) onto a projection plane of which the centerpoint of the projection is O.

Next we introduce the manner in which we can build this kind of mathematical model. The projection centerpoint O is the new pole of the pseudo-spherical coordinate system, and OA is the central meridian of the pseudo-geographical coordinate system $(\Delta\lambda_0' = 0)$. In the pseudo-geographical system a coordinate of point O is $\varphi_0' = 90° - Z_A$. The distance from pole A of the pseudo-geographical coordinate system to an arbitrary point P is S, which can be expressed by an angle:

$$k = \frac{S}{R} \cdot \frac{180}{\pi} \qquad (12.1.6)$$

Then in the pseudo-geographical coordinate system the coordinates of point P are:

$$\varphi' = 90° - k, \quad \Delta\lambda' = \beta - a \qquad (12.1.7)$$

where a is the azimuth of great-circle arc AP at point A, and β is the azimuth of line AO at point A.

In a pseudo-spherical coordinate system for which OA is the central meridian and O is the pole, the pseudo-spherical coordinates of point P are (Z', a'), and the formulae for their calculation are:

$$\left.\begin{array}{l} \cos Z' = \sin\varphi'\sin\varphi_0' + \cos\varphi'\cos\varphi_0'\cos\Delta\lambda' \\ \sin Z'\cos a' = \cos\varphi_0'\sin\Delta\lambda' - \sin\varphi_0'\cos\varphi'\cos\Delta\lambda' \\ \sin Z'\sin a' = \cos\varphi'\sin\Delta\lambda' \end{array}\right\} \qquad (12.1.8)$$

$$\varphi_0' = 90° - Z_A$$

$$\cos Z_A = \sin\varphi_0\sin\varphi + \cos\varphi_0\cos\varphi\cos(\lambda - \lambda_0) \qquad (12.1.9)$$

$$\tan\beta = \frac{\cos\varphi_0\sin(\lambda_0 - \lambda_A)}{\cos\varphi_A\sin\varphi_0 - \sin\varphi_A\cos\varphi_0\cos(\lambda_0 - \lambda_A)} \qquad (12.1.10)$$

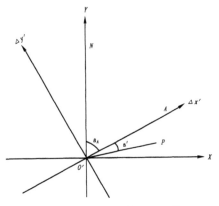

Figure 12.2 A rectangular coordinate system corresponding to Figure 12.1

On the projection plane we can build a rectangular coordinate system $x'O'y'$ (see Figure 12.2) for which O' is the origin and the image of OA is the ordinate axis. For a projection of the azimuth, the formulae for the coordinates of the images of the great and small circles are:

$$x' = \rho\sin a', \quad y' = \rho\cos a' \qquad (12.1.11)$$

where $\rho = f(Z')$ is the radius of the azimuthal projection.

If we build a rectangular coordinate system $xO'y$ using O' as origin and the image of ON as the ordinate axis, then we have:

$$\left.\begin{array}{l} y = \rho\cos(a_A + a') = y'\cos a_A - x'\sin a_A \\ x = \rho\sin(a_A + a') = y'\sin a_A + x'\cos a_A \end{array}\right\} \qquad (12.1.12)$$

where:

$$\tan a_A = \frac{\cos\varphi_A\sin(\lambda_A - \lambda_0)}{\cos\varphi_0\sin\varphi_A - \sin\varphi_0\cos\varphi_A\cos(\lambda_A - \lambda_0)}$$

and a_A is the azimuth of line OA at point O.

Using the above method we can get directly the image of a great circle and small circle on the projection plane instead of inversely solving from geographical coordinates (φ, λ).

Examples

Example 1   Given the geographical coordinates of the centerpoint A of a small circle on the spherical surface $\varphi_A = 30°$, $\lambda_A = 105°$, calculate the geographical coordinates of a small circle for which the radius is 1 000 km, using an average radius of curvature at point A as the spherical radius, i.e., $R = 6\,367.5181$ km.

With equation (12.1.2), we obtain the geographical coordinates as follows:

$$R = 6\,367.5181 \text{ km}, \quad S = 1\,000 \text{ km}, \quad k = 8.5953281459$$

Table 12.1   Small-circle coordinates

φ	21°00'06".7185	25°	30°	35°	38°59'53".2815
λ_1	105°	96°33'34".1457	94°36'22".6038	96°07'15".4254	105°
λ_2	105°	113°26'25".854	115°23'37".396	113°52'44".574	105°

Example 2   Given the geographical coordinates of point A on the sphere as $\varphi_A = 30°$, $\lambda_A = 105°$, and the azimuth passing through point A as $a = 30°$, calculate the geographical coordinates of its great circle. Using equation (12.1.5), the calculated result is given in Table 12.2.

Table 12.2   Great-circle coordinates

λ	90°	105°	135°	165°	180°
φ	2°17'34.3193"	30°	56°18'35.7568"	63°40'13.7873"	64°20'13.0414"

12.1.2   Ellipses and hyperbolas on the sphere and their projections

Formulae for ellipses and hyperbolas on the surface of the sphere

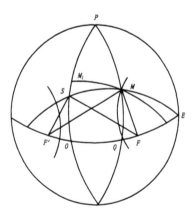

Figure 12.3   Spherical polar coordinate system and ellipses and hyperbolas having the same foci

Spherical rectangular coordinates   Referring to spherical rectangular coordinates, as shown in Figure 12.3, assume that $F(0, \varepsilon)$, $F'(0, -\varepsilon)$ are two fixed points symmetrical about O. Then the locus of points for which the sum of distances is constant from the moving point to the two fixed points is a spherical ellipse, and the locus of points for which the difference of distances is constant is a spherical hyperbola (Yang 1983b).

Assuming that point M is the intersecting point of the spherical ellipse and hyperbola having the same foci, its spherical rectangular coordinates are $OM_1 = \eta$, $OQ = \lambda'$.

Let $MF = \rho$, $MF' = \rho'$, and $\rho' + \rho = 2\alpha$, $\rho' - \rho = 2\delta$, $OS = \beta$. First we calculate the equation of the spherical ellipse.

From the spherical rectangular triangle OSF, we can get:

$$\cos \alpha = \cos \beta \cos \varepsilon$$

i.e.,

$$\cos \beta = \frac{\cos \alpha}{\cos \varepsilon} \tag{12.1.13}$$

and through a simple rearrangement, it becomes:

$$\tan^2 \varepsilon = \frac{\tan^2 \alpha - \tan^2 \beta}{1 + \tan^2 \beta} \tag{12.1.14}$$

From the spherical rectangular triangle MQF' and MQF, we obtain:

$$\left.\begin{aligned} \cos \rho' &= \cos \varphi' \cos(\varepsilon + \lambda') \\ \cos \rho &= \cos \varphi' \cos(\varepsilon - \lambda') \end{aligned}\right\} \tag{12.1.15}$$

where φ' and λ' are the coordinates of M in a pseudo-geographical coordinate system for which P is the new pole.

Taking into account that $\rho' + \rho = 2\alpha$, $\rho' - \rho = 2\delta$, we can obtain:

$$\frac{1}{2}(\cos \rho + \cos \rho') = \cos \alpha \cos \delta$$

and from equation (12.1.15) we obtain:

$$\frac{1}{2}(\cos \rho + \cos \rho') = \cos \varphi' \cos \varepsilon \cos \lambda' = \frac{\cos \varepsilon \cos \lambda'}{\sqrt{1 + \tan^2 \varphi'}}$$

Thus we have:

$$\cos \alpha \cos \delta = \frac{\cos \varepsilon \cos \lambda'}{\sqrt{1 + \tan^2 \varphi'}} \tag{12.1.16}$$

From equation (12.1.13) we have:

$$\cos \beta \cos \delta = \frac{\cos \alpha \cos \delta}{\cos \varepsilon} = \frac{\cos \lambda'}{\sqrt{1 + \tan^2 \varphi'}} \tag{12.1.17}$$

and considering equation (12.2.12), we get:

$$\tan \varphi' = \cos \lambda' \tan \eta \tag{12.1.18}$$

Substituting equation (12.1.18) into (12.1.17) yields:

$$\cos \beta \cos \delta = \frac{\cos \lambda'}{\sqrt{1 + \cos^2 \lambda' \tan^2 \eta}}$$

Squaring both sides of this equation and taking the reciprocals, we find:

$$\frac{1}{\cos^2 \beta \cos^2 \delta} = \frac{1}{\cos^2 \lambda'} + \tan^2 \eta$$

i.e.,

$$(1 + \tan^2 \beta)(1 + \tan^2 \delta) = 1 + \tan^2 \lambda' + \tan^2 \eta \tag{12.1.19}$$

From equation (12.1.15) we get:

$$\cos \rho' - \cos \rho = -2 \sin \alpha \sin \delta = -2 \cos \phi' \sin \varepsilon \sin \lambda'$$
$$\cos \rho' + \cos \rho = 2 \cos \alpha \cos \delta = 2 \cos \phi' \cos \varepsilon \cos \lambda'$$

In the above two equations, dividing the first by the second gives:

$$\tan \delta = \frac{\tan \varepsilon \tan \lambda'}{\tan \alpha} \qquad\qquad (12.1.20)$$

Squaring both sides of this equation and combining with equation (12.1.14) gives:

$$\tan^2 \delta = \frac{\tan^2 \lambda'}{\tan^2 \alpha} \cdot \frac{\tan^2 \alpha - \tan^2 \beta}{1 + \tan^2 \beta}$$

Substituting the above equation into (12.1.19), after arranging we get:

$$1 + \tan^2 \beta + \tan^2 \lambda' - \frac{\tan^2 \beta}{\tan^2 \alpha} \cdot \tan^2 \lambda' = 1 + \tan^2 \lambda' + \tan^2 \eta$$

i.e.,

$$\tan^2 \eta = \tan^2 \beta - \frac{\tan^2 \beta \tan^2 \lambda'}{\tan^2 \alpha}$$

Thus we have:

$$\frac{\tan^2 \lambda'}{\tan^2 \alpha} + \frac{\tan^2 \eta}{\tan^2 \beta} = 1 \qquad\qquad (12.1.21)$$

where:

$$\tan^2 \beta = \frac{\tan^2 \alpha - \tan^2 \varepsilon}{1 + \tan^2 \varepsilon}$$

Equation (12.1.21) is the equation for a spherical ellipse of which the spherical triangle coordinates η and λ' are parameters. Now we continue to calculate the equation for a spherical hyperbola. From equation (12.1.13) and considering (12.1.18) we get:

$$\cos \alpha \cos \delta = \frac{\cos \varepsilon \cos \lambda'}{\sqrt{1 + \tan^2 \phi'}} = \frac{\cos \varepsilon \cos \lambda'}{\sqrt{1 + \cos^2 \lambda' \tan^2 \eta}}$$

Thus we have:

$$(1 + \tan^2 \alpha)(1 + \tan^2 \delta) = (1 + \tan^2 \varepsilon)(1 + \tan^2 \lambda' + \tan^2 \eta) \qquad\qquad (12.1.22)$$

From equation (12.1.20) we get:

$$\tan^2 \alpha = \frac{\tan^2 \lambda'}{\tan^2 \delta} \tan^2 \varepsilon \qquad\qquad (12.1.23)$$

Incorporating the above equation into (12.1.22) we get:

$$\frac{\tan^2 \lambda}{\tan^2 \delta} \tan^2 \varepsilon + \tan^2 \delta = \tan^2 \lambda' + \tan^2 \eta + \tan^2 \varepsilon + \tan^2 \varepsilon \tan^2 \eta$$

i.e.,

$$\frac{\tan^2 \lambda'}{\tan^2 \delta}(\tan^2 \varepsilon - \tan^2 \delta) = (\tan^2 \varepsilon - \tan^2 \delta) + (1 + \tan^2 \varepsilon)\tan^2 \eta$$

Let:

$$\tan^2 \gamma = \frac{\tan^2 \varepsilon - \tan^2 \delta}{1 + \tan^2 \varepsilon} \qquad (12.1.24)$$

we obtain:

$$\frac{\tan^2 \lambda'}{\tan^2 \delta} - \frac{\tan^2 \eta}{\tan^2 \gamma} = 1 \qquad (12.1.25)$$

This is the equation for a spherical hyperbola, in which spherical triangle coordinates η and λ' are parameters.

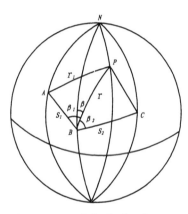

Figure 12.4 Spherical polar coordinate system and spherical hyperbola

Spherical hyperbola equation in which the spherical polar coordinates are parameters As is shown in Figure 12.4, the three station points are respectively $A\ (\varphi_A, \lambda_A)$, $B\ (\varphi_B, \lambda_B)$, $C\ (\varphi_C, \lambda_C)$.

The distance from the active station B to the passive station A is $S_1 = 2c_1$, the distance from B to the passive station C is $S_2 = 2c_2$, the azimuths of spherical arcs $B\hat{A}$ and $B\hat{C}$ are β_1 and β_2, and the differences in distance from every station to the target point are, respectively:

$$r_1 - r = 2a_1 \quad \text{and} \quad r_2 - r = 2a_2 \qquad (12.1.26)$$

Now let us calculate the geographic coordinates (φ, λ) of point P.

From the spherical triangles APB and BPC, according to the cosine formula we get:

$$\left.\begin{array}{l} \cos r_1 = \cos 2c_1 \cos r + \sin 2c_1 \sin r \cos \psi_1 \\ \cos r_2 = \cos 2c_2 \cos r + \sin 2c_2 \sin r \cos \psi_2 \end{array}\right\} \qquad (12.1.27)$$

From equation (12.1.26) we get:

$$r_1 = r + 2a_1 \quad \text{and} \quad r_2 = r + 2a_2$$

Introducing them into (12.1.27) yields:

$$\left.\begin{array}{l} \cos 2a_1 - \cos 2c_1 = \tan r\,(\sin 2a_1 + \sin 2c_1 \cos \psi_1) \\ \cos 2a_2 - \cos 2c_2 = \tan r\,(\sin 2a_2 + \sin 2c_2 \cos \psi_2) \end{array}\right\} \qquad (12.1.28)$$

Formula (12.1.28) is the equation for two spherical hyperbolas having the same foci, their focal axis lengths being $B\hat{A} = 2c_1$ and $B\hat{C} = 2c_2$.

Noting that:

$$\psi_1 = 360 + \beta - \beta_1 \quad \text{and} \quad \psi_2 = \beta_2 - \beta \qquad (12.1.29)$$

equation (12.1.28) can be written as:

$$\cos 2a_1 - \cos 2c_1 = \tan r\,[\sin 2a_1 + \sin 2c_1 \cos(\beta - \beta_1)] \qquad (12.1.30)$$

$$\cos 2a_2 - \cos 2c_2 = \tan r\,[\sin 2a_2 + \sin 2c_2 \cos(\beta_2 - \beta)] \qquad (12.1.31)$$

We can get (r, β) from the above equations. For this purpose we divide (12.1.30) by (12.1.31). Then:

$$\frac{\cos2a_1 - \cos2c_1}{\cos2a_2 - \cos2c_2} = \frac{\sin2a_1 + \sin2c_1(\cos\beta_1\cos\beta + \sin\beta_1\sin\beta)}{\sin2a_2 + \sin2c_2(\cos\beta_2\cos\beta + \sin\beta_2\sin\beta)} \tag{12.1.32}$$

The left side of the above equation is a constant; let

$$k = \frac{\cos2a_1 - \cos2c_1}{\cos2a_2 - \cos2c_2} = \frac{\sin(c_1 + a_1)\cdot\sin(c_1 - a_1)}{\sin(c_2 + a_2)\cdot\sin(c_2 - a_2)} \tag{12.1.33}$$

then equation (12.1.32) can be written:

$$(k\cdot\sin2c_2\cdot\sin\beta_2 - \sin2c_1\cdot\sin\beta_1)\cdot\sin\beta + (k\cdot\sin2c_2\cdot\cos\beta_2$$
$$- \sin2c_1\cdot\cos\beta_1)\cdot\cos\beta + (k\cdot\sin2a_2 - \sin2a_1) = 0 \tag{12.1.34}$$

Introducing the symbols:

$$\left.\begin{array}{l} m = (k\cdot\sin2c_2\cdot\sin\beta_2 - \sin2c_1\cdot\sin\beta_1) \\ n = (k\cdot\sin2c_2\cdot\cos\beta_2 - \sin2c_1\cdot\cos\beta_1) \\ l = (k\cdot\sin2a_2 - \sin2a_1) \end{array}\right\} \tag{12.1.35}$$

we get:

$$m\sin\beta + n\cos\beta + l = 0 \tag{12.1.36}$$

From the above equation we have:

$$\sin\beta + \frac{n}{m}\cos\beta = -\frac{l}{m}$$

Letting $\tan\delta = \dfrac{n}{m}$, then we have:

$$\sin(\beta + \delta) = -\frac{l}{m}\cos\delta \tag{12.1.37}$$

Hence we can get azimuth β. From equation (12.1.31) we obtain the following equation by means of (12.1.30):

$$\tan r = \frac{\cos2a_1 - \cos2c_1}{\sin2a_1 + \sin2c_1\cdot\cos\psi_1} = \frac{\cos2a_2 - \cos2c_2}{\sin2a_2 + \sin2c_2\cdot\cos\psi_2} \tag{12.1.38}$$

Using the above equation we can obtain the arc length $\widehat{BP} = r$.

If given the value of r, we get the arc length $r_1 = \widehat{AP}$ and $r_2 = \widehat{CP}$ by using equations (12.1.26).

From this, with spherical triangle BNP, we can obtain:

$$\left.\begin{array}{l} \sin\varphi = \sin\varphi_B\cos r + \cos\varphi_B\sin r\cos\beta \\ \tan(\lambda - \lambda_B) = \dfrac{\sin r\sin\beta}{\cos\varphi_B\cos r - \sin\varphi_B\sin r\cos\beta} \end{array}\right\} \tag{12.1.39}$$

or:

$$\sin(\lambda - \lambda_B) = \sin r\frac{\sin\beta}{\sin\varphi} \tag{12.1.40}$$

Introducing the focal reference and the eccentric ratio of the spherical hyperbola, i.e.,

$$
\left.
\begin{aligned}
P_1 &= \frac{\cos 2a_1 - \cos 2c_1}{\sin 2a_1}, \quad e_1 = \frac{\sin 2c_1}{\sin 2a_1} \\[2mm]
P_2 &= \frac{\cos 2a_2 - \cos 2c_2}{\sin 2a_2}, \quad e_2 = \frac{\sin 2c_2}{\sin 2a_2}
\end{aligned}
\right\}
\tag{12.1.41}
$$

then equation (12.1.38) can be written:

$$
\left.
\begin{aligned}
\tan r' &= P_1 : [e_1 \cdot \cos(\beta - \beta_1) + 1] \\
\tan r'' &= P_2 : [e_2 \cdot \cos(\beta_2 - \beta) + 1]
\end{aligned}
\right\}
\tag{12.1.42}
$$

Given the distance r of the spherical surface or azimuth β, from the above we can get the geographical coordinates of every point of the left and right group of hyperbolas. We will develop a method of calculating latitude from a given longitude along the spherical hyperbolas. According to equation (12.1.26) and Figure 12.4, for point P we have:

$$
r_1 = r + 2a_1, \quad r = r_1 - 2a_1
\tag{12.1.43}
$$

Calculating the cosine of the two sides of the above equation, we get:

$$
\left.
\begin{aligned}
\cos r_1 &= \cos r \cos 2a_1 - \sin r \sin 2a_1 \\
\cos r &= \cos r_1 \cos 2a_1 + \sin r_1 \sin 2a_1
\end{aligned}
\right\}
\tag{12.1.44}
$$

Multiplying the above two equations by $\cos r_1$ and $\cos r$, respectively, and adding together, we get:

$$
\cos^2 r_1 + \cos^2 r - 2\cos r_1 \cdot \cos r \cdot \cos 2a_1 = \sin^2 2a_1
\tag{12.1.45}
$$

Applying the cosine formula to spherical triangles APN and BPN, we have:

$$
\left.
\begin{aligned}
\cos r_1 &= \sin \varphi_A \cdot \sin \varphi + \cos \varphi_A \cdot \cos \varphi \cdot \cos(\lambda_P - \lambda_A) \\
\cos r &= \sin \varphi_B \cdot \sin \varphi + \cos \varphi_B \cdot \cos \varphi \cdot \cos(\lambda_P - \lambda_B)
\end{aligned}
\right\}
\tag{12.1.46}
$$

Noticing that:

$$
\sin^2 \varphi = \frac{1}{2}(1 - \cos 2\varphi) \quad \text{and} \quad \cos^2 \varphi = \frac{1}{2}(1 + \cos 2\varphi)
$$

substituting (12.1.46) into (12.1.45), and rearranging we get:

$$
m \sin 2\varphi + n \cos 2\varphi + l = 0
\tag{12.1.47}
$$

where:

$$
\begin{aligned}
m &= 2[ac + bd - (ad + bc) \cdot e] \\
n &= [-(a^2 + b^2) + (c^2 + d^2) + 2(ab - cd)e] \\
l &= [(a^2 + b^2) + (c^2 + d^2) - 2(ab + cd)e - 2f^2]
\end{aligned}
$$

and:

$$
\begin{aligned}
a &= \sin \varphi_A, \quad b = \sin \varphi_B, \quad c = \cos \varphi_A \cdot \cos(\lambda_B - \lambda_A) \\
d &= \cos \varphi_B \cdot \cos(\lambda_P - \lambda_B), \quad e = \cos 2a_1, \quad f = \sin 2a_1
\end{aligned}
$$

From equation (12.1.47) we get:

$$
\tan \delta = \frac{n}{m} \quad \text{and} \quad \sin(2\varphi + \delta) = -\frac{l}{m} \cos \delta
\tag{12.1.48}
$$

It can be seen from this that for the same spherical hyperbolas we can obtain latitude by only changing the value of d.

As for the latitude of points on symmetrical spherical hyperbolas, they can be calculated by an equation such as the following:

$$\sin(2\varphi + \delta) = \sin[180° - (2\varphi + \delta)] \tag{12.1.49}$$

Examples *Example 1.* Given the active station B: $\varphi_B = 55°58'58''.0$, $\lambda_B = 38°48'08''.9$, a passive station A, $S_1 = 2c_1 = 1°23'49''.1$, $\beta_1 = 289°49'03''.9$, a passive station C, $S_2 = 2c_2 = 1°28'58''.2$, $\beta_2 = 75°12'04''.0$, a targeted point P for which the difference in distances are $(r_1 - r) = 2a_1 = 27\ 215.8$ m and $(r_2 - r) = 2a_2 = 38\ 994.9$ m, respectively, and a sphere of radius $R = 6\ 386\ 200.373$ m. Now calculate the geographical coordinates (φ, λ) of the target point P. Based on equations (12.4.20)–(12.4.26). The result is as follows: For the target point, $\varphi = 57°21'06''.7$, $\lambda = 38°51'16''.8$.

Example 2. Given the active station B: $\varphi_B = 55°58'58''$, $\lambda_B = 38°48'09''$, a passive station A at a distance $S_1 = 2c_1 = 1°23'49''$, with $\beta_1 = 289°49'04''$ its azimuth; the difference in distance from active and passive stations to the target point is $(r_1 - r) = 2a_1 = 0°14'39''$, calculate the geographical coordinates (φ_P, λ_P) of the position lines according the given azimuths β. From equation (12.1.42), let $\psi = \beta - \beta_1$, i.e., $\beta = \beta_1 + \psi$. The results of the calculations are given in Table 12.3.

Table 12.3 Results from Example 2

ψ	0°	10°	30°	71°21'36''	90°
φ_P	56°10'27''.7213	56°16'11''.2454	56°28'39''.0088	57°21'06''.5649	59°35'48''.3779
λ_P	37°49'42''.1108	37°53'24''.3269	38°02'32''.694	38°51'16''.8062	41°23'33''.4478

Example 3. Given the active station B: $\varphi_B = 55°58'58''.0$, $\lambda_B = 38°48'08''.9$, the passive station A: $\varphi_A = 56°26'01''.5$, $\lambda_A = 36°25'30''.1$, and the difference in distance to the target point $2a_1 = 0°14'39''$, then we can calculate the corresponding latitude φ according the given longitudes of the position lines.

From equations (12.1.47) and (12.1.48) we can get the results as shown in Table 12.4.

Table 12.4 Results from Example 3

λ_P	38°02'32''.7	38°51'16''.8	41°23'33''.4
φ_P	56°28'28''.0356	57°21'02''.2676	59°35'47''.1383

The image of the spherical ellipse and hyperbola on the gnomonic projections

Ellipses and hyperbolas having the same foci on the sphere still have the same foci on gnomonic projections of the sphere As shown in Figure 12.3, on the horizontal (or equatorial) gnomonic projection (using the pseudo-geographical coordinate system) for which the point O is the point of tangency,

$$x' = \tan \lambda', \quad y' = \tan \eta = \tan \varphi' \sec \lambda' \tag{12.1.50}$$

with the radius of the sphere taken as $R = 1$, let

$$a = \tan \alpha, \quad b = \tan \beta, \quad c = \tan \gamma, \quad d = \tan \delta$$

Then from equations (12.1.21) and (12.1.25) we get:

$$\frac{x'^2}{a^2} + \frac{y'^2}{b^2} = 1 \tag{12.1.51}$$

and:

$$\frac{x'^2}{c^2} - \frac{y'^2}{d^2} = 1 \tag{12.1.52}$$

Equations (12.1.51) and (12.1.52) provide plane ellipses and hyperbolas having the same foci on the gnomonic projection, which is what we are trying to obtain.

The projections of hyperbolas having the same foci on the sphere remain hyperbolas having the same foci on any gnomonic projection As shown in Figure 12.4, on oblique gnomonic projections in which point B (φ_B, λ_B) is the center of projection, the polar coordinates of point P (φ, λ) are:

$$\rho' = \tan r, \quad a' = \beta_1 + \psi' \quad \text{or} \quad \rho'' = \tan r, \quad a'' = \beta_2 - \psi''$$

From equation (12.1.42) we get:

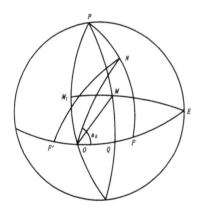

Figure 12.5 Spherical geographic coordinate system and ellipses and hyperbolas having the same foci

$$\rho' = \frac{P_1}{1 + e_1 \cos \psi'} \tag{12.1.53}$$

$$\rho'' = \frac{P_2}{1 + e_2 \cos \psi''} \tag{12.1.54}$$

Hence on the plane of gnomonic projections we get equations for plane hyperbolas on which $B'A'$, $B'C'$ constitute the polar axis, and B' is the polar point. This shows that gnomonic projections of hyperbolas having the same foci on the sphere remain hyperbolas having the same foci.

Calculating formulae for ellipses and hyperbolas having the same foci As shown in Figure 12.5, assume the coordinates of the foci are F' (φ_1, λ_1), F (φ_2, λ_2).

1. Fixing the coordinates of the new polar point O at (φ_0, λ_0), the focal distance at 2, radius R at 1, and the azimuth of OF at a_0, we can obtain the spherical focal distance 2ε:

$$\cos 2\varepsilon = \sin \varphi_1 \sin \varphi_2 + \cos \varphi_1 \cos \varphi_2 \cos(\lambda_2 - \lambda_1) \tag{12.1.55}$$

From spherical triangles $NF'F$, $NF'O$, and NOF we can get the calculating formulae for φ_0, λ_0, a_0:

$$\left. \begin{array}{l} \sin F' = \dfrac{\sin(\lambda_2 - \lambda_1)}{\sin 2\varepsilon} \cos \varphi_2 \\[2mm] \sin \varphi_0 = \sin \varphi_1 \cos \varepsilon + \cos \varphi_1 \sin \varepsilon \cos F' \end{array} \right\} \tag{12.1.56}$$

$$\sin(\lambda_0 - \lambda_1) = \frac{\sin F'}{\cos\varphi_0}\sin\varepsilon \tag{12.1.57}$$

$$\sin a_0 = \frac{\sin(\lambda_1 - \lambda_0)}{\sin\varepsilon}\cos\varphi_2 \tag{12.1.58}$$

2. Calculating the coordinates (x, y) of triangles of the projections.

If OP' is the y-axis, Of is the x-axis, and the spherical radius is $R = 1$, then we can get the formula for coordinates of the oblique gnomonic projections:

$$x = \tan Z \cos(a_0 - a), \quad y = \tan Z \sin(a_0 - a) \tag{12.1.59}$$

where:

$$\cos Z = \sin\varphi_0 \sin\varphi + \cos\varphi_0 \cos\varphi \cos(\lambda - \lambda_0)$$

$$\tan a = \frac{\cos\varphi\sin(\lambda - \lambda_0)}{\cos\varphi_0 \sin\varphi - \sin\varphi_0 \cos\varphi\cos(\lambda - \lambda_0)}$$

3. Drawing the network of ellipses and hyperbolas on the projection plane.

For the ellipses we have:

$$\frac{x^2}{a^2} + \frac{y^2}{b^2} = 1$$

where:

$$a^2 = \tan^2 a, \quad b^2 = \tan^2\beta = \frac{\tan^2 a - \tan^2\varepsilon}{1 + \tan^2\varepsilon}$$

For

$$x = \pm\frac{a}{b}\sqrt{b^2 - y^2}$$

given a series of values of y, we can get the corresponding values of $\pm x$.

If the network for the spherical ellipses is following α_0, $\alpha_0 + \Delta\alpha$, $\alpha_0 + 2\Delta\alpha$, ..., we can get a series of $a_0, a_1, a_2, \ldots$ and $b_0, b_1, b_2, \ldots$. From this we can draw a network of ellipses.

For the hyperbolas we have:

$$\frac{x^2}{c^2} - \frac{y^2}{d^2} = 1$$

where:

$$d^2 = \tan^2\delta, \quad c^2 = \tan\gamma = \frac{\tan^2\varepsilon - \tan^2\delta}{1 + \tan^2\varepsilon}$$

For

$$x = \pm\frac{c}{d}\sqrt{d^2 + y^2}$$

given a series of values of y, we can get the corresponding values of $\pm x$.

If the spherical hyperbola network is according to δ_0, $\delta_0 + \Delta\delta$, $\delta_0 + 2\Delta\delta$, ... we can get a series of c_0, c_1, c_2, ... and d_0, d_1, d_2, ... From this we can draw a network of hyperbolas.

The image of spherical hyperbolas on the Mercator projection

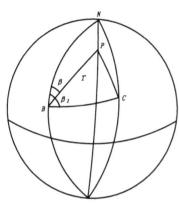

Figure 12.6 Spherical polar coordinate system and spherical hyperbola

As shown in Figure 12.6, the two station points are B (φ_B, λ_B), and C (φ_C, λ_C), respectively, and the distances are $BC = 2c$, $BP = r$, and $CP = r + 2a$.

From equation (12.4.25) we obtain the equations for spherical hyperbolas which are referenced to spherical polar coordinates (r, β):

$$\tan r = \frac{\cos 2a - \cos 2c}{\sin 2a + \sin 2c \cos(\beta_2 - \beta)} \quad (12.1.60)$$

where β_2 is the azimuth of line BC and β is the azimuth of line BP.

When the value of $\cos 2a - \cos 2c$ is very small, it should be replaced with $2 \sin(c + a)\sin(c - a)$. Then equation (12.1.60) can be written:

$$\tan r = \frac{2 \sin(c + a)\sin(c - a)}{\sin 2a + \sin 2c \cos(\beta_2 - \beta)} \quad (12.1.61)$$

In equation (12.1.61) the base line $2c$ and the azimuth β_2 are constants, and the formula for calculation is:

$$\left.\begin{array}{l} \cos 2c = \sin\varphi_B \sin\varphi_C + \cos\varphi_B \cos\varphi_C \cos(\lambda_C - \lambda_B) \\[2mm] \tan\beta_2 = \dfrac{\cos\varphi_C \cos(\lambda_C - \lambda_B)}{\cos\varphi_B \sin\varphi_C - \sin\varphi_B \cos\varphi_C \cos(\lambda_C - \lambda_B)} \end{array}\right\} \quad (12.1.62)$$

If given the spherical polar coordinates of the original point B (φ_B, λ_B) and the spherical coordinates (r, β), we can obtain geographical coordinates (φ, λ) according to the formula for transformation of spherical coordinates, i.e.,

$$\left.\begin{array}{l} \sin\varphi = \sin\varphi_B \cos r + \cos\varphi_B \sin r \cos\beta \\[2mm] \tan(\lambda - \lambda_B) = \dfrac{\sin r \sin\beta}{\cos\varphi_B \cos r - \sin\varphi_B \sin r \cos\beta} \end{array}\right\} \quad (12.1.63)$$

For hyperbolas, the difference in distance $2a$ is constant. Hence from equation (12.1.61) we can calculate the polar radius r using the polar angle β as an input variable and get (r_i, β_i). Then we can obtain (φ_i, λ_i) from equation (12.1.63). Substituting (φ_i, λ_i) into the coordinate formulae for the Mercator projection,

$$x = r_0 \lambda, \quad y = r_0 \ln \tan\left(45° + \frac{\varphi}{2}\right) \quad (12.1.64)$$

we can obtain the image of a hyperbola on the Mercator projection. Taking the difference $2a$ in distance as the increment variable we can get a group of images of hyperbolas.

12.2   DETERMINING THE POSITION OF A TARGET

12.2.1   Determining the position of a target using the gnomonic projection

Any great circle on any gnomonic projection is portrayed as a straight line. If the azimuths of the target point surveyed and determined synchronously from two fixed stations are given respectively as a_1, a_2, then we only need to draw two straight lines through the two fixed station points according to azimuths A_1, A_2, of the projection, and the intersecting point is the position of the target point which we want to know (Yang 1983b). Or we can use a method of calculation to get the position using the observed azimuths (a_1, a_2) on the gnomonic projection plane. In the next section we introduce them in order.

Locating a target on the gnomonic projection using graphic intersections of the observed azimuths (Yang 1983b)

As shown in Figure 12.7, on the sphere the direction angle of great circle AM which passes through point A is β, the azimuth is α, and the direction angle of the meridian which passes through point A is β_m on the gnomonic projection plane (see Figure 12.8). Relatively the direction angle of great circle $A'M'$ is β', the azimuth is A, and the direction angle of the meridian is β'_m.

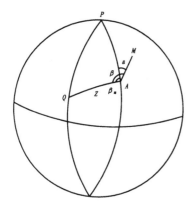

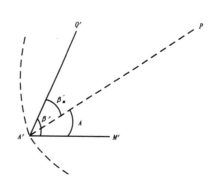

Figure 12.7   The direction angle which passes through point A on the sphere

Figure 12.8   The imaging of gnomonic projection

For the gnomonic projection we have:

$$\tan \beta' = \cos Z \tan \beta \tag{12.2.1}$$

or:

$$\tan (\beta'_m + A) = \cos Z \tan (\beta_m + a) \tag{12.2.2}$$

From spherical triangle PQA we have:

$$\frac{\sin(90° - \varphi_0)}{\sin \beta_m} = \frac{\sin Z}{\sin(\lambda_A - \lambda_0)}$$

i.e.,

$$\sin \beta_m = \frac{\cos\varphi_0 \sin(\lambda_A - \lambda_0)}{\sin Z} \tag{12.2.3}$$

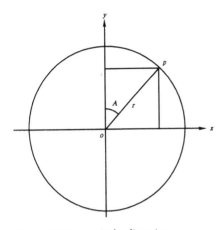

Figure 12.9 A circle direction-correcting nomograph

Thus we obtain the formula for calculation of azimuth A on the gnomonic projection:

$$A = \arctan\left[\cos Z_A \tan(\beta_m + a)\right] - \beta'_m$$

(12.2.4)

where:

$$\sin\beta_m = \frac{\cos\varphi_0 \sin(\lambda_A - \lambda_0)}{\sin Z_A}$$

$$\tan\beta'_m = \cos Z_A \tan\beta_m$$

$$\cos Z_A = \sin\varphi \sin\varphi_A$$
$$+ \cos\varphi_0 \cos\varphi_A \cos(\lambda_A - \lambda_0)$$

At azimuth a from fixed point Q (φ_Q, λ_Q), we can calculate points to azimuth A with the above formula, that is, construct the great circle on the map.

For the convenience of graphical construction, we may calculate a series of azimuths A_i to fixed point Q and corresponding to a_i using the above formula, that is, we can plot a kind of circle as a direction-correcting nomograph (see Figure 12.9).

The calculating formulae for the direction-correcting nomograph are:

$$x = r\sin A, \quad y = r\cos A$$

(12.2.5)

Locating a target on the gnomonic projection by analytical intersection with the observed azimuth

Transverse gnomonic projection As shown in Figure 12.10, we can express point P (φ, λ) on a spherical surface by using spherical rectangular coordinates $OR = \eta$, $OQ = \lambda$.

In the transverse (or equatorial) gnomonic projection, the meridians and great circles passing through the equator E,W will all be straight lines which are horizontal and vertical (perpendicular to each other after projection; see Figure 12.11). Each great circle

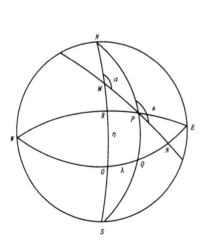

Figure 12.10 Spherical rectangular coordinate system and great circle lines

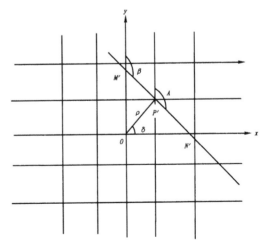

Figure 12.11 Transverse gnomonic projection grid and the imaging of great circle lines

spaced equally along a meridian on the sphere is also spaced from the center of the projection increasing gradually in proportion to the tangent function.

The formulae for coordinates of the transverse gnomonic projection are:

$$x = \rho \sin \delta = R \tan \lambda, \quad y = \rho \cos \delta = R \tan \eta \tag{12.2.6}$$

Assume that the azimuth of point P of great circle MPN on the sphere is a, and the azimuth of point M is a; then on the projection plane the corresponding azimuths are A and β.

According to spherical triangle XMP we have:

$$\cos \alpha = -\cos \lambda \cos (180° - a) + \sin \lambda \sin (180° - a) \cos (90° - \varphi)$$

i.e.,

$$\cos \alpha = \cos \lambda \cos a + \sin \varphi \sin \lambda \sin a \tag{12.2.7}$$

We also have:

$$\tan \beta = \cos m \tan \alpha \tag{12.2.8}$$

where $m = OM$ is the polar distance of point M.

From spherical triangle XMP we have:

$$\cos m = \frac{\sin a}{\sin \alpha} \cos \varphi \tag{12.2.9}$$

Substituting equation (12.2.9) into (12.2.8) yields:

$$\tan \beta = \frac{\sin a}{\cos \alpha} \cos \varphi \tag{12.2.10}$$

From the spherical triangle MON we have:

$$\tan n = -\sin m \tan \alpha \tag{12.2.11}$$

where n is the equatorial arc ON which is cut by the arc MPN.

From the spherical triangle XPR we have:

$$\cos \lambda = \cot (90° - \varphi) \tan (90° - \eta)$$

i.e.,

$$\tan \eta = \tan \varphi \sec \lambda \tag{12.2.12}$$

Using equation (12.2.12) we can get the spherical rectangle coordinates (η, λ) based on the geographical coordinates (φ, λ), whereas we have:

$$\tan \varphi = \cos \lambda \tan \eta \tag{12.2.13}$$

If given geographical coordinates (φ, λ) and the azimuth a of point P, from equations (12.2.7) and (12.2.10) we can obtain the azimuth a of the great circle MPN along the central meridian and its corresponding azimuth β on the projection plane.

If given φ, a, and α, from equations (12.2.9) and (12.2.11) we can get arcs m, n which are cut by the great circle MPN on the central meridian and equator.

Calculating the position of a target based on two observed bearings

Assume that as the spherical straight line MN (see Figure 12.10) passes the arbitrary point P, it meets the spherical ordinate axis at azimuth a. The line segments cut on the spherical coordinate axis are m and n, respectively. The spherical straight line is still a straight line on the projection plane xOy, intersecting all meridians at angle β.

When the radius of the sphere is $R = 1$, on the projection plane the corresponding values of the spherical arcs m and n are:

$$a = \tan m, \quad b = \tan n \tag{12.2.14}$$

Assume that the coordinates in the spherical rectangle for point P are (η, λ); then their plane coordinates are:

$$x = \tan \lambda, \quad y = \tan \eta \tag{12.2.15}$$

Hence we can get the equation for the projection of the spherical straight line MN on the projection plane:

$$\frac{\tan \eta}{\tan m} + \frac{\tan \lambda}{\tan n} = 1 \tag{12.2.16}$$

If the spherical straight line passes through a point P_1' for which the coordinates are $x_1 = \tan \lambda_1$, $y_1 = \tan \eta_1$, then its projection equations are:

$$\tan \lambda - \tan \lambda_1 = k(\tan \eta - \tan \eta_1) \tag{12.2.17}$$

where:

$$k = \tan \beta = -\frac{\tan n}{\tan m}$$

If passing through points P_1' and P_2' of which coordinates are:

$$\left.\begin{array}{ll} x_1 = \tan \lambda_1, & y_1 = \tan \eta_1 \\ x_2 = \tan \lambda_2, & y_2 = \tan \eta_2 \end{array}\right\} \tag{12.2.18}$$

then the projection equation for the spherical straight line is:

$$\tan \lambda - \tan \lambda_1 = \frac{\tan \lambda_1 - \tan \lambda_2}{\tan \eta_1 - \tan \eta_2}(\tan \eta - \tan \eta_1) \tag{12.2.19}$$

In the following we will discuss ways to determine the coordinates of the position points by using the spherical straight line projection equations.

1. Given arcs m_1, n_1 and m_2, n_2 cut by two spherical straight lines on the spherical coordinate axis, we want to get the spherical coordinates η, λ of the target. From equation (12.2.16) we have:

$$\frac{\tan \eta}{\tan m_1} + \frac{\tan \lambda}{\tan n_1} = 1, \quad \frac{\tan \eta}{\tan m_2} + \frac{\tan \lambda}{\tan n_2} = 1 \tag{12.2.20}$$

Simultaneously solving the equations in (12.2.20), we get:

$$\left.\begin{array}{l} \tan \eta = \dfrac{\tan m_1 \tan m_2 (\tan n_1 - \tan n_2)}{\tan m_2 \tan n_1 - \tan m_1 \tan n_2} \\[3mm] \tan \lambda = \dfrac{\tan n_1 \tan n_2 (\tan m_2 - \tan m_1)}{\tan m_2 \tan n_1 - \tan m_1 \tan n_2} \end{array}\right\} \tag{12.2.21}$$

Noting from Figure 12.9 that:

$$\tan m = x_P - y_P \cot \beta, \quad \tan n = y_P - x_P \tan \beta \tag{12.2.22}$$

we obtain:

$$\tan m_1 = \tan \eta_1 - \tan \lambda_1 \cot \beta_1, \quad \tan m_2 = \tan \eta_2 - \tan \lambda_2 \cot \beta_2$$
$$\tan n_1 = \tan \lambda_1 - \tan \eta_1 \tan \beta_1, \quad \tan n_2 = \tan \lambda_2 - \tan \eta_2 \tan \beta_2 \qquad (12.2.23)$$

We can get the geographical coordinates (φ, λ) of target point P by using equations (12.2.23), (12.2.21), and (12.2.13).

2. Given the angular coefficients $k_1 = \tan \beta_1$ and $k_2 = \tan \beta_2$, respectively, of two spherical straight lines on the projection plane, we shall seek the spherical rectangular coordinates (η, λ) of the target. From equation (12.2.17) we have:

$$k_1 \cdot \tan \eta - \tan \lambda = k_1 \cdot \tan \eta_1 - \tan \lambda_1$$
$$k_2 \cdot \tan \eta - \tan \lambda = k_2 \cdot \tan \eta_2 - \tan \lambda_2 \qquad (12.2.24)$$

Then the coordinates of the intersecting point are:

$$\tan \eta = \frac{(k_2 \tan \eta_2 - k_1 \tan \eta_1) - (\tan \lambda_2 - \tan \lambda_1)}{k_2 - k_1}$$
$$\tan \lambda = \frac{(k_2 \tan \lambda_1 - k_1 \tan \lambda_2) + k_1 k_2 (\tan \eta_2 - \tan \eta_1)}{k_2 - k_1} \qquad (12.2.25)$$

We can get the geographical coordinates (φ, λ) of the target point P by using equations (12.2.12), (12.2.25), and (12.2.13).

3. Given the spherical coordinates of points along the two spherical straight lines (η_1, λ_1) and (η_2, λ_2), respectively, we get the spherical rectangle coordinates (η, λ) of the target. According to equation (12.2.17) we can write:

$$\tan \beta_1 = \frac{\tan \lambda - \tan \lambda_1}{\tan \eta - \tan \eta_1}, \quad \tan \beta_2 = \frac{\tan \lambda - \tan \lambda_2}{\tan \eta - \tan \eta_2} \qquad (12.2.26)$$

From the above equation we have:

$$\sin \beta_1 (\tan \eta - \tan \eta_1) - \cos \beta_1 (\tan \lambda - \tan \lambda_1) = 0$$
$$\sin \beta_2 (\tan \eta - \tan \eta_2) - \cos \beta_2 (\tan \lambda - \tan \lambda_2) = 0$$

Through a simple rearranging we get:

$$\tan \eta \sin \beta_1 - \tan \lambda \cos \beta_1 = \tan \rho_1$$
$$\tan \eta \sin \beta_2 - \tan \lambda \cos \beta_2 = \tan \rho_2 \qquad (12.2.27)$$

where:

$$\tan \rho_1 = \tan \eta_1 \sin \beta_1 - \tan \lambda_1 \cos \beta_1$$
$$\tan \rho_2 = \tan \eta_2 \sin \beta_2 - \tan \lambda_2 \cos \beta_2 \qquad (12.2.28)$$

Solving equation (12.2.27) we can get the formula for calculating spherical rectangular coordinates of the target P as follows:

$$\tan \eta = \frac{\tan \rho_1 \cos \beta_2 - \tan \rho_2 \cos \beta_1}{\sin(\beta_1 - \beta_2)}$$
$$\tan \lambda = \frac{\tan \rho_1 \sin \beta_2 - \tan \rho_2 \sin \beta_1}{\sin(\beta_1 - \beta_2)} \qquad (12.2.29)$$

Thus we can get the geographical coordinates (β, λ) of the target by using equations (12.2.7), (12.2.10), (12.2.28), and (12.2.13).

Examples *Example 1.* Given the geographical coordinates $\varphi_0 = 40°$, $\lambda_0 = 100°$ of the centerpoint Q of a gnomonic projection, calculate the value of the azimuth in several directions of fixed point A $(39°55', 116°23')$ on the gnomonic projection plane.

From equation (12.2.4) we can obtain the values of azimuths as shown in Table 12.5.

Table 12.5 Azimuths for gnomonic projection

a	0	30°	60°	90°	120°
A	0	30°39'39".8267	60°47'24".4296	90°16'17".667	119°37'19".229

Example 2. Given the geographical coordinates and azimuths of two stations $\varphi_1 = 52°18'10"$, $\lambda_1 = 30°31'20"$, $a_1 = 356°22'$ and $\varphi_2 = 43°49'30"$, $\lambda_2 = 69°02'40"$, $a_2 = 348°45'$, calculate the geographical coordinates (φ, λ) of the target at their intersection. The calculated geographical coordinates of the target are $\varphi = 80°20'33".3$, $\lambda = 20°13'32".1$.

12.2.2 Determining the position of a target on the stereographic projection

Characteristics of the stereographic projection

From equations (2.2.1) and (2.2.6) we know that the coordinate formulae of the stereographic projection take the form:

$$x = 2R\tan\frac{Z}{2}\sin a, \quad y = 2R\tan\frac{Z}{2}\cos a \tag{12.2.30}$$

From map projection theory we know that on the stereographic projection arbitrary circles on the sphere are still circles. We illustrate the characteristic as follows.

As shown in Figure 12.12, if a circle with O as its center and k as its radius has for point P the spherical coordinates of (Z, a), then from spherical triangle QOP we have:

$$\cos k = \cos Z \cos Z_0 + \sin Z \sin Z_0 \cos(a_0 - a) \tag{12.2.31}$$

Developing the $\cos(a_0 - a)$ on the right side of the above equation,

$$\cos k = \cos Z \cos Z_0 + \sin Z \sin Z_0 \cos a \cos a_0 + \sin Z \sin Z_0 \sin a \sin a_0 \tag{12.2.32}$$

Introducing the following symbols:

$$\left.\begin{array}{ll} A = \sin Z_0 \cos a_0, & B = \sin Z_0 \sin a_0 \\ C = \cos Z_0, & D = \cos k \end{array}\right\} \tag{12.2.33}$$

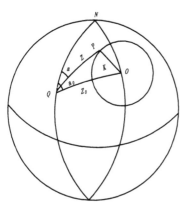

Figure 12.12 Spherical polar coordinate system and spherical circle

Substituting (12.2.33) into (12.2.32) yields:

$$A \sin Z \cos a + B \sin Z \sin a + C \cos Z = D$$

(12.2.34)

Substituting (12.2.30) into the above and rearranging gives:

$$\left(y - \frac{2RA}{C+D} \right)^2 + \left(x - \frac{2RB}{C+D} \right)^2 = \frac{4R^2(A^2 + B^2 + C^2 - D^2)}{(C+D)^2}$$

(12.2.35)

It is clear that on the sphere the formula for projection of the circle we obtained is the formula of a circle, and the coordinates of its center and radius are:

$$y_0 = \frac{2RA}{C+D}, \quad x_0 = \frac{2RB}{C+D}, \quad K = \frac{2R\sqrt{A^2 + B^2 + C^2 - D^2}}{C+D}$$

(12.2.36)

Substituting (12.2.33) into the above we obtain:

$$x_0 = \frac{2R \sin Z_0 \sin a_0}{\cos Z_0 + \cos k}, \quad y_0 = \frac{2R \sin Z_0 \cos a_0}{\cos Z_0 + \cos k}, \quad K = \frac{2R \sin k}{\cos Z_0 + \cos k}$$

(12.2.37)

From the above equation we know that when $Z_0 = 0$:

$$x_0 = 0, \quad y_0 = 0, \quad K = 2R \tan \frac{k}{2}$$

(12.2.38)

that is, the center of the circle on the projection plane coincides with the new polar point. But the projection of the center of the circle on the spherical surface does not coincide with the center of the circle on projection plane. They are located at the same perpendicular circle for which the azimuth is a_0.

This characteristic of the stereographic projection can be applied to the position of a target point.

Locating a target using the observed distance

If we know that the distances from two fixed points A (φ_A, λ_A) and B (φ_B, λ_B) to a target P are S_A, S_B, we can calculate the geographic coordinates (φ_P, λ_P) of the target. Assuming that on the stereographic projection plane of which point A is the new pole and the coordinates of the target are (x, y), then we can get the projection of the spherical circle for which point A is the center of the circle and the radius is S_A:

$$x^2 + y^2 = \rho^2$$

(12.2.39)

where:

$$\rho = 2R \tan \frac{S_A}{2R}$$

We can get the equation of the spherical circle of the projection plane for which the center of the circle is B and the radius is S_B:

$$(x - x_0)^2 + (y - y_0)^2 = K^2$$

(12.2.40)

where:

$$x_0 = \frac{2R\sin Z_B \sin a_B}{\cos Z_B + \cos k}, \quad y_0 = \frac{2R\sin Z_B \sin a_B}{\cos Z_B + \cos k}$$

$$K = \frac{2R\sin k}{\cos Z_B + \cos k}, \quad k = \frac{S_B}{R}$$

Substituting equation (12.2.39) into (12.2.40) we have:

$$\rho^2 - 2x_0 x - 2y_0 y + x_0^2 + y_0^2 = K^2$$

i.e.,

$$x_0 x + y_0 y = C \tag{12.2.41}$$

where:

$$C = (x_0^2 + y_0^2 + \rho^2 - K^2)/2$$

From equation (12.2.41) we have:

$$x = \frac{C - y_0 y}{x_0} \tag{12.2.42}$$

Substituting equation (12.2.41) into (12.2.39) we get:

$$(x_0^2 + y_0^2)y^2 - 2Cy_0 y + (C^2 - x_0^2 \rho^2) = 0 \tag{12.2.43}$$

Calculating with the above equation we have:

$$y = \frac{Cy_0 \pm \sqrt{C^2 y_0^2 - (x_0^2 + y_0^2)(C^2 - x_0^2 \rho^2)}}{x_0^2 + y_0^2}$$

i.e.,

$$y = \frac{Cy_0 \pm \sqrt{x_0^2 \rho^2 (x_0^2 + y_0^2) - C^2 x_0^2}}{x_0^2 + y_0^2} \tag{12.2.44}$$

We can get the coordinates (x, y) of point P on the stereographic projection by using equations (12.2.44) and (12.2.42).

From inverse solution formulae (4.1.35) and (4.1.36) for the stereographic projection, we have:

$$Z = 2\arctan\left(\frac{\sqrt{x^2 + y^2}}{2R}\right), \quad a = \arctan\left(\frac{x}{y}\right) \tag{12.2.45}$$

and:

$$\left.\begin{array}{l} \sin\varphi = \sin\varphi_A \cos Z + \cos\varphi_A \sin Z \cos a \\[2mm] \tan(\lambda - \lambda_A) = \dfrac{\sin Z \sin a}{\cos\varphi_A \cos Z - \sin\varphi_A \sin Z \cos a} \end{array}\right\} \tag{12.2.46}$$

We can get the geographical coordinates (φ, λ) of the target by using the above formulae.

Examples Given two fixed stations A: $\varphi_1 = 30°$, $\lambda_1 = 100°$ and point B: $\varphi_2 = 40°$, $\lambda_2 = 110°$, if the observed values of the distances to the target are $S_1 = 2\ 262.989089$ km and

$S_2 = 1\ 178.692741$ km, respectively, then we can get the geographical coordinates (φ, λ) of the target. Given the average radius of curvature at $\varphi = 35°$ as the radius of the earth, i.e. $R = 6\ 370.8922$ km, according to the above formulae we can get results as follows:

$\varphi = 50°$, $\lambda = 104°59'59''.999$

$\varphi = 36°59'19''.7578$, $\lambda = 123°00'04''.265$

Comparing the two pairs of results with the values according to the section from another station or the approximate azimuth, for example the third station C: $\varphi_3 = 35°$, $\lambda_3 = 115°$, we can state that the coordinates of the target point by station B and C are as follows:

$\varphi = 50°00'00''.0003$, $\lambda = 105°00'00''.001$;

$\varphi = 44°13'33''.0228$, $\lambda = 96°52'18''.0034$.

So the geographical coordinates of the target are $\varphi = 50°$, $\lambda = 105°$.

CHAPTER THIRTEEN

Spatial Information Positioning Systems

13.1 INTRODUCTION

Spatial information refers to data that take space coordinates as their reference system. Nowadays observing the earth's surface from above, and the information superhighway, provide a vast amount of information that urgently needs to be analyzed, manipulated, simulated and managed, so that mankind can understand and grasp temporal–spatial processes and the functioning of the earth and of society.

All activities of mankind take place within geographic space. Spatial information positioning systems (SIPS) are formed of spatial information (such as digital maps and remote sensing images), and registration models (referring to map projections) that form the basis and framework for the study of spatial information. Spatial information positioning systems make the applied information geographically coherent and render point positions in a plane coordinate system amenable to measurement.

A spatial information positioning system based on a digital map is called a mapping positioning system for spatial information. A spatial information positioning system based on a remote sensing image is called an image positioning system for spatial information. And a spatial information positioning system based on both a digital map and a remote sensing image is called a mapping and image positioning system for spatial information.

All of the above-mentioned spatial information positioning systems can be integrated using positioning techniques and can be applied to special models which can serve society. These new concepts have made spatial information an important knowledge industry, and have given spatial information analysis, spatial models, and simulations, a viable and reliable foundation. This has great theoretical and practical significance for the industrialization of spatial information.

A spatial information positioning system (SIPS) is formed of the following items (Yang *et al.* 1997):

- digital map
- remote sensing image
- registration model
- positioning technique

- data processing
- automatic building of map mathematical foundation
- integration and application.

13.2 MAPPING AND IMAGE POSITIONING SYSTEM

13.2.1 Mapping positioning system for spatial information

Traditional analog maps reflect all the complications of the natural and social features, and represent these by a mapping system using symbols, colors, characters, etc. Because of the properties of the plane coordinate system and the mathematical basis of the mapping science, it converts differences from a scene or photo, and makes them into a mapping system with positioning that supports spatial information. In a mapping positioning system, the registration model includes a geodetic coordinate system and a map projection to provide a mathematical basis for accurate positioning.

A digital map based on the development of computer technology is the geographic basis of spatial information positioning. The so-called digital map refers to the aggregation of digital spatial information that is digitized from an analog map, uses a certain data structure, and is then stored on some media such as tape, disk, or CD-ROM after processing. A mapping positioning system for spatial information, when integrated using a digital map and a registration model, is the carrier of spatial information and the basis of positioning. It has been widely used as the geographical basis in geographical information systems (GIS) and many special information systems. Since it is a uniform and standardized framework, and with metric properties – that is, the map projection system – it is quite possible to interchange, co-register and share all kinds of special spatial information, at all levels and for all regions. Furthermore, the data can then be used for comprehensive analysis and evaluation.

Mapping positioning systems of spatial information based on digital maps has become one important part of GIS. All aspects of GIS, whether the acquiring of data, preprocessing, storing and recording of information, or the processing, application and output of data, require a spatial positioning framework, i.e., a common geographical coordinate system and a plane coordinate system. The digital map includes a vector mapping system and raster mapping. The mathematical transformation for a vector map is a corresponding single-valued transformation, but the raster image is not a single-valued transformation. That is, the size, in area, of a raster element in an original piece of map is not equal to that in the resulting map. Compared with an analog map, a digital mapping system has more flexibility and adaptability, and a wider application area. Combined with positioning technology, it can make a static mapping system into a dynamic mapping system.

13.2.2 Image positioning systems for spatial information

With the unceasing development of remote sensing platforms and remote sensing instruments, remote sensing has entered a new stage of rapid and timely provision of a vast amount of data pertaining to the earth. The spatial resolution, spectral resolution and temporal resolution of remote sensing images have been greatly improved. The processes and methods of processing remote sensing images are mature.

Remote sensing images contain considerable spatial information and are an important data source for geographical information systems. They are also an important part of a spatial positioning system. But there are practical problems. These are because realizing a system with a specified precision is usually very difficult and time consuming. Only remote sensing images corresponding to a particular map projection and having a uniform scale can yield an image having spatial coordinates and possessing value for applications. Nowadays the commonly used method is to acquire satellite imagery and then to use brute force methods for initial rectification and alignment of the image to a particular map by geometrical rectification. That is, one uses a polynomial fit to more than 20 control points in one section of the image. Here we suggest a new method of building a mathematical foundation for a satellite image. It includes geometrical control and a spatial projection system to fit a satellite image.

The spatial projection system for a satellite image is one kind of optimized mathematical model for the fitting of a satellite image. It has these properties:

1. It takes into consideration the effect of the satellite motion, the earth's rotation and orbital precession. It can simulate well the spatial model of the physical processes of satellite imaging and the one to one analytical relationship between pixel elements and ground control points. It has the merit of matching with the numerical fitting method now applied in image processing.

2. The spatial projection system can force the strip images in the same flight orbit to be in a unified coordinate system. Hence only a small number of ground control points are needed for building the projection system.

3. Within an extent of $\pm 1°$ from the locus of the sub-satellite track of the satellite, the spatial projection preserves the conformality property and it is thus quite suitable for satellite image mapping when used in conjunction with the Gauss–Krüger projection transform, the UTM projection, and so on. It can meet the needs of all agencies using satellite images. The satellite images built upon this mathematical foundation provide an image positioning system rich in spatial possibilities.

Combining a mapping positioning system with an image positioning system by data fusing technology for spatial information and positioning places them into the same geographical framework and this results in a mapping and image spatial information system with a vast range of application prospects.

13.3   REGISTRATION MODEL (MAP PROJECTION SYSTEM)

The positioning function of a mapping and image system is decided using a rigorous mathematical basis. For this reason, we can regard a map projection system as a spatial information positioning system. The map projection system includes a projection coordinate system, origin positioning, ellipsoid (or surface) fitting, a grid system and a map projection transformation system. We introduce them as follows.

13.3.1   Map projection coordinate system

There is no essential distinction between a digital map projection coordinate system and an analog map projection coordinate system. They all a require a mathematical basis for spatial information positioning. The mathematical basis of an analog map stresses the construction of the graticule of meridians and parallels and a rectangular coordinate system. These are taken as a framework within which to fill in the map content. On the other hand the mathematical structure of a digital map stresses the positioning of map elements as points (including a point's position and its attributes) in a specified projection coordinate system. When this is to be a single worldwide system special considerations are required (Tobler and Chen 1986).

For a large-scale digital map, one usually uses a topographic scale map projection such as the Gauss–Krüger projection. This is convenient to produce an abstract of the map detail to be applied to a new map at a nearby scale. But a map projection transformation must be performed when converting the map to an adjacent zone or to another map projection. For remote sensing images it is suitable to use the mathematical spatial projection.

13.3.2 Position line

For the theory and application of position lines, see Chapter 9. For software systems for positioning and navigation of position lines, see Appendix 4.

13.3.3 Ellipsoid (or sphere) positioning

For a small-scale digital map we should process the acquired data, transform them into geographical coordinates by an inverse map projection, and store them in terms of latitude and longitude. That is, take locations on the earth ellipsoid as a carrier of spatial information. The advantage of using geographical coordinate positions is that it is easy to join together pieces into a large region or to a global scale. And the position of a point will not change with the selection of a projection system. As a result, we can format the output into different map projections, if we provide several commonly used map projections in the system.

13.3.4 Map grid system

Map grids include the geographic graticule and rectangular grids. They are usually used in classifying remotely sensed images, for regional database querying and information indexing. The geographic graticule has the advantage of easily matching adjoining areas and the graticule is carried along with the selection of projection system. But the density of the graticule is not uniform on the earth. That is, the density is greater in high latitude regions and less in low latitude regions. A rectangular grid has the advantage of uniformly distributed locations. But the actual terrestrial position corresponding to a grid location changes with the map projection selected. If the Gauss–Krüger projection is used, the problem of incompatible grids will appear at the edge of adjacent zones.

13.3.5 Map projection transformation system

When a new map based on one projection is edited for use on an analog map on another projection, one usually forces the old data to fit the new projection framework, thus obtaining the map projection transformation. But in computer-aided mapping, by process-ing the acquired data positions in advance, building a mathematical model, and using appropriate software, the projection transformation for producing another digital map on a different kind of projection can be effected. For this reason, in a digital spatial informa-tion system, the extraction and application of spatial information depends completely on a map projection transformation software system. Accordingly, in order to convert to a different scale or map projection, the transformation software system should include: a software subsystem for computer-aided construction on the requisite mathematical basis; a system package for acquiring, rectification and processing of geographic information; a software package for the Gauss–Krüger projection coordinate transformation and soft-ware for other commonly used map projection transformations; and a software package for the numerical transformation of map projections.

The software package for the Gauss–Krüger projection coordinate transformation in-cludes: the direct and inverse transformation of Gauss–Krüger projection; 3° and 6° zone coordinate transformations for the Gauss–Krüger projection; Gauss–Krüger projection coordinate transformation between local coordinate systems and a national coordinate system; and constant coefficient coordinate transformations among the Gauss–Krüger projection, the Mercator projection and the conformal conic projection used for the 1:1 000 000 scale map.

The commonly used map projection transformation software package includes: direct and inverse coordinate transformation between the normal and oblique or transverse azimuthal projections with different characteristics; direct and inverse coordinate trans-formation between the normal and oblique or transverse cylindrical projection with differ-ent characteristics; direct and inverse coordinate transformation between the normal and oblique or transverse conical projection with different characteristics; direct and inverse coordinate transformation between three kinds of pseudo-projection and the polyconical projection; and finally other direct and inverse projection coordinate transformations.

The software package for the numerical transformation of map projections includes: numerical transformation of intersection point coordinates on a densified geographic graticule; direct and inverse numerical transformation of non-conformal projections; and direct and inverse numerical transformation of conformal projections.

13.4 DATA PROCESSING OF A DIGITAL MAP

13.4.1 Error analysis of a digital map

The position of points on a map (or graphic) acquired from an analog map usually take the form of vector data or raster data. It is quite important to design a fixed data structure such that the acquired data can be used with flexibility, and comparability, and that the topological structure be maintained. The so-called data structure is the arrangement of data and the relationship among its elements. The data structures employed for mapped (or graphed) information, maintaining the spatial information, can generally be divided into two types: vector structure and raster structure.

The errors resulting from vector and scan digitization depend on the following factors: width of the element; complexity of the element; resolution of digitization; density of elements, and so on. The data acquired by digitizing a map according to one of the data structures generally have the following kinds of errors (Yang and Yang 1995):

1. Errors resulting from the deformation of a map sheet. In normal atmospheric tempera-
   tures, the size of paper stretches by up to 1.6%. In printing the length may extend by
   1.5% and the width by 2.5%.

2. Errors resulting from map orientation. The digitization coordinate system is not con-
   cordant with the projection coordinate system.

3. Errors resulting from digital sampling.

4. Errors resulting from digital manipulations.

The above errors belong to systematic error, accidental error and gross error. The existence of all kinds of errors makes digital map data elements of maps difficult to match
when they are transformed to a different map; they make the digital results from different
sources difficult to link accurately; and make it difficult to join adjacent sheets. Hence the
digitized data should be rectified or processed before storage or application.

13.4.2 Data processing model

The linear similarity transformation

$$x = a_1x' + a_2y' + a_0, \quad y = a_1y' - a_2x' + b_0 \tag{13.4.1}$$

The values of a_1, a_2, a_0, b_0 can be found using two common points. When the number of
points is greater than two, use the method of least squares.

The affine transformation

$$x = a_1x' + a_2y' + a_0, \quad y = b_1x' + b_2y' + b_0 \tag{13.4.2}$$

The values of a_1, a_2, a_0, b_1, b_2, b_0 can be found using three common points. If the number
of points is greater than three, use the method of least squares.

The homographic transformation model

$$x = \frac{a_1x' + a_2y' + a_0}{c_1x' + c_2y' + 1}, \quad y = \frac{b_1x' + b_2y' + b_0}{c_1x' + c_2y' + 1} \tag{13.4.3}$$

The values of a_1, a_2, a_0, b_1, b_2, b_0, c_1, c_2 can be found using four common points. If the
number of points ≥ 4, solve the system by the method of least squares.

Double polynomial of degree 1

$$x = a_1x' + a_2y' + a_3x'y' + a_0, \quad y = b_1x' + b_2y' + b_3x'y' + b_0 \tag{13.4.4}$$

The values of a_1, a_2, a_3, a_0, b_1, b_2, b_3, b_0 can be found using four common points. If the
number of points is greater than four, solve the system with the method of least squares.

The least-squares model

The mathematical model for the linear similarity transformation If digital coordinates
(x_i', y_i') and their corresponding projection coordinates (x_i, y_i) are known, and the
number of points (m) is greater than two, we can solve for the coefficients of (13.4.1)
according to method of least squares. That is, the following conditions hold:

$$\varepsilon = \sum_{i=1}^{m} [(a_1 x'_i + a_2 y'_i + a_0 - x_i)^2 + (a_1 y'_i - a_2 x'_i + b_0 - y_i)^2] = \min$$

Hence we get the following matrix formula:

$$\begin{vmatrix} \sum x'^2_i + \sum y'^2_i & 0 & \sum x'_i & \sum y'_i \\ 0 & \sum x'^2_i + \sum y'^2_i & \sum y'_i & -\sum x'_i \\ \sum x'_i & \sum y'_i & m & 0 \\ \sum y'_i & -\sum x'_i & 0 & m \end{vmatrix} \cdot \begin{vmatrix} a_1 \\ a_2 \\ a_0 \\ b_0 \end{vmatrix} = \begin{vmatrix} \sum x'_i x_i + \sum y'_i y_i \\ \sum y'_i x_i - \sum x'_i y_i \\ \sum x_i \\ \sum y_i \end{vmatrix}$$

(13.4.5)

We get the values of a_1, a_2, a_0, b_0 by solving the above system of equations.

Mathematical model of affine transformation If digital coordinates (x'_i, y'_i) and their corresponding projection coordinates (x_i, y_i) are known, and the number of points (m) is greater than three, then we can solve for the coefficients of (13.4.2) according to the method of least squares. That is, the minimization conditions are:

$$\varepsilon = \sum_{i=1}^{m} (a_1 x'_i + a_2 y'_i + a_0 - x_i)^2 = \min, \quad \varepsilon' = \sum_{i=1}^{m} (b_1 x'_i + b_2 y'_i + b_0 - y_i)^2 = \min$$

Hence we get the following matrix formulae:

$$\begin{vmatrix} \sum x'^2_i & \sum y'_i x'_i & \sum x'_i \\ \sum x'_i y'_i & \sum y'^2_i & \sum y'_i \\ \sum x'_i & \sum y'_i & m \end{vmatrix} \cdot \begin{vmatrix} a_1 \\ a_2 \\ a_0 \end{vmatrix} = \begin{vmatrix} \sum x_i x'_i \\ \sum x_i y'_i \\ \sum x_i \end{vmatrix}$$

(13.4.6)

$$\begin{vmatrix} \sum x'^2_i & \sum y'_i x'_i & \sum x'_i \\ \sum x'_i y'_i & \sum y'^2_i & \sum y'_i \\ \sum x'_i & \sum y'_i & m \end{vmatrix} \cdot \begin{vmatrix} b_1 \\ b_2 \\ b_0 \end{vmatrix} = \begin{vmatrix} \sum y_i x'_i \\ \sum y_i y'_i \\ \sum y_i \end{vmatrix}$$

We can get the values of a_1, a_2, a_0, b_1, b_2, b_0 by solving the above equations one at a time.

13.4.3 Map accuracy evaluation and data processing methods

Topographic map accuracy evaluation and data processing methods

Topographic map accuracy evaluation The linear similarity transformation consists of a translation, a rotation, and adjustment in size, thus maintaining the similarity of graphics. Thus it can be used for accuracy evaluation relating points of a geographic graticule and the square grid used on topographic maps.

Data processing methods for a topographic map

1. Evaluate the point position accuracy using the linear similarity transformation model. If it satisfies the accuracy demands, use the model to transform the data.

2. If it does not meet the conditions, use the affine transformation model or use a double polynomial model of first degree to transform the data. This may satisfy the adjustment.

A data processing method for a small-scale map

Analytical method

1. Data preprocessing (coordinate the projection coordinate system): use the linear similarity transformation model to adjust the projection coordinate system and evaluate the position accuracy of points.

2. Data processing: find the geographic coordinates of each point using the inverse transformation model of the map projection.

Numerical method

1. Block transformation based on the geographic graticule. The data structure should be taken into account before block digitizing and data processing.

2. Block transformation based on a rectangular grid. Digitizing and data processing are independent of each other. Therefore data processing is very convenient. The numerical transformation model introduced in Chapter 5 can be used here.

Numerical–analytical method   If the analytical expression of the map projection is given but some parameters of the projection are unknown, we can get the parameters with a numerical method according to the projection's properties and then process data with the analytical method.

13.4.4 Application examples

Here we present a practical example to show the digital processing of a world map using the numerical–analytical method.

The world map uses the modified Gore's projection. The direct solution of the equation is:

$$\left.\begin{array}{l} x = R\lambda \left[c_1 - c_2 \sin^2 \left[45° tg \left(\dfrac{\varphi}{2} \right) \right] \right] \\[2em] y = R(1 + \cos\varphi_0) tg \dfrac{\varphi}{2} \end{array}\right\} \tag{13.4.7}$$

where $\varphi_0 = \pm 45°$ is given, but the value of the radius of the earth R and the projection parameters c_1, c_2 are unknown.

From (13.4.7) we can obtain the inverse coordinate formula:

$$\left.\begin{array}{l} \varphi = 2\,arctg\,[y/1.70711R] \\[1.5em] \lambda = \dfrac{x}{\left\{ R \left[c_1 - c_2 \sin^2 \left(45° \dfrac{y}{1.70711R} \right) \right] \right\}} \end{array}\right\} \tag{13.4.8}$$

Systems for the processing of digital world map data are composed of five modules (see Appendix 4). We introduce each module's function:

1. Module for acquiring digital data and for generating the arrangement of feature points on the map.

   ▪ Interpreting feature point data: convert the coordinate file of points on a digital graticule of meridians and parallels into the data of groups of numbers.

- Generating the distribution map of feature points: generate the sketch map of graticule of meridians and parallels and label the number of feature points.

- Interpreting the data of each feature on the map: convert the data format to form a data file for easy processing.

2. Module for map feature recognition and reconstructing.

- Determining the radius of the earth R: determine the radius of the earth R according to the digital coordinates on the central meridian.

- Determining the projection parameters c_1, c_2: determine parameter c_1 according to the digital coordinates on the equator; determine parameter c_2 according to the standard parallel at 45° N and S.

- Modifying the direct coordinate transformation model for Gore's projection: calculate the acquired data x, y from the given geographical coordinates φ, λ.

- Modifying the inverse coordinate transformation model for Gore's projection: calculate geographical coordinates φ, λ from the given projection coordinates x, y.

3. Module for the coordinating of the projection coordinate system and the precision of point positions.

- Adjusting of the projection coordinate system: using a linear similarity transformation, transform all the digital coordinates into the coordinates of the projection coordinate system according to the method of least squares.

- Adjusting the precision of point positions: calculate the location error of feature points on the projection surface; calculate the same error of feature points on the original surface.

4. Module for the uniting of the projection coordinate system and the adjustment of point positions.

- Uniting of the projection coordinate system: using an affine transformation, transform all the digital coordinates into the coordinates of the projection coordinate system according to the least squares method.

- Adjustment of point positions: calculate the error of feature points on the projection surface; calculate the error of feature points on original surface.

5. Module for the unification of the projection coordinate system and the solving for geographical coordinates with an inverse solution transformation.

- Uniting of projection coordinate systems: use an affine transformation or a linear similarity transformation to convert all the digital coordinates into the coordinates of the projection coordinate system taking into account the positional error of feature points.

- Solving for geographical coordinates with an inverse solution transformation: inversely solve for geographical coordinates with the coordinates in a unified projection coordinate system.

The system is applicable to the processing of digital world map data. It can generate pure geographical coordinates for the building and updating of map databases, and for automatic cartography as well as for the construction of a GIS.

13.5   AUTOMATICALLY SETTING UP MAP MATHEMATICAL FOUNDATION

13.5.1   General method

Historically, the creation of the mathematical foundation of a map was realized through the following steps: calculate the coordinates of a series of points at fixed intervals of longitude and latitude from the projection coordinate formula, then plot points with a coordinatograph; finally the corresponding points were connected and the graticule of meridians and parallels was formed.

With the application of computers and automatic cartographic systems to the practice of cartography, this classical manual operating method for creating the mathematical foundation of a map has been completely replaced by the automatic method using hardware such as computers and automatic drawing instruments and the associated software system. They are generally referred to as computer-assisted cartography.

The computer-assisted setting up of a map's mathematical basis is similar to the method of manual plotting of a graticule of meridians and parallels. It is obtained by the following steps: calculate the coordinates of a series of points on meridians and parallels at fixed intervals of longitude and latitude from the projection coordinate formula; output the coordinate information of the points in a plotting format; and finally plot the graticule of meridians and parallels using an automatic drawing instrument.

Given map projection formula:

$$x = f_1(\varphi, \lambda), \quad y = f_2(\varphi, \lambda) \tag{13.5.1}$$

Then the parametrized parallel equation is:

$$x = f_1(\varphi_i, \lambda), \quad y = f_2(\varphi_i, \lambda) \tag{13.5.2}$$

where φ_i denotes different parallels, $i = 1, 2, 3, \ldots n$.

And the parametrized meridian equation is:

$$x = f_1(\varphi, \lambda_i), \quad y = f_2(\varphi, \lambda_i) \tag{13.5.3}$$

In the case of a parallel $\varphi = \varphi_1$ and a difference in longitude of $\Delta\lambda$, we can obtain a series of coordinates (x_i, y_i) such as (φ_1, λ_0), $(\varphi_1, \lambda_0 + \Delta\lambda)$, $(\varphi_1, \lambda_0 + 2\Delta\lambda), \ldots$ from equation (13.5.3), and then plot the parallel φ_1 by connecting two points into a line, and in turn plot every parallel $\varphi_2, \varphi_3, \ldots$.

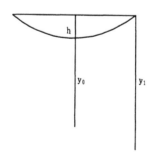

Similarly, in the case of a meridian $\lambda = \lambda_1$ and difference in latitude of $\Delta\varphi$, we can obtain a series of coordinates (x_i, y_i) from equation (13.5.3), then plot the meridian λ_1, and in turn plot each meridian $\lambda_2, \lambda_3, \ldots$.

This way of automated plotting of meridians and parallels uses straight lines instead of a curve. To ensure the smoothness of the plotted meridians and parallels and the precision of point positions, the maximum of vector length between the arc and the chord should be less than 0.1 mm at map scale.

Generally, the relation of the interval of difference of longitudes and latitudes and the vector length can be read off from existing coordinate tables; the vector length of a parallel with fixed difference of longitude can be calculated from the following equation (see Figure 13.1):

Figure 13.1   The relation of vector length h and ordinate axis

$$h = y_1 - y_0 \tag{13.5.4}$$

In addition, we can derive a widely used calculating equation for the vector length of a parallel for a conical projection. For conical projections, the calculating equation of vector length of a tangent parallel is:

$$h = R \cot \varphi_0 \left(1 - \cos \frac{\alpha \Delta \lambda}{2} \right) = R \frac{\Delta \lambda^2}{16} \sin 2\varphi_0 \tag{13.5.5}$$

Let $\varphi_0 = 45°$, $R = 6\,378$ mm (reduced to a $1:1\,000\,000$ map scale), then from the above equation we obtain:

$$h = 0.12\Delta \lambda^2 \text{ (mm)} \tag{13.5.6}$$

where the unit for $\Delta \lambda$ is in degrees.

13.5.2 The automatic plotting of the map graticule of meridians and parallels

The computational flow is as follows:

Data preparation → Calculating coordinates of the points of a parallel family → Plotting parallels → Calculating coordinates of the points of a meridian family → Plotting meridians

For the projection of limit points, infinite points and double points, as well as the case when a pole point becomes a line and a closed curve on the original surface becomes an open curve or discontinuous curve after being projected, special software is required. Let us take an oblique conical projection as an example; since its new polar point is $B_0 \neq 0$, if we expand the conical surface along a central meridian passing through the North Pole, then the projection parallels of $B_i > B_0$ will become discontinuous curves, the projection parallels of $B_i < -B_0$ will become closed curves and the projection parallels of $-B_0 < B_i > B_0$ will become open curves. Figures 13.2 and 13.3 separately show the world graticule of meridians and parallels of a transverse conical projection and an oblique conical projection. They have been automatically produced with the appropriate software.

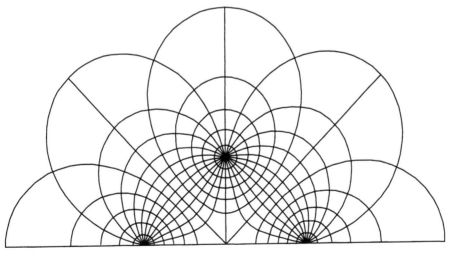

Figure 13.2 The world graticule of meridians and parallels of a transverse conformal conic projection

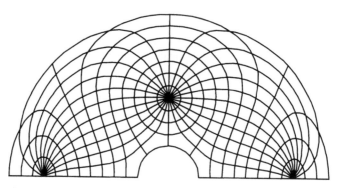

Figure 13.3 The world graticule of meridians and parallels of an oblique equidistant conic projection

13.5.3 Automatic drawing of thematic mathematical feature map (Yang 1986c; Appendix 4)

The graticule of meridians and parallels on a map represents the principal mathematical elements of the map's mathematical basis. The main task of map projection research is just the theory and method of transforming the graticule of meridians and parallels on an ellipsoid (or sphere) to the map plane. With the development of modern science and technology, there have been breakthroughs in the field of classical research and methods of map projection. Modern science and technology demand maps able to provide more spatial position information such as point positions, distance, azimuth and locus. The information provided by the map graticule alone is insufficient. Hence, the classical mathematical elements of the graticule of meridians and parallels cannot meet the needs of practical applications. As a result, computer-aided position line drawing and its measurement, i.e., thematic mathematical elements of a map, have become a new component of map projection.

Creating thematic mathematical elements does not result in a new map. It only adds thematic mathematical elements to the original map to meet the requirements of spatial positioning information by various departments. This method is suited to maps using all types of map projections. And it is quite easy to plot thematic mathematical elements on transparent film which is overlaid on the original map for use. This is the introduction of multiple map plane layers on a mathematical foundation. It has opened up a vast range of prospects for the application of map projections to productive practice. Now we introduce the automatic drawing of two kinds of thematic mathematical elements.

Automatic drawing of hyperbolic navigation lattices

To meet the need of radio navigation, a hyperbolic navigation lattice should be added to a Mercator projection map. The hyperbolic navigation lattice is just a kind of thematic mathematical element.

Adding a hyperbolic projection lattice to a Mercator map uses double projection, that is, conformal projection of the surface of the ellipsoid onto the surface of the sphere according to the second Gauss projection method, then projection of a spherical hyperbolic grid onto the Mercator plane.

The automatic drawing of a hyperbolic lattice also involves the calculation of parameters for controlling curve smoothness, the calculation of the width of the hyperbolic

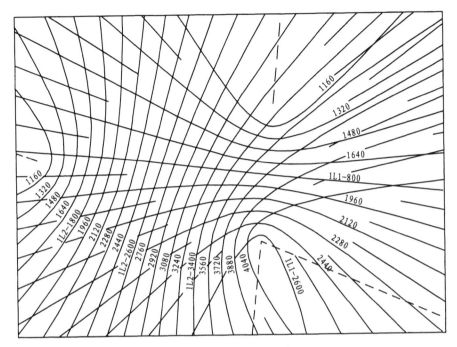

Figure 13.4 Hyperbolic lattice on the Mercator projection map

zones, the conversion of difference of distance between annotations and great circles, the calculation of the beginning value and the end value for annotation, the calculation of the position of annotations, etc.

The design sequence for the automatic drawing of a hyperbolic lattice is: take the polar angle β as an independent variable, using the spherical hyperbolic formula expressed in spherical polar coordinates (S, β) to solve a series of polar coordinates $(S_i, \beta_i) \rightarrow (\varphi_i, \lambda_i)$ $\rightarrow (x_i, y_i)$ of points on the hyperbola. Then automatically draw a hyperbola.

Figure 13.4 shows the reduced automatically drawn graph of a 1:1 000 000 hyperbolic lattice.

Automatic drawing of distance and direction measurements as thematic mathematical elements on a map

As shown in Figure 13.5, given a central point O (φ_0, λ_0) from which to project, then the distance and the azimuthal track from a fixed point A (φ_A, λ_A) on the spherical surface to an arbitrary point K are equal to the small circle's line at centerpoint A and the great circle arc at centerpoint A.

Since a projection unavoidably distorts, distance and azimuthal measurement on a map will result in errors; on some maps these will be large. If we plot on a projection surface the equidistant line and the orthodrome from a fixed point to an arbitrary point, then we can easily, quickly and accurately measure distances and azimuths on the map.

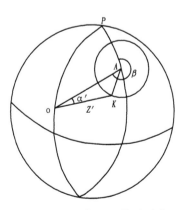

Figure 13.5 The small circle's line at centerpoint A and the great circle arc at centerpoint A

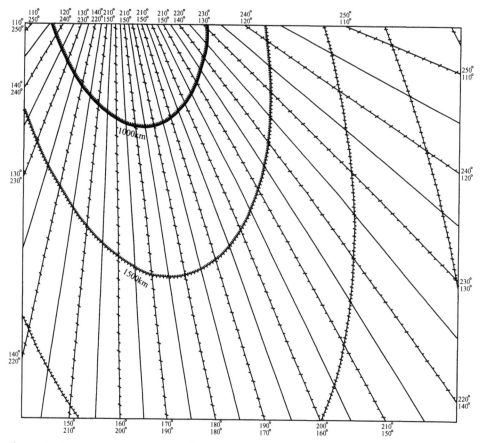

Figure 13.6 The measurement of distances and angles map for normal gnomonic projection

The principle of automatic drawing of the projection graticule for the measurement of distances and angles is: to project spherical small circle lines with fixed radii and spherical great circle lines with fixed spacing onto the projected surface, then finely divide and annotate the lines of equidistance and the orthodromes according to the requirements.

Our software for automatic drawing of the thematic mathematical element of distance and azimuths on a map for the measurement of distances and angles consists of three parts: the software for automatic drawing the graticule for distance measuring and azimuthal measuring; the software for automatic drawing the finer divisions of the graticule; and the software for the automatic annotations of the lines of equidistance and orthodromes.

This software can be used to add the graticule for distance measuring and azimuth measuring to a fixed point on some special map, or just to produce a special map for the fixed point. This distance-orthodromic graticule can improve the effectiveness of existing maps.

We have automatically drawn 13 sheets of this graticule for the 1:12 500 000 map on the normal gnomonic projection. The size of a sheet is 78.79 cm × 68.14 cm. Figure 13.6 is a reduction of one of them.

Appendices

Inverse Transformation of Algebraic Series

▮

In Section 3.7.1, we introduced series such as:

$$y = x + \varepsilon \cdot f(y) \tag{A1.1}$$

where ε is a small parameter. The corresponding Lagrange series is:

$$y = x + \frac{\varepsilon}{1} f(x) + \frac{\varepsilon^2}{2} \frac{d}{dx} f^2(x) + \frac{\varepsilon^3}{6} \frac{d^2}{dx^2} f^3(x) + \dots \tag{A1.2}$$

The more general Lagrange series for function $u = \varphi(y)$ is as follows:

$$u = \varphi(x) + \frac{\varepsilon}{1} \varphi'(x) f(x) + \frac{\varepsilon^2}{2} \frac{d}{dx} \varphi'(x) f^2(x) + \frac{\varepsilon^3}{6} \frac{d^2}{dx^2} \varphi'(x) f^3(x) + \dots \tag{A1.3}$$

A1.1 INVERSE TRANSFORMATION OF THE ALGEBRAIC POWER SERIES OF DIMENSION ONE

Suppose:

$$y = a_1 x + a_2 x^2 + a_3 x^3 + a_4 x^4 + a_5 x^5 \tag{A1.4}$$

where a_i are coefficients and $|x| < 1$.

Then its inverse transformation formula is:

$$x = b_1 y + b_2 y^2 + b_3 y^3 + b_4 y^4 + b_5 y^5 \tag{A1.5}$$

Substituting (A1.1.5) into (A1.1.4), we get:

$$
\begin{aligned}
y = {} & a_1(b_1 y + b_2 y^2 + b_3 y^3 + b_4 y^4 + b_5 y^5) + a_2(b_1 y + b_2 y^2 + b_3 y^3 + b_4 y^4 + b_5 y^5)^2 \\
& + a_3(b_1 y + b_2 y^2 + b_3 y^3 + b_4 y^4 + b_5 y^5)^3 + a_4(b_1 y + b_2 y^2 + b_3 y^3 + b_4 y^4 + b_5 y^5)^4 \\
& + a_5(b_1 y + b_2 y^2 + b_3 y^3 + b_4 y^4 + b_5 y^5)^5
\end{aligned}
$$

After rearranging the above equation we obtain:

$$
\begin{aligned}
y = {} & a_1 b_1 y + (a_1 b_2 + a_2 b_1^2) y^2 + (a_1 b_3 + 2a_2 b_1 b_2 + a_3 b_1^3) y^3 \\
& + (a_1 b_4 + a_2 b_2^2 + 2a_2 b_1 b_3 + 3a_3 b_1^2 b_2 + a_4 b_1^4) y^4 \\
& + (a_1 b_5 + 2a_2 b_1 b_4 + 2a_2 b_2 b_3 + 3a_3 b_1^2 b_3 + 3a_3 b_1 b_2^2 + 4a_4 b_1^3 b_2 + a_5 b_1^5) y^5
\end{aligned}
$$

Comparing the above two equations with the coefficients of like terms, we have:

$$a_1 b_1 = 1, \quad \therefore \quad b_1 = \frac{1}{a_1}$$

$$a_1 b_2 + a_2 b_1^2 = 0, \quad \therefore \quad b_2 = -\frac{a_2}{a_1^3}$$

$$a_1 b_3 + 2 a_2 b_1 b_2 + a_3 b_1^3 = 0, \quad \therefore \quad b_3 = \frac{2 a_2^2}{a_1^5} - \frac{a_3}{a_1^4}$$

$$a_1 b_4 + a_2 b_2^2 + 2 a_2 b_1 b_3 + 3 a_3 b_1^2 b_2 + a_4 b_1^4 = 0, \quad \therefore \quad b_4 = -\frac{5 a_2^3}{a_1^7} + \frac{5 a_2 a_3}{a_1^6} - \frac{a_4}{a_1^5}$$

$$a_1 b_5 + 2 a_2 b_1 b_4 + 2 a_2 b_2 b_3 + 3 a_3 b_1^2 b_3 + 3 a_3 b_1 b_2^2 + 4 a_4 b_1^3 b_2 + a_5 b_1^5 = 0$$

$$b_5 = \frac{14 a_2^4}{a_1^9} - \frac{21 a_2^2 a_3}{a_1^8} + \frac{6 a_2 a_4 + 3 a_3^2}{a_1^7} - \frac{a_5}{a_1^6}$$

Thus, we obtain the inverse transformation formula of (A1.4) as follows:

$$x = b_1 y + b_2 y^2 + b_3 y^3 + b_4 y^4 + b_5 y^5 \tag{A1.6}$$

where:

$$b_1 = \frac{1}{a_1}, \quad b_2 = -\frac{a_2}{a_1^3}, \quad b_3 = \frac{2 a_2^2}{a_1^5} - \frac{a_3}{a_1^4},$$

$$b_4 = -\frac{5 a_2^3}{a_1^7} + \frac{5 a_2 a_3}{a_1^6} - \frac{a_4}{a_1^5}$$

$$b_5 = \frac{14 a_2^4}{a_1^9} - \frac{21 a_2^2 a_3}{a_1^8} + \frac{6 a_2 a_4 + 3 a_3^2}{a_1^7} - \frac{a_5}{a_1^6}$$

When $a_2 = 0$ in formula (A1.4), each coefficient mentioned above is:

$$b_1 = \frac{1}{a_1}, \quad b_2 = 0, \quad b_3 = -\frac{a_3}{a_1^4}, \quad b_4 = -\frac{a_4}{a_1^5}, \quad b_5 = \frac{3 a_3^2}{a_1^7} - \frac{a_5}{a_1^6} \tag{A1.7}$$

When $a_2 = a_4 = 0$ in formula (A1.4), each coefficient mentioned above is:

$$b_1 = \frac{1}{a_1}, \quad b_2 = 0, \quad b_3 = -\frac{a_3}{a_1^4}, \quad b_4 = 0, \quad b_5 = \frac{3 a_3^2}{a_1^7} - \frac{a_5}{a_1^6} \tag{A1.8}$$

For example, suppose the abscissa of the Gauss–Krüger projection is:

$$x = a_1 l + a_3 l^3 + a_5 l^5 + \dots \tag{A1.9}$$

where:

$$a_1 = N \cos B, \quad a_3 = \frac{N}{6} \cos^3 B (1 - \tan^2 B + \eta^2)$$

$$a_5 = \frac{N}{120} \cos^5 B (5 - 18 \tan^2 B + \tan^4 B)$$

Substituting the coefficients above into (A1.8), we obtain:

$$l = \frac{x}{N \cos B} - \frac{(1 - \tan^2 B + \eta^2) x^3}{6 N^3 \cos B} + \frac{(5 - 2 \tan^2 B + 9 \tan^4 B) x^5}{120 N^5 \cos B} \tag{A1.10}$$

For the series of even powers:

$$y = a_0 + a_2 x^2 + a_4 x^4 + a_6 x^6 + \dots \tag{A1.11}$$

The case 0 may occur in formula (A1.6) when $a_1 = a_3 = 0$. Hence, letting $y - a_0 = y'$, $a_2 x^2 = u$, we have:

$$u = y' - \frac{a_4}{a_2^2} u^2 - \frac{a_6}{a_2^3} u^3 - \dots \tag{A1.12}$$

If we let:

$$\varepsilon = -\frac{a_4}{a_2^2}, \quad f(u) = u^2 + \frac{a_6}{a_2 a_4} u^4 + \dots$$

Applying the Lagrangian series (A1.2), we get:

$$u = a_2 x^2 = y' - \frac{a_4}{a_2^2} y'^2 - \frac{a_6}{a_2^3} y'^3 + \frac{2a_4^2}{a_2^4} y'^3 + \frac{2a_4^2}{a_2^4} y'^3 + \dots \tag{A1.13}$$

or:

$$x = \sqrt{\frac{u}{a_2}} = b_{1/2} y'^{1/2} + b_{3/2} y'^{3/2} + b_{5/2} y'^{5/2} + \dots \tag{A1.14}$$

where:

$$b_{1/2} = \frac{1}{\sqrt{a_2}}, \quad b_{3/2} = -\frac{a_4}{2a_2^2 \sqrt{a_2}}, \quad b_{5/2} = \frac{7a_4^2}{8a_2^4 \sqrt{a_2}} - \frac{a_6}{2a_2^3 \sqrt{a_2}}$$

For example, the cosine power series is:

$$\cos x = 1 - \frac{x^2}{2} + \frac{x^4}{24} - \frac{x^6}{720} + \dots \tag{A1.15}$$

Its inverse transformation is:

$$x = \sqrt{2}(1 - \cos x)^{1/2} + \frac{\sqrt{2}}{12}(1 - \cos x)^{3/2} + \frac{3\sqrt{2}}{60}(1 - \cos x)^{5/2} + \dots \tag{A1.16}$$

or:

$$x = 2\sin\frac{x}{2} + \frac{1}{3}\sin^3\frac{x}{2} + \frac{3}{2}\sin^5\frac{x}{2} + \dots$$

Letting $x = \frac{\pi}{2}$, from the above equation we get:

$$\frac{\pi}{2} = \sqrt{2} + \frac{\sqrt{2}}{12} + \frac{3\sqrt{2}}{60} + \dots \tag{A1.17}$$

A1.2   INVERSE TRANSFORMATION OF THE ALGEBRAIC POWER SERIES OF DIMENSION TWO

Suppose the two-variable function is:

$$x = a_{10}u + a_{20}u^2 + a_{02}v^2 + a_{30}u^3 + a_{12}uv^2 + a_{40}u^4 + a_{22}u^2v^2 + a_{04}v^4 \tag{A1.18}$$

$$y = b_{01}v + b_{20}u^2 + b_{02}v^2 + b_{30}u^3 + b_{12}uv^2 + b_{40}u^4 + b_{22}u^2v^2 + b_{04}v^4 \tag{A1.19}$$

Then, its inverse transformation formula is (Butekovich 1964):

$$u = \frac{1}{a_{10}}x - \frac{a_{20}}{a_{10}^3}x^2 - \frac{a_{02}}{a_{10}b_{01}^2}y^2 + \left(\frac{2a_{20}^2}{a_{10}^5} - \frac{a_{30}}{a_{10}^4}\right)x^3$$

$$+ \left(\frac{2a_{20}a_{02}}{a_{10}^3b_{01}^2} + \frac{2a_{02}b_{11}}{a_{10}^2b_{01}^3} - \frac{a_{12}}{a_{10}^2b_{01}^2}\right)xy^2 + \left(-\frac{5a_{20}^3}{a_{20}^7} + \frac{5a_{20}a_{30}}{a_{10}^6} - \frac{a_{40}}{a_{10}^5}\right)x^4$$

$$+ \left(-\frac{6a_{20}^2a_{02}}{a_{10}^5b_{01}^2} - \frac{6a_{20}a_{02}b_{11}}{a_{10}^3b_{01}^4} + \frac{3a_{02}a_{30}}{a_{10}^4b_{01}^2} + \frac{3a_{20}a_{12}}{a_{10}^4b_{01}^2} + \frac{2a_{02}b_{21}}{a_{10}^3b_{01}^4} + \frac{2a_{12}b_{11}}{a_{10}^3b_{01}^3}\right.$$

$$\left. - \frac{a_{22}}{a_{10}^3b_{01}^2}\right)x^2y^2 + \left(-\frac{a_{20}a_{02}^2}{a_{10}^3b_{01}^4} - \frac{2a_{02}^2b_{11}}{a_{10}^2b_{01}^5} + \frac{a_{02}a_{12}}{a_{10}^2b_{01}^4} + \frac{2a_{02}b_{03}}{a_{10}b_{01}^5} - \frac{a_{04}}{a_{10}b_{01}^4}\right)y^4$$

(A1.20)

$$v = \frac{1}{b_{01}}y - \frac{b_{11}}{a_{10}b_{01}^2}xy + \left(\frac{a_{20}b_{11}}{a_{10}^3b_{01}^2} + \frac{b_{11}^2}{a_{10}^2b_{01}^3} - \frac{b_{21}}{a_{10}^2b_{01}^2}\right)x^2y + \left(\frac{a_{02}b_{11}}{a_{10}b_{01}^4} - \frac{b_{03}}{b_{01}^4}\right)y^3$$

$$+ \left(-\frac{2a_{20}^2b_{11}}{a_{10}^5b_{01}^2} - \frac{2a_{20}b_{11}^2}{a_{10}^4b_{01}^3} + \frac{a_{30}b_{11}}{a_{10}^4b_{01}^2} + \frac{2a_{20}b_{21}}{a_{10}^4b_{01}^2} - \frac{b_{31}}{a_{10}^4b_{01}^2} - \frac{b_{11}^3}{a_{10}^3b_{01}^4}\right.$$

(A1.21)

$$\left. + \frac{2b_{11}b_{21}}{a_{10}^3b_{01}^3}\right)x^3y + \left(-\frac{2a_{20}a_{02}b_{11}}{a_{10}^3b_{01}^4} - \frac{4a_{02}b_{11}}{a_{10}^2b_{01}^5} + \frac{5a_{02}b_{21}}{a_{10}^2b_{01}^4} + \frac{a_{12}b_{11}}{a_{10}^2b_{01}^4}\right)xy^3$$

Constant Coefficient Tables of Zone Transformation for the Gauss–Krüger Projection

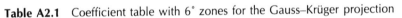

Table A2.1 Coefficient table with 6° zones for the Gauss–Krüger projection

No.	B_0	Coordinate of point 0		i	a_i	b_i
1	2°	x_0	22.14587111	1	0.999993306732	0.3658065747E-2
		y_0	33.39153100	2	−0.45299202 E-6	0.82554819E-4
		X_0	22.14587111	3	−0.68020540E-8	−0.15993464E-8
		Y_0	−33.39153100	4	0.47961634E-12	−0.18054002E-10
2	6°	x_0	66.43918005	1	0.999939977931	0.10956079030E-1
		y_0	33.22963328	2	−0.13499393E-5	0.82133596E-4
		X_0	66.43918005	3	−0.66428356E-8	−0.47855337E-8
		Y_0	−33.22963328	4	0.14352963E-11μ	−0.17912782E-10
3	10°	x_0	110.73715964	1	0.999834366798	0.18199840717E-1
		y_0	32.90659273	2	−0.22199812E-5	0.81294035E-4
		X_0	110.73715964	3	−0.63272409E-8	−0.79352240E-8
		Y_0	−32.90659273	4	0.23718804E-11	−0.17632062E-10
4	14°	x_0	155.04282554	1	0.999678545570	0.25353519216E-1
		y_0	32.42391689	2	−0.30458069E-5	0.80041890E-4
		X_0	155.04282554	3	−0.58628675E-8	−0.11024985E-7
		Y_0	−32.42391689	4	0.32405744E-11	−0.17207790E-10
5	18°	x_0	199.35904799	1	0.999475570487	0.32381786334E-1
		y_0	31.78386141	2	−0.38110588E-5	0.78385601E-4
		X_0	199.35904799	3	−0.52599311E-8	−0.14033263E-7
		Y_0	−31.78386141	4	0.40339953E-11	−0.16623036E-10
6	22°	x_0	243.68849719	1	0.999229419803	0.39250008288E-1
		y_0	30.98942321	2	−0.45006582E-5	0.76336321E-4
		X_0	243.68849719	3	−0.45311203E-8	−0.16939713E-7
		Y_0	−30.98942321	4	0.47521986E-11	−0.15979995E-10

Table A2.1 *cont'd*

No.	B_0	Coordinate of point 0		i	a_i	b_i
7	26°	x_0	288.03359214	1	0.998944915199	0.45924431550E-1
		y_0	30.04433097	2	-0.51011063E-5	0.73907498E-4
		X_0	288.03359214	3	-0.36924951E-5	-0.19725060E-7
		Y_0	-30.04433097	4	0.53486104E-11	-0.15229373E-10
8	30°	x_0	332.39645391	1	0.998627624893	0.52372362321E-1
		y_0	28.95303312	2	-0.56007647E-5	0.71114758E-4
		X_0	332.39645391	3	-0.27621354E-8	-0.22373228E-7
		Y_0	-28.95303312	4	0.58406612E-11	-0.14344081E-10
9	34°	x_0	376.77886411	1	0.998283752981	0.58562331622E-1
		y_0	27.72068286	2	-0.59900797E-5	0.67975720E-4
		X_0	376.77886411	3	-0.17591830E-8	-0.24872004E-7
		Y_0	-27.72068286	4	0.62171379E-11	-0.13437917E-10
10	38°	x_0	421.18222944	1	0.997920016479	0.64464248704E-1
		y_0	26.35312011	2	-0.62617325E-5	0.64509652E-4
		X_0	421.18222944	3	-0.70468786E-9	-0.27210560E-7
		Y_0	-26.35312011	4	0.64471761E-11	-0.12500223E-10
11	42°	x_0	465.60755366	1	0.997543512630	-.70049548935E-1
		y_0	24.85685018	2	-0.64107919E-5	0.60737143E-4
		X_0	465.60755366	3	0.38010483E-9	-0.29378522E-7
		Y_0	-24.85685018	4	0.65475402E-11	-0.11448619E-10
12	46°	x_0	510.05541587	1	0.997161579990	0.75291320562E-1
		y_0	23.23901902	2	-0.64347732E-5	0.5680064E-4
		X_0	510.05541587	3	0.14736656E-8	-0.31371428E-7
		Y_0	-23.23901902	4	0.65334404E-11	-0.10457412E-10
13	50°	x_0	554.52595820	1	0.996781653881	0.80164421784E-1
		y_0	21.50738459	2	-0.63336199E-5	0.52361063E-4
		X_0	554.52595820	3	0.25542803E-8	-0.33183141E-7
		Y_0	-21.50738459	4	0.63633542E-11	-0.93853813E-11
14	54°	x_0	599.01888057	1	0.996411122226	0.84645589601E-1
		y_0	19.67028469	2	-0.61097381E-5	0.47803408E-4
		X_0	599.01888057	3	0.3601412E-8	-0.34813602E-7
		Y_0	-19.67028469	4	0.61111116E-11	-0.83937301E-11
15	58°	x_0	643.23798931	1	0.999014190959	0.44391965261E-1
		y_0	8.86963269	2	-0.14464712E-5	0.21683423E-4
		X_0	643.23798931	3	0.11496013E-8	-0.18237187E-7
		Y_0	-8.86963269	4	0.14637180E-11	-0.41069370E-11

Table A2.2 Coefficient table with 3° zones for the Gauss–Krüger projection

No.	B_0	Coordinate of point 0		i	a_i	b_i
1	2°	x_0	22.12296702	1	0.999998328971	0.1827757903E-2
		y_0	16.69001710	2	−0.11320594E-6	0.41291790E-4
		X_0	22.12296702	3	−0.17017025E-8	−0.78020767E-9
		Y_0	−16.69001710	4	0.11696199E-12	−0.88132764E-11
2	6°	x_0	66.37091253	1	0.999985015106	0.5474331925E-2
		y_0	16.60920758	2	−0.33737994E-6	0.41085241E-4
		X_0	66.37091253	3	−0.16621615E-8	−0.23357640E-8
		Y_0	−16.60920758	4	0.35047659E-12	−0.87584106E-11
3	10°	x_0	110.62485433	1	0.999958647060	0.9094126406E-2
		y_0	16.44796041	2	−0.55487544E-6	0.40673388E-4
		X_0	110.62485433	3	−0.15840555E-8	−0.38767211E-8
		Y_0	−16.44796041	4	0.59330318E-12	−0.86384233E-11
4	14°	x_0	154.88866345	1	0.999919738864	0.12669438729E-1
		y_0	16.20701850	2	−0.76139258E-6	0.40058690E-4
		X_0	154.88866345	3	−0.14687771E-8	−0.53921667E-8
		Y_0	−16.20701850	4	0.81001871E-12	−0.84696694E-11
5	18°	x_0	199.16602338	1	0.999869049549	0.16182790611E-1
		y_0	15.88749395	2	−0.95286173E-6	0.39244802E-4
		X_0	199.16602338	3	−0.13189254E-8	−0.68742629E-8
		Y_0	−15.88749395	4	0.10079714E-11	−0.82406859E-11
6	22°	x_0	243.46035979	1	0.9998075674057	0.19617016726E-1
		y_0	15.49086535	2	−0.11255283E-5	0.38236585E-4
		X_0	243.46035979	3	−0.11375806E-8	−0.83134592E-8
		Y_0	−15.49086535	4	0.11868284E-11	−0.79677930E-11
7	26°	x_0	287.77477471	1	0.999736490821	0.22955348800E-1
		y_0	15.01897387	2	−0.12760102E-5	0.37039871E-4
		X_0	287.77477471	3	−0.92885876E-9	−0.97009351E-8
		Y_0	−15.01897387	4	0.13402612E-11	−0.76811890E-11
8	30°	x_0	332.11198657	1	0.999657205390	0.26181500774E-1
		y_0	14.47401826	2	−0.14013934E-5	0.35661610E-4
		X_0	332.11198657	3	−0.69684702E-9	−0.11030509E-7
		Y_0	−14.47401826	4	0.14578893E-11	−0.73022699E-11
9	34°	x_0	376.47427695	1	0.999571255874	0.29279748353E-1
		y_0	13.85854846	2	−0.14992639E-5	0.34109625E-4
		X_0	376.47427695	3	−0.44646597E-9	−0.12295098E-7
		Y_0	−13.85854846	4	0.15488166E-11	−0.69116934E-11
10	38°	x_0	420.86344551	1	0.999480317211	0.32235004678E-1
		y_0	13.17545774	2	−0.15677596E-5	0.32392586E-4
		X_0	420.86344551	3	−0.18297896E-9	−0.13487090E-7
		Y_0	−13.17545774	4	0.16216472E-11	−0.64919181E-11
11	42°	x_0	465.28077402	1	0.999386159610	0.35032896265E-1
		y_0	12.42797317	2	−0.16056109E-5	0.30519963E-4
		X_0	465.28077402	3	0.88931528E-10	−0.14605102E-7
		Y_0	−12.42797317	4	0.16410206E-11	−0.60190741E-11

Table A2.2 *cont'd*

No.	B_0	Coordinate of point 0		i	a_i	b_i
12	46°	x_0	509.72699917	1	0.999290617084	0.37659828668E-1
		y_0	11.61964435	2	-0.16121593E-5	0.28501867E-4
		X_0	509.72699917	3	0.36328628E-9	-0.15641376E-7
		Y_0	-11.61964435	4	0.16413537E-11	-0.55386806E-11
13	50°	x_0	554.20229717	1	0.999195549011	0.401030534744E-1
		y_0	10.75433006	2	-0.15873468E-5	0.26348890E-4
		X_0	554.20029717	3	0.63513283E-9	-0.16594533E-7
		Y_0	-10.75433006	4	0.16102674E-11	-0.51172399E-11
14	54°	x_0	598.70627744	1	0.9991028054237	0.42350726890E-1
		y_0	9.83618308	2	-0.15317326E-5	0.24072243E-4
		X_0	598.70627744	3	0.89912788E-9	-0.17460415E-7
		Y_0	-9.83618308	4	0.15407675E-11	-0.45241588E-11
15	58°	x_0	643.23798931	1	0.999014190959	0.44391965261E-4
		y_0	8.86963269	2	-0.14464712E-5	0.21683423E-4
		X_0	643.23798931	3	0.11496013E-8	-0.18237187E-7
		Y_0	-8.86963269	4	0.14637180E-11	-0.41069370E-11

Coordinate Units 10 km = 10^4 m.

Examples of Numerical Transformation

A3.1 CONVENTIONAL AND LEAST-SQUARES POLYNOMIAL APPROXIMATIONS

Example 1: Direct numerical transformation for bivariate cubic approximating polynomials

A map showing several discrete points for the Mercator and the conformal conic projections is shown in Figure A3.1 and the coordinates are known as in Table A3.1. A coordinate transformation from the Mercator projection to the conformal conic projection is applied. The coordinates in Tables A3.1 and A3.2 are from Tables 47 and 7 in Yang (1979b).

Using (6.1.13) as the program, select the coordinates of 10 common points as in Table A3.1, according to the elimination method for principal elements; the value of coefficients a_i, b_i are obtained as shown in Table A3.3. Using these coefficients and formula (6.1.10), coordinates for the Mercator projection in Table A3.2 are converted to the coordinates of transformed points as shown in Table A3.4. Comparing the computed results in Table A3.4 with the check values of Table A3.2, the deviation of coordinates is seen to be less than 0.002 cm; that is, the precision would satisfy mapping requirements perfectly.

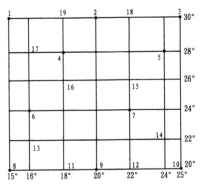
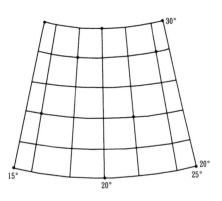

Figure A3.1 Distributive map of common points and checkpoints in the transformed area

333

Table A3.1 The coordinates of the common points

No.	B	l	y_M	x_M	y_c	x_c
1	30°	15°	348.225	166.982	259.803	141.982
2	30°	20°	348.225	222.643	268.243	188.734
3	30°	25°	348.225	278.303	279.057	234.994
4	28°	18°	322.898	200.378	243.076	174.081
5	28°	24°	273.608	267.171	255.486	231.092
6	24°	16°	273.608	178.114	196.254	162.083
7	24°	22°	225.846	244.907	208.017	221.982
8	20°	15°	225.846	166.982	150.546	158.840
9	20°	20°	225.846	222.643	159.989	211.143
10	20°	25°	225.846	278.303	172.086	262.896

Table A3.2 The coordinates of the checkpoints

No.	B	l	y_M	x_M	y_c	x_c
11	20°	18°	225.846	200.378	155.891	190.280
12	20°	22°	225.846	244.907	164.510	231.918
13	22°	16°	249.557	178.114	174.322	165.697
14	22°	24°	249.557	267.171	191.185	247.164
15	26°	22°	298.041	244.907	229.547	217.064
16	26°	18°	298.041	200.378	221.480	178.094
17	28°	16°	322.898	178.114	239.718	154.922
18	30°	22°	348.225	244.907	272.285	207.304
19	30°	18°	348.225	200.378	264.581	170.085

Table A3.3 The value of the coefficients a_i, b_i

a_i		b_i	
i = 0, 1, 2, 3, 4	i = 5, 6, 7, 8, 9	i = 0, 1, 2, 3, 4	i = 5, 6, 7, 8, 9
−101.822789	5.177691551E-04	5.513249808E-01	−1.216096761E-05
1.176462255	1.299514318E-07	−6.838182612E-03	−3.444473361E-08
7.4749412E-03	6.704350788E-08	1.17435471	3.791690456E-07
−5.320500501E-04	−3.692826561E-07	2.899227858E-05	7.635762011E-08
−4.18518308E-05	−8.995900958E-09	−1.062337665E-03	−1.402809059E-07

Table A3.4 The coordinates of the transformed points

No.	y_T	x_T	No.	y_T	x_T
11	155.8926652	190.2789979	16	221.479775	178.0936078
12	164.5098241	231.9182147	17	239.7180662	154.9228053
13	174.3222631	165.6956719	18	272.284225	207.3050673
14	191.1868958	247.1637636	19	264.5814784	170.08351
15	229.5466789	217.065545			

Example 2: Direct numeric transformation for bivariate cubic polynomials using the least-squares method

A map showing the distribution of several common points for the Mercator and conformal conic projections is shown in Figure A3.2 and the coordinates are as in Table A3.5. Coordinate transformation from the Mercator projection to the conformal conic projection is computed.

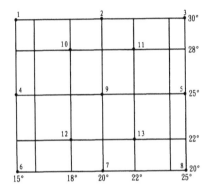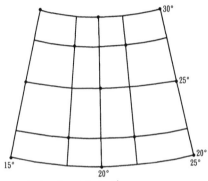

Figure A3.2 Distributive map of common points and checkpoints in the transformed area

Table A3.5 The coordinates of the common points

No.	B	l	y_M	x_M	y_C	x_C
1	30	15	348.225	166.982	259.803	141.982
2	30	20	348.225	222.643	268.243	188.734
3	30	25	348.225	278.303	279.507	234.994
4	25	15	285.774	166.982	205.581	150.348
5	25	25	285.774	278.303	225.969	248.842
6	20	15	225.846	166.982	150.546	158.840
7	20	20	225.846	222.643	159.989	211.143
8	20	25	225.846	278.303	172.086	262.896
9	25	20	285.774	222.643	214.518	199.855
10	28	18	322.898	200.378	243.076	174.081
11	28	22	322.898	244.907	250.962	212.174
12	22	18	249.557	200.378	177.913	186.188
13	22	22	249.557	244.907	186.347	226.931

Using (6.1.21) as the program, select the coordinates of 13 points from Table A3.5, and according to the elimination method for principal elements; the values of coefficients a_i, b_i are obtained as shown in Table A3.6 (origin translated to point no. 9 when computing). Using these coefficients and formula (6.1.10), coordinates of the Mercator projection in Table A3.2 are converted to the coordinates of the transformed points as shown in Table A3.7. Comparing these computed results with the check values in Table A3.2, the deviation of the coordinates is seen to be less than 0.001 cm.

Table A3.6 The values of the coefficients a_i, b_i

a_i		b_i	
$i = 0, 1, 2, 3, 4$	$i = 5, 6, 7, 8, 9$	$i = 0, 1, 2, 3, 4$	$i = 5, 6, 7, 8, 9$
−6.922086206E-04	4.060846272E-04	6.465176635E-04	−8.40141414E-05
8.851432946E-01	5.226695925E-07	−1.832225143E-01	−2.688476117E-08
1.832373933E-01	7.815110657E-08	8.851527084E-01	3.692755243E-07
−4.05623902E-04	−3.692376125E-07	8.3926367768E-05	7.637222146E-08
−1.679907524E-04	−2.93497217E-08	−8.116079914E-04	−1.220634725E-07

Table A3.7 The coordinates of the transformed points

No.	y_T	x_T	No.	y_T	x_T
11	155.891718	190.2800529	15	229.5472491	217.064725
12	164.511216	231.9171853	16	221.479776	178.0933068
13	174.3216429	165.6960652	17	239.7187561	154.9218358
14	191.1859582	247.1644421	18	272.2856288	207.304003
			19	264.5805056	170.0845822

A3.2 REFINING PARALLELS AND MERIDIANS

Example 3: Refining the coordinates of parallels for an oblique pseudo-azimuthal projection by cubic spline interpolation

The coordinates of several points along parallels for an oblique pseudo-azimuthal projection are as shown in Table A3.8. According to (6.2.11) and (6.2.8), select coordinates of nodes, using the accelerated method to obtain the first derivatives $\{m_i\}$ at the nodes, then use (6.2.1) to refine the coordinates. Comparing the computed results in Tables A3.9–A3.12 with the look-up values of Table 6 in Yang (1979b), the deviation of the coordinates is less than 0.001 cm.

Table A3.8 Coordinates at parallel 0° for an oblique azimuthal projection

I	0°	1°	2°	3°	4°	5°
λ	5°	10°	15°	20°	25°	30°
x	12.004	14.895	19.534	26.065	34.683	45.591
y	610.113	668.048	725.950	783.943	842.064	900.272

Table A3.9 The values of the first derivative at the nodes

I	0	1	2	3	4	5
m_i	2.422813333	1.946190351	1.508025263	1.111108597	7.495403508E-01	0.40873

Table A3.10 Refining the y coordinate

y	12.004	14.895	19.534	26.065	34.683	45.591
y_i	12.44632474	15.67938444	20.68304946	27.61501771	36.67534369	
	12.95613823	16.53426492	21.90905806	29.25014282	38.76092265	
	13.53388935	17.46085318	23.21364194	30.97205907	40.94112977	
	14.18002698	18.46036095	24.59841721	32.78245022	43.21735792	

Table A3.11 The values of the first derivative at the nodes

I	1	2	3	4	5	6
m_i	11.64402667	11.63546403	11.61151719	11.5868672	11.57801403	11.60327667

Table A3.12 Refining the x coordinate

x	610.113	668.048	725.95	783.943	842.064	900.272
x_i	621.711854	679.625838	737.539024	795.557280	853.701284	
	622.030303	680.203977	738.132552	796.176861	854.341617	
	623.888282	681.783398	739.730867	797.801402	855.984107	
	656.4693553	714.3650792	772.3342557	830.4305618	888.6278652	

A3.3 CONVENTIONAL AND LEAST-SQUARES CONFORMAL POLYNOMIAL APPROXIMATIONS

Example 4: Numerical transformation from the Mercator projection to the conformal conic projection

The transformed region is shown in Figure A3.3; the coordinates of the expansion point 0 and four discrete points are as shown in Tables A3.13 and A3.14. Following (7.1.5), select the coordinates of the expansion point 0 and four discrete points from

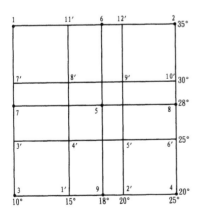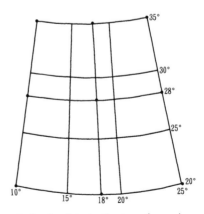

Figure A3.3 Distributive map of common points and checkpoints in the transformed area

Table A3.13 The coordinates of common points

No.	B	l	y_M	x_M	y_C	x_C
0	28°	18°	322.898	200.378	243.076	174.081
1	35°	10°	323.945	201.321	244.918	175.314
2	35°	25°	324.945	202.303	245.737	176.254
3	20°	10°	325.846	203.321	246.785	177.124
4	20°	25°	225.846	278.303	172.086	262.896

Table A3.14 Coordinates of checkpoints

No.	B	l	y_M	x_M	y_C	x_C
1	20°	15°	225.846	166.982	150.546	158.840
2	20°	20°	226.846	167.643	151.989	159.143
3	25°	10°	227.774	168.321	152.180	160.450
4	25°	15°	228.774	169.982	153.581	161.348
5	25°	20°	229.774	170.643	154.518	162.855
6	25°	25°	230.774	171.303	155.969	163.842
7	30°	10°	231.225	172.321	156.759	164.861
8	30°	15°	232.225	173.982	157.803	165.982
9	30°	20°	233.225	174.643	158.243	167.734
10	30°	25°	234.225	175.303	159.057	168.994
11	35°	15°	235.945	176.982	160.609	169.680
12	35°	20°	413.945	222.643	321.556	177.698

Table A3.13; then the values of the coefficients a_i, b_i are computed by the elimination method for the principal elements. Now the coordinate transformation can be effected by formula (5.1.36).

To compare the coordinates in Tables A3.13 and A3.14, see Tables 47 and 7 in Yang (1979b).

Table A3.15 The values of the coefficients a_i, b_i

I	1	2	3	4
a_i	8.589610456E-01	-3.937283848E-04	1.204794938E-07	-3.045562737E-11
b_i	-1.595957952E-01	7.314752736E-05	-2.240586908E-08	4.744945641E-12

Table A3.16 The coordinates of the transformed points

No.	x_T	y_T	No.	x_T	y_T
1	150.5459933	158.8402148	7	253.7579474	94.86108663
2	151.9878693	159.1435226	8	254.8025623	95.9826594
3	152.1794862	160.4512822	9	255.242602	96.7345895
4	153.5799033	161.3490647	10	256.0558672	97.9944769
5	154.5171792	162.8557868	11	257.6085071	98.6806123
6	225.9678754	248.8411278	12	321.5551393	177.698635

Example 5: Least-squares numerical transformation from the Mercator projection to the conformal conic projection

A map of the distribution of the discrete points for the transformed region is shown in Figure A3.3; for the coordinates of expansion point 0 and discrete points 1, 2, 3, 4 see Table A3.13. Coordinates of the other discrete points are as shown in Table A3.17.

According to (7.1.16), select the coordinates of the expansion point 0 and nine discrete points from Tables A3.13 and A3.17; then the values of coefficients a_i, b_i are computed by the elimination method for the principal elements as shown in Table A3.18. Coordinates of the transformed points are computed according to (5.1.36) as shown in Table A3.19.

Table A3.17   The coordinates of common points

No.	B	l	y_M	x_M	y_C	x_C
5	28°	18°	322.898	200.378	243.076	174.081
6	35°	18°	323.945	201.378	244.108	175.140
7	28°	10°	324.898	202.321	245.000	176.089
8	28°	25°	325.898	203.303	246.892	177.515
9	20°	18°	225.846	200.378	155.891	190.280

Table A3.18   The values of the coefficients a_i, b_i

I	1	2	3	4
a_i	8.589630203E-01	-3.937234819E-04	1.204665654E-07	-2.997152463E-11
b_i	-1.595956278E-01	7.314841665E-05	-2.24163596E-08	4.594893155E-12

Table A3.19   The coordinates of the transformed points

No.	y_T	x_T	No.	y_T	x_T
1	150.5458695	158.8402377	7	253.7580195	94.86091257
2	151.9877559	159.1434907	8	254.8026166	95.9825912
3	152.1793692	160.4510717	9	255.2426481	96.7346435
4	153.5798299	161.349007	10	256.0558767	97.9946355
5	154.5171074	162.855815	11	257.6087129	98.6804993
6	225.9677641	248.8412664	12	321.5553785	177.6987315

A3.4   THE FINITE ELEMENT METHOD OF NUMERICAL TRANSFORMATION FOR A CONFORMAL PROJECTION

Example 6: Transformation from the Mercator projection to a conformal projection

The transformed region is shown in Figure A3.4. The triangular elements in the figure are numbered 1–24; 1–6 are numbers of the internal transformation points; 7–20 are the numbers of boundary points. Nodes 1–20 of the Mercator projection are obtained from

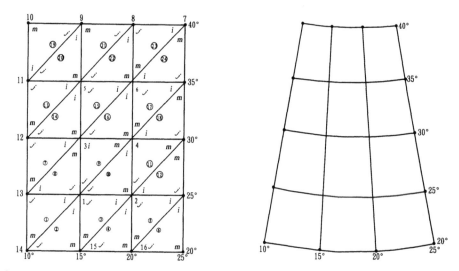

Figure A3.4 The transformed region and triangle units

Table 47 in Yang (1979b). Coordinates of points 7–20 for the conformal conic are obtained from Table 7 of the same reference book. Using the finite element procedure for the calculation, the data of the example are as follows:

2:DATA 1, 13, 14, 1, 14, 15, 2, 1, 15, 2, 15, 16, 18, 2, 16, 18, 16, 17

3:DATA 3, 12, 13, 3, 13, 1, 4, 3, 1, 4, 1, 2, 4, 2, 19, 2, 18, 19

4:DATA 5, 11, 12, 5, 12, 3, 6, 5, 3, 6, 3, 4, 20, 6, 4, 20, 2, 19

5:DATA 9, 10, 11, 9, 11, 5, 8, 9, 5, 8, 5, 6, 7, 8, 6, 7, 6, 20

6:DATA 285.774, 166.982, 285.774, 222.643, 252, 166.982, 348.225, 222.643

7:DATA 413.945, 166.982, 413.945, 222.643, 856, 278.303, 483.856, 222.643

8:DATA 483.856, 166.982, 483.856, 111.321, 945, 111.321, 348.225, 111.321

9:DATA 285.774, 111.321, 225.846, 111.321, 225.846, 166.982, 225.846, 222.643

10:DATA 225.846, 278.303, 285.774, 278.303, 348.225, 278.303, 413.945, 278.303

11:DATA 348.400, 207.518, 374.851, 166.666, 367.398, 125.381, 362.060, 83.769

12:DATA 307.918, 89.314, 253.759, 94.861, 199.180, 100.450, 143.785, 106.124

13:DATA 150.546, 158.840, 159.989, 211.143, 172.086, 262.896, 225.969, 248.842

14:DATA 279.057, 234.994, 331.737, 221.254

Illustration: in the program the single parameter K0 represents the number of triangular elements; N is the number of the transformation points; L is the number of boundary points. The array (3, K0) stores the total coding number of every triangle element's vertices; X(N + L), Y(N + L) store the node's coordinates; XT(L), YT(L) store the new projection coordinates of the boundary points.

The order of the DATA is N(3, K0), X(N + L), Y(N + L), XT(L), YT(L). The results are obtained as shown in Table A3.20. Comparing the above computing results with the values of the table, the deviation of the coordinates is less than 0.001 cm.

Table A3.20 The coordinates of the transformed points

No.	1	2	3	4	5	6
y_T	205.5810102	214.5186288	259.8042958	268.2443987	313.6103409	321.5569191
x_T	150.3484996	199.855605	141.9822599	188.7341845	133.6801729	177.6984745

A3.5 THE THIRD TYPE OF COORDINATE TRANSFORMATION FOR A CONFORMAL PROJECTION

Example 7: Coordinate transformation of the points for the graticule of meridians and parallels along the map border

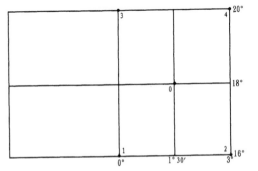

Figure A3.5 The region for Gauss–Krüger projection

Chinese topographic maps adopt the Gauss–Krüger projection. The requirement is to draw a square grid for the 1:100 000 or larger scale map. Then the coordinates of intersections between the graticule of meridians and parallels along the map border, and thus the coordinates of the lines, can be found.

Now we use the numerical method to solve the third type of coordinate transformation for the Gauss–Krüger projection along the margins. The third type of coordinate transformation for points on the marginal lines of a sheet of a topographic map, shown in Figure A3.5, is applied.

Given values of (B, x), to obtain (l, y) The Gauss–Krüger coordinates of discrete points shown in Figure A3.6 are known, as in Table A3.21. The values (B, x) of northern and southern parallel on the margin are known, as in Table A3.22.

Table A3.21 The coordinates of common points

No.	B	l	y	x
0	18°	1°30′	1 991 660.2	158 874.9
1	16°	0	1 769 674.8	0
2	16°	3°	1 771 994.4	321 233.9
3	20°	0	2 212 405.7	0
4	20°	3°	2 215 219.6	314 057.2

Table A3.22 The values (B, x) of the northern and southern parallel on the margin

No.	1	2	3	4	5	6
B	16	16	18	18	20	20
x	107 040.8	214 109.5	105 911.2	211 848.6	104 653.0	209 330.5

According to the coordinates of the expansion point 0 and the nine common points in Table A3.21, using (8.3.1) we obtain the values of coefficients a_i, b_i as shown in Table A3.23. From (8.3.6) and (8.3.2) we obtain, l, y as shown in Table A3.24.

Table A3.23 The values of the coefficients a_i, b_i

i	1	2	3	4
a_i	606.9704773	−93.8999979	−82.47130676	34.99539634
b_i	−4.911566955	−6.47154529	3.66981332	1.051429855

Table A3.24 The values (l, y) of the northern and southern parallel on the margin

No.	1	2	3	4	5	6
l	5.959999775E-01	2.00000012	1.000000292	2.000000193	1.000000253	2.000000217
y	176.9932323	177.070515	199.1303072	199.2160351	221.2718041	221.3655602

Given values of (l, y), to obtain (B, x) The values (l, y) of the eastern and western meridians on the margin are known as in Table A3.25. According to the coordinates of the expansion point 0 and the four common points in Table A3.21, use (8.3.2) to obtain the values of coefficients a_i, b_i as shown in Table A3.26. From (8.3.2), (3.7.18) and (8.3.11), we obtain the value of (B, x) as shown in Table A3.27.

Illustration: the unit of the solution coordinates y, x is 10 km; the above transform results agree with the check values.

Table A3.25 The values (l, y) of the eastern and western meridians on the margin

No.	1	2	3	4	5	6
l	0	0	2	2	3	3
y	1 991 017.5	2 101 705.6	1 881 428.0	2 102 902.8	1 882 788.5	2 104 400.7

Table A3.26 The values of the coefficients a_i, b_i

i	1	2	3	4
a_i	1.647418194E-03	4.190491468E-07	8.201829812E-10	4.840318127E-13
b_i	1.333080248E-05	3.912068633E-08	3.074451037E-11	6.239797299E-14

Table A3.27 The values of (B, x) of the eastern and western meridian on the margin

No.	1	2	3	4	5	6
B	18.00000019	18.59599983	17.00000015	19.00000012	17.00000014	18.59599993
x	−0.00000969	−0.00000539	21.30113402	21.06214782	31.95847671	31.59958974

A3.6   COORDINATE TRANSFORMATION FOR THE GAUSS–KRÜGER PROJECTION BETWEEN ADJACENT ZONES

Example 8

Given the Gauss–Krüger coordinates of a point in the western zone $B = 28°$, $l = 3°20'$ in a 6° zone system as $y = 3\ 102\ 979.1905$, $x = 327\ 982.5948$ m, compute its coordinates in the adjacent eastern zone (6° zone system).
   From column $B_0 = 30°$ in Table A2.1 we look up DATA as follows:

   46:DATA 332.39645391, 28.95303312, 332.39645391, –28.95303312

   48:DATA 0.998627624893, –0.56007674E-5, –0.27621354E-8, 0.58406612E-11

   49:DATA 0.52372362321E-1, 0.71114758E-4, –0.22373228E-7, –0.14344081E-10

The result is: $y_T = 3\ 101\ 364.4713$ $x_T = -262\ 356.0705$.
   Coordinate transformation from the east to the west zone needs x_0, X_0 and b_i for the inverse symbol.

Example 9

Given the Gauss–Krüger coordinates of a point in the western zone $B = 28°$, $l = 1°40'$ in the 3° zone system as $y = 3\ 099\ 616.5660$, $x = 163\ 952.2283$ m, compute its coordinates in the adjacent eastern zone (3° zones).
   From column $B_0 = 30°$ in Table A2.2 we look up DATA as follows:

   46:DATA 332.11198657, 14.47401826, 332.11198657, –14.47401826

   48:DATA 0.999657205390, –0.14013934E-5, –0.69684702E-9, 0.14578893E-11

   49:DATA 0.26181500774E-1, 0.35661610E-4, –0.11030509E-7, –0.73022699E-11

The result is: $y_T = 3\ 099\ 213.4049$, $x_T = -131\ 158.0319$. The accuracy of the above result is within 1–2 mm.
   The following examples are zone transformations of constant coefficients for several discrete points. The maps of several expansion points and discrete points for two adjacent 6° zones and two adjacent 3° zones are shown in Figures 9.3 and 9.4.

Example 10

Given the Gauss–Krüger coordinates of a point in the eastern zone $B = 28°$, $l = -1°20'$ in 6° zones as $y = 3\ 099\ 213.4061$ m, $x = -131\ 158.0320$ m, compute its coordinates in the adjacent western zone (3° zone system).
   Using the program of zone transformation to compute constant coefficients:

   $y_0 = 332.1119867$, $x_0 = -14.47401825$, $Y_0 = 332.1119867$, $X_0 = 14.47401825$

Table A3.28   The values of the coefficients a_i, b_i

i	1	2	3	4
a_i	9.996572055E-01	-1.401395549	-6.963953833E-10	1.470533138E-12
b_i	-2.61815002e-02	-3.566152827E-05	1.1030603E-08	7.343996444E-12

Calculation result: $y_T = 3\ 099\ 616.5660$, $x_T = 163\ 952.2284$.

Illustration: transformation from the western to the eastern zone WE = 1; from eastern to western zone WE = −1; transformation between adjacent 6° zones KL = 0; transformation between adjacent 3° zones (or 6° ↔ 3°) KL = 1.5.

A3.7 COORDINATE TRANSFORMATION OF ADJACENT GRID LINES

Using the program of coordinate transformations of adjacent grid lines, the example is computed as follows.

Example 11

To solve the value of coordinate transformation between adjacent grid lines for the 1:25 000 scale sheet L-51-97-C-3. The sheet range is B:45° to 45°05′; l: −3° to 2°52′30″.

Table A3.29 The values of the coordinate transformation between adjacent grid lines

B	45°	B	45°05′	B	45°	B	45°05′
l	−2°52′30″	l	−3°	l	−3°	l	2°52′30″
x_1'	737000	x_1'	737 000	y_1'	4 990 000	y_1'	4 990 000
y	4 989 055.598	y	4 998 674.423	x	263 455.4092	x	273 640.5579
x_i	263 885.5284	x_i	264 599.9218	y_i	4 990 001.614	y_i	4 989 245.644
	264 888.2675		265 602.665		4 991 004.366		4 990 248.277
	265 890.995		266 605.3967		4 992 007.119		4 991 250.911
	266 893.7108		267 608.1168		4 993 009.873		4 992 253.546
	267 896.415		268 610.8252		4 994 012.629		4 993 256.183
	268 899.1075		269 613.522		4 995 015.386		4 994 258.82
	269 901.7884		270 616.2072		4 996 018.144		4 995 261.46
	270 904.4577		271 618.8808		4 997 020.904		4 996 264.1
	271 907.1154		272 621.5427		4 998 023.665		4 997 266.742
	272 909.7614		273 624.1931		4 999 026.427		4 998 269.385

Illustration: the values of DATA in the main program can be checked in Table A2.1; y_0, x_0, $Y_0(y_0')$, $X_0(x_0')$ and a_i, b_i can be looked up in the column $B_0 = 46°(= 12 \times 4 − 2)$; x_1', y_1' are the initial values of the vertical and horizontal grid lines, respectively.

Transformation for the Mathematical Elements of Topographic Maps

A4.1 TRANSFORMATION OF GEODETIC COORDINATE SYSTEMS AND POSITION DIFFERENCE OF SHEET CORNERS

The mathematical elements of a topographic map consist mainly of a geodetic coordinate system, a map projection, a map sheet division, an initial meridian, an angle system and the mathematical elements outside the margin, etc. The influential mathematical properties of topographic maps are the geodetic coordinate system and the map projection. The so-called transformation of the mathematical elements of a topographic map is to transform the topographic map to a different coordinate system and map projection to bring them to a uniform system.

Every point on the ground has a unique value of longitude and latitude in a single geodetic coordinate system. The essence of different geodetic coordinate systems is that the single ground point has different numerical values of longitude and latitude.

Suppose the geodetic coordinates of a point in the new system is (B', L') and that in the old system is (B, L) then:

$$\Delta B = B' - B, \quad \Delta L = L' - L \tag{A4.1.1}$$

If the old geodetic coordinate system is being changed into the new one there exists:

$$B' = B + \Delta B, \quad L' = L + \Delta L \tag{A4.1.2}$$

Therefore the essence of the transformation of different geodetic coordinate systems is to obtain the adjustment values ΔB, ΔL from the old geodetic coordinate system.

Establishing a one-to-one relation of the surface of the reference ellipsoid and the projection plane according to a particular projection method forms the mathematical basis of a topographic map. The relation represented by mathematical formula is that:

$$x = f_1(B, L), \quad y = f_2(B, L) \tag{A4.1.3}$$

where B, L are coordinates in the old geodetic coordinate system.

In the new projection plane there are:

$$x' = F_1(B', L'), \quad y' = F_2(B', L') \tag{A4.1.4}$$

Hence to transform map data from different coordinate systems to that of a single coordinate system one needs to reestablish the margin lines on the old map, that is to move the
geographic mesh; and to redraw the grid from the previously used projection coordinates
according to the new margin points, that is to move the vertical and horizontal coordinate
lines. This is also equivalent to moving old map data in order to merge them with new
map data. So the essence of the transformation of the mathematical elements of a topographic map is that of merging the old and new map data together by reference points and
then to redefine the margin lines and grid lines by the new coordinate system so that the
ground points have unique longitudes and latitudes and unique coordinate values in the
same projection plane coordinate system.

A4.2   POSITION DIFFERENCE OF SHEET CORNERS

The transformation from one kind of cartographic document to another can be realized
with the help of a photographic print and this can be pieced together as in conventional
cartographic work. This transformation is essentially a kind of similarity transformation.
Therefore the basics of realizing a transformation of the mathematical elements of a
topographic map in the old system to the new one is to regard a topographic map in
different coordinate systems as being similar within a certain range. Thus we can perform
the transformation of the mathematical elements of a topographic map by moving the
sheet corners and redrawing the grid.

The adjustment values of geodetic coordinates on a topographic map have constant
values ΔB, ΔL and the modified values of the plane coordinates of the corresponding
point in the projection plane are dx, dy. For Gauss–Krüger projection, if selecting the first
term of the differential then:

$$dx = M \cdot \frac{\Delta B''}{\rho''}, \quad dy = N\cos B \cdot \frac{\Delta L''}{\rho''} \tag{A4.1.5}$$

where $\rho'' = 206\ 265''$.

The variation in value of plane coordinates that arises from the change of a geodetic
coordinate system on the topographic map of the old system is generally regarded as a
constant. Therefore when a map figure in the new system is similar to that in the old
system, the sheet corner of the map in the new system can be found directly on the map
of the old system and redrawing of the square grid by the coordinates that are calculated
by the projection formulae of the new system. Then the transformation from one kind of
mathematical element of a topographic map to another can be achieved.

Next we discuss the computation formula for determining the sheet corners of a new
system on the map of the old system.

In order to determine the position of sheet corners in the new system the position
difference from the sheet corners of the old map to the new one needs to be obtained,
generally for the topographic maps that have sheets partitioned by parallels and meridians.

Suppose the geodetic coordinate of the old sheet corner is (B, L). The geodetic coordinate of this point corresponding to the new system is $(B + \Delta B, L + \Delta L)$. The position of
the sheet corner of the new system corresponding to the old system is $(B - \Delta B, L - \Delta L)$
since the geodetic coordinate of the new sheet corner is also (B, L).

Hence the position differences from the old sheet corner to the new one on the map of
the old system are:

$$\Delta x_{corner} = x_{new\ corner}(B - \Delta B, L - \Delta L) - x_{old\ corner}(B, L)$$
$$= x(B, L) - dx - x(B, L) = -dx$$
$$\Delta y_{corner} = y_{new\ corner}(B - \Delta B, L - \Delta L) - y_{old\ corner}(B, L)$$
$$= y(B, L) - dy - y(B, L) = -dy$$
$$\therefore \quad \Delta x_{corner} = -dx, \quad \Delta y_{corner} = -dy \tag{A4.1.6}$$

Formula (A4.1.6) is the general formula of calculating the position difference of sheet corners.

In consideration of the formula (A4.1.5) we have:

$$\Delta y_{corner} = -M\frac{\Delta B''}{\rho''}, \quad \Delta x_{corner} = -N\cos B\frac{\Delta L''}{\rho''} \tag{A4.1.7}$$

The above formula indicates that sheet corners move to north and east when the values of $\Delta x_{corner}, \Delta y_{corner}$ are positive; sheet corners move to south and west when $\Delta x_{corner}, \Delta y_{corner}$ are negative. This is equivalent to noting that sheet corners move to north and east when $\Delta B, \Delta L$ are negative and sheet corners move to south and west when $\Delta B, \Delta L$ are positive.

For the map sheets bounded by a rectangle, generally the position of new sheet corners can be obtained directly on the map of the old system instead of obtaining the position difference of sheet corners, that is:

$$\left. \begin{array}{l} x_{corner} = x(B - \Delta B, L - \Delta L) = x(B, L) - dx \\ y_{corner} = y(B - \Delta B, L - \Delta L) = y(B, L) - dy \end{array} \right\} \tag{A4.1.8}$$

The following steps can carry out the transformation of the mathematical elements of a topographic map when the correction values $\Delta B, \Delta L$ are known.

First, determine the position of the sheet corner of new system by formula (A4.1.6), (A4.1.7) or (A4.1.8). Second, draw the coordinates of sheet corners that are calculated by the projection formula of the new system by the common method of piecing the photographic prints together. Thus the transformation of the mathematical elements of a topographic map from the old system to the new one can be realized.

Next we discuss the formula for calculating the position difference of sheet corners according to the plane coordinate difference $\Delta x, \Delta y$ of a discrete point in the new and old systems.

Given the geodetic coordinate (B, L) of a new sheet corner and the geodetic coordinate $(B + \Delta B, L + \Delta L)$ of the old sheet corner corresponding to the new system, then the position difference of sheet corners calculated by the projection formulae in the new system is:

$$\left. \begin{array}{l} \Delta x_{corner} = x'_{new\ corner}(B, L) - x'_{old\ corner}(B + \Delta B, L + \Delta L) \\ \Delta y_{corner} = y'_{new\ corner}(B, L) - y'_{old\ corner}(B + \Delta B, L + \Delta L) \end{array} \right\} \tag{A4.1.9}$$

We also know the plane coordinate difference of the same point on the ground in the new and old systems is:

$$\left. \begin{array}{l} \Delta x = x'_{new}(B + \Delta B, L + \Delta L) - x_{old}(B, L) \\ \Delta y = y'_{new}(B + \Delta B, L + \Delta L) - y_{old}(B, L) \end{array} \right\} \tag{A4.1.10}$$

where x', y' are the coordinates calculated by the projection formulae in the new system; $\Delta x, \Delta y$ are the variation in the values of squares transformed from the old system to the new system. It is equivalent to the position difference of the origin of the coordinates. Its movement rule is that the origin of the coordinates moves to the south and west when

$\Delta x, \Delta y$ are positive; the origin of coordinates moves to the north and east when $\Delta x, \Delta y$ are negative.

Substituting (A4.1.9) into (A4.1.10) we have:

$$\left.\begin{array}{l} \Delta x_{corner} = x'_{new\ corner}\,(B, L) - [x_{old\ corner}\,(B, L) + \Delta x] = \Delta X - \Delta x \\ \Delta y_{corner} = y'_{new\ corner}\,(B, L) - [y_{old\ corner}\,(B, L) + \Delta y] = \Delta Y - \Delta y \end{array}\right\}$$

Therefore:

$$\Delta x_{corner} = \Delta X - \Delta x, \quad \Delta y_{corner} = \Delta Y - \Delta y \qquad\qquad (A4.1.11)$$

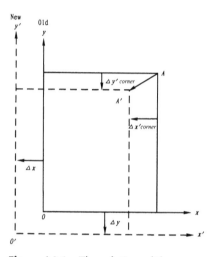

Figure A4.1 The relation of the position difference of a sheet corner and the variation in the value of a square grid line

where ΔX, ΔY are the coordinate differences of the new and old sheet corners arising from the computation formulae of the projection. Δx, Δy are the variation in the values of the grid line.

The above formula presents the relation of the position difference of a sheet corner and the variation in the value of a square grid line, or the position difference of the origin of the coordinates (see Figure A4.1).

It can be approximately regarded that $\Delta Y = \Delta S_m$, $\Delta X = \Delta S_n$; when the first terms of the projection formulae are S_m and S_n then (A4.1.11) can be written in the form of:

$$\begin{array}{l} \Delta y_{corner} = \Delta S_m - \Delta y, \\ \Delta x_{corner} = \Delta S_n - \Delta x \end{array} \qquad (A4.1.12)$$

where ΔS_m, ΔS_n are the differences of arc length of parallels and meridians on different ellipsoids.

So the following steps in cartographic work can carry out the transformation of the mathematical elements of a topographic map:

- First, to measure the plane coordinates of discrete points to obtain the difference of the coordinates Δx, Δy.

- Second, to obtain the difference of coordinates ΔX, ΔY or ΔS_m, ΔS_n of sheet corners.

- Finally, to obtain the position differences Δx_{corner}, Δy_{corner} of sheet corners by formula (A4.1.11) or (A4.1.12).

The preparatory work of cartography can be carried out by conventional methods after the sheet corners of the new system are determined on the map of the old system.

The Position Line and Position Navigation Software System

The position line is a contour line that is composed of points on the earth's surface that have equal values of some geometric parameter. The Position Line and Position Navigation Software System was constructed based on the research results in the theory of position lines. The system is multifunctional and all-purpose. It can be used for digital mapping, computation of position lines, and graphical positioning, all in an integrative technology.

This software system includes the following parts: (a) software for digital map data processing and graph plotting; (b) software for the digital mapping of the administrative divisions of China and the world; (c) software for the solution of the geodetic problem; (d) software for radio-range orientation and navigation; (e) software for GPS applications; (f) software for auto-plotting grids of distance and/or angle-measures; (g) software for projection transformation; and (h) software for fast display of thematic mathematical elements. The software systems are designed rationally and are easy to operate. They can perform many different functions and serve many purposes.

Integrated with a digital map, the system can be used to display and trace radio fixes and navigation in real time. It can quickly display the range line, direction line and track of a target on a background map of China and a world map. One can read the instantaneous distance and position angle from the centerpoint to the target point and provide surveying and mapping users with a screen display and graphically output various thematic mathematical elements on a map, including such items as a planned sea route map, flying range map, distance-angle measuring grids and hyperbolic navigation lines, etc.

Now we present a description of each software module, with its coded abbreviations. To learn more about the software package of this system see: *Software for a Space Information Positioning Model and Its Application System – Achievements in Scientific Research* (Yang Qihe 1994; in Chinese).

A5.1 SOFTWARE FOR DIGITAL MAP DATA PROCESSING AND GRAPHICAL PLOTTING

ZDTP

Functions: generation of digital coordinate files and conversion of data formats; correction of data errors; joining of polygon chains, node matching, automatic construction of topological relations; extraction, indexing and display of polygons.

SJCL

Functions: for digitizing, error correction and geographical coordinate adjustment of variously scaled maps based on different mathematical foundations; especially for the data handling of world maps, maps of China and of topographic map series.

Solutions: include the analytical method, polynomial method, least-squares polynomial method, the orthomorphic polynomial method, etc.

JWX

Functions: cutting window-sized rectangle sheets based on all kinds of mathematical basis projections; automatic plotting of the geographic mesh and automatic lettering.

A5.2 SOFTWARE FOR THE SOLUTION OF GEODETIC PROBLEMS

A geodesic is the shortest line between two points on the earth ellipsoid. If we take the earth ellipsoid as sphere, then the geodesic will be a great circle. The geodetic elements are the geodetic longitude L and geodetic latitude B of a point on the ellipsoid, the geodesic length S between two points, and the direct and inverse geodetic azimuths A1, A2. The process of finding some unknown geodetic elements from other known geodetic elements is called a solution of the geodetic problem. It is very complicated.

CZ001 – Solution of geodetic problem on the sphere

Direct solution: input B1, L1, A1, S, output B2, L2.
Inverse solution: input B1, L1, B2, L2, output S, A1, A2.

CZ002 – Solution of geodetic problem on the ellipsoid

Direct solution: input B1, L1, A1, S, output B2, L2.
Inverse solution: input B1, L1, B2, L2, output S, A1, A2.

A5.3 SOFTWARE FOR RADIO-RANGE ORIENTATION AND NAVIGATION

Radio-range orientation

The radio wave transmitted by a radio station travels at the speed of light along the shortest route. This route is just a geodesic. Hence, geodetic position line is widely used in radio-range orientation.

Functions: fast and accurate calculation of radio orientation of a target intersection, ranging the target intersection, orientation and ranging target intersections, hyperbolic curve target intersections and determination of the best position for multi-station orientation and positioning on a sphere or ellipsoid. The geographic coordinates of target points can be printed, and their position and path can be displayed on the background of a map of China or on a world map. These functions are included in the following programs:

- AZ1 – calculating and displaying of target intersections by orientation, ranging and hyperbolic curves on a sphere.

- AZ2 – calculating and displaying of target intersections of orientation, ranging and hyperbolic curve on an ellipsoid.

Radio-range navigation

When at sea, for the sake of navigation, ships usually sail along a rhumb line. This is a straight line on a Mercator projection map. But this voyage is not the shortest; the shortest route is along the great circle. In practice, both paths are used in conjunction with each other.

BZ1

Functions: display the actual sailing route updated from a planned shipping path and implement auto-navigation using the radio.
   Input: the positions of start and end points B1, L1, B2, L2.
   Output: display the real sailing track on an electronic map.

A5.4   GPS APPLICATION SOFTWARE

GPS is the United States' new generation of a global positioning system. Its broad application possibilities and development potentialities have aroused general interest:

- GPS1 – computation of an ephemeris for the two-body problem and for considering the effect of perturbation.
- GPS2 – transformation between a geocentric coordinate system and a geographical coordinate system.
- GPS3 – GPS positioning, navigation and sailing track display in real time.
- GPS4 – the projecting of the sub-satellite point on all kinds of map projection planes, when the shape of earth is taken as an ellipsoid and the effect of perturbation is considered.

A5.5   SOFTWARE FOR AUTO-PLOTTING GRIDS OF DISTANCE AND ANGLE MEASUREMENT

Based on the three commonly used classes of projections – conical projections, azimuthal projections, and cylindrical projections – the software can automatically plot grids of distance, direction or both from a given arbitrary central position, in arbitrary increments of graduation, and provide a map with border, geographic mesh, and lettering. Hence it is suitable for the measurement of a fixed area around an arbitrary central point on the map. The software includes the following.

TEST1

Functions: displaying the map grid of distance and angle measurement that uses a conical projection as the mathematical basis.
   Input: prompt with a multi-level menu for the function selection of: (a) displaying an equidistant line, loxodrome or grid of distance and angle measurement; (b) for the

projection type, such as conformal, equivalent or equidistant; (c) for the selecting of the geographical coordinate of the central point position; (d) for the determining of azimuth, distance increments and graduation value.

TEST2

Displaying a grid map of distance and angle measurement that takes an azimuthal projection as the mathematical basis. The input is as above.

TEST3

Displaying a grid map of distance and angle measurement that takes a cylindrical projection as the mathematical basis. Its input is just as above.

A5.6   SOFTWARE FOR RAPID DISPLAY OF THEMATIC MATHEMATICAL ELEMENT

AMEM

Functions: fast display of the range, direction, track and sub-satellite track from a fixed point to a target point on a background of map of China or a world map; fast acquisition of distance and azimuth from a fixed point to an arbitrary point; plotting of thematic elements such as a planned shipping route on a map, a flight radius map, track map of the sub-stellar point of satellite.
   Parameter input: follow menu prompts.

A5.7   SOFTWARE FOR MAP PROJECTION TRANSFORMATION

ZL04

Functions: to realize transformations among all kinds of maps based on different mathematical foundations. These maps are aeronautical charts, nautical charts, topographical maps, radio-orientation maps, and other navigation and positioning maps. Many map projections, such as azimuthal projections, cylindrical projections, conical projections and the Gauss–Krüger projection are involved.
   Solutions: analytical method, numerical method, analytical–numerical method.

A5.8   SOFTWARE FOR THE DIGITAL MAP OF ADMINISTRATIVE REGIONS OF CHINA AND THE WORLD

CWDM

Digitizing using the fundamental materials of the 1:2 500 000 Map of Chinese Administrative Divisions and the 1:33 000 000 Map of the World Administrative Divisions. This software produces and processes digital forms of individual elements, such as boundary lines, main rivers, route networks, and settlement places. These maps are digitally stored and can be applied for graphical display and plotting.

Bibliography of Chinese Literature on Map Projections and Other References

CHINESE LITERATURE

Bu Kongshu, 1957, Ditu touying xue [*The Study of Map Projections*], Taiwan, College of Surveying.

Cheng Yang, 1985, [Complex functions and conformal projection], *Acta Geodetica et Cartographica Sinica*, 14(1): 51–60 [with English abstract].

1989, On numerical method to develop conformal projections with special character, Intenational Cartographic Association, 14th World Conference, August 17–24, Budapest, *Abstracts-Resumés*, 622 [Abstract only].

1990a, [On conformal projection maintaining the length of any desired curve on the earth ellipsoid true to scale], *Acta Geodetica et Cartographica Sinica*, 19(2): 110–119 [with English abstract].

1990b, [A study of conformal projection transformation by using finite element method], in Assn. of Chinese Surveyors and Cartographers [*Collection of Outstanding Articles of National Young Surveyors and Cartographers*], 112–119.

1991, [On conformal space projection], *Acta Geodetica et Cartographica Sinica*, 20(1): 36–45 [with English abstract].

Dang Songshi, 1960, [Comments about functions for cartographic projections. (tr. from Russ. title)]: *Acta Geodetica et Cartographica Sinica*, 4(1): 44–48 [Chinese with Russian abstract. Reprinted (without abstract) in Wu and Hu, 1983 [*Map Projection Papers*], as [Comments on map projection functions], 173–178.

Ding Jiabo, 1982, [A complement to the analytical transformation between conformal projections], *Acta Geodetica et Cartographica Sinica*, 11(1): 46–50.

1993, [The transforming of zones of Gauss projection by using latitude of pedal], *Acta Geodetica et Cartographica Sinica*, 22(3): 212–217.

1994, [The projection of Geodesic curvature and application on Mercator chart], *Acta Geodetica et Cartographica Sinica*, 23(2): 155–158.

1995, [The extension of conditions for conformal, equal-area and equidistant projection], *Cartography*, 1995(1): 15–16.

Fang Bingyan, 1979, [*Tables for Calculating Map Projections*], Peking, China Surveying and Mapping Press.

1982, An analytical method for researching polyconical projections for world maps, presented at the 11th International Cartographic Conference, Warsaw.

1983, [A polyconic projection with unequally spaced parallel lines and its applications] in *Map Projection Papers*, 225–275.

Fang Jun, 1934, *Ti-t'u t'ou-ying-hsueh* [*The Study of Map Projections*], Peking, Science Press, 1952 ed., 1957–58 ed. (2 vols).

1943, *Lambert Projection Tables for China*, Chungking, National Geological Survey of China, Cartographic records 3 [Chinese and English].

1948–49, On the Lambert conformal projection as applied in China, *Empire Survey Review*, 1948, 9(70): 357–363; 1949, 10(71): 30–71.

1949, Transformations between the Lambert conformal and Gauss–Krueger (transverse Mercator) projections, *Empire Survey Review*, 10(74): 181–186.

1950, Azimuth and distance corrections for the geodetic lines on the Lambert conformal projection, *Empire Survey Review*, 10(75): 218–227.

1955, [On Gauss–Krüger projection transformation zone], *Acta Geophysica Sinica*, 4(1): 35–45.

1957a, The equation of the geodetic line on conformal projection and the direction and distance corrections for them, *Acta Geodetica et Cartographica Sinica*, 1(2): 166–169 [English].

1957b, [Formulae and tables for conversion of the geographical latitude to the authalic and isometric latitudes], *Acta Geographica Sinica*, 23(4): 379–388 [with English abstract].

1957c, [*Map Projection*] Book One, Peking, Science Press.

1958, [*Map Projection*] Book Two, Peking, Science Press.

Feng Kejun, 1983, [An elementary study on the rules of the density of the map graticule] in Wu and Hu [*Map Projection Papers*], 335–345.

Gong Jianwen, 1964, [Graphic interpretation methods in seeking cylindrical map projections with pre-defined distortion, title based on reprint], *Acta Geodetica et Cartographica Sinica*, 7(1). Reprinted in Wu and Hu, 1983 [*Map Projection Papers*], 96–104.

1989, [*Map Cartometry*], Peking, China Surveying and Mapping Press.

Gong Jianwen and Hu Yuju, 1992, [*Exercises for Map Projections*], China Surveying and Mapping Press.

Hu Jingxing, 1990, [On probing into the graph topology in map projection], *Cartography*, 1990(19): 3–7.

1994, [A family of new pseudocylindrical equal-area projections], *Chinese Yearbook of Cartography*, 1994: 25–27.

1996, [The topological model and classification framework of map projections], *Cartography*, 40: 5–8.

Hu Peng, 1981 [An analysis of the pseudo-azimuthal projection], Wuhan College of Geodesy, Photogrammetry and Cartography, *Journal*, (1): 98–121 [with English abstract].

1982, [The semi-numerical method for the map projection transformation], *Acta Geodetica et Cartographica Sinica*, 11(3): 191–202 [with English abstract].

Hu Peng and Hu Yuju, 1988, [A research on map projection and its classification], *Acta Geodetica et Cartographica Sinica*, 17(4): 286–294 [with English abstract].

Hu Yuju, 1957a, [Some comments on 'The principles and properties of Mercator projection'], *Bulletin of Surveying and Mapping*, (2).

1957b, [On the determination of the constants of conic projections], *Bulletin of Surveying and Mapping*, (4).

1958a, [Study on the projections for the provinces' atlas of China], *Journal of Survey-ing and Mapping*, (2).

1958b, [Characteristics of some famous projections in the USSR], *Bulletin of Survey-ing and Mapping*, (3).

1962, [On graphic interpretation methods in seeking azimuthal projection, presumed title from reprint], *Acta Geodetica et Cartographica Sinica*, 5(2). Reprinted in Wu and Hu, 1983, in [*Map Projection Papers*], 44–57.

1964a, [*Mathematical Cartography*], Publishing House of Industry.

1964b, [A special world map projection network for measuring and drawing], *Bulletin of Surveying and Mapping* (4).

1980, [Contents and tasks of mathematical cartography], *Proceedings of all-China Cartographic Conference*, Publishing House of Surveying and Mapping.

1987, [Cartographic projection system for variable scale maps], Wuhan Technical University of Surveying and Mapping, *Journal*, 12(2): 47–54.

Hu Yuju and Gong Jianwen, 1974, [*Map and Projections for Small-Scale Maps*], part 1, Chinese Cartographic Publishing House.

1981, [*Mathematical Cartography*], Bureau of Surveying, Ministry of Geology.

1992a, [*Map Projections*], 2nd ed., Publishing House of Surveying and Mapping.

1992b, [*Album of Map Projections*], 2nd ed., Publishing House of Surveying and Mapping.

Hu Yuju and Wang Jinren, 1962, [Nomograms for measuring distance and azimuth on equal-area conic projections], *Acta Geodetica et Cartographica Sinica*, 5(1).

Hu Yuju and Wu Bochun, 1983, [A thematic world map projection and its measurement and annotation methods] in Wu and Hu [*Map Projection Papers*], 363–376.

Hu Yuju and Zhou Yujun, 1958, [Conformal or equal-spaced conical projection with total-equal area], *Acta Geodetica et Cartographica Sinica*, 2(3): 195–202; (also *Journal of Surveying and Mapping of WTUSM*, (3) [with English abstract]. Re-printed (without abstract) in Wu and Hu, 1983 [*Map Projection Papers*] as [Conformal or equidistant conic projection with true total mapped area], 105–111.

Hua Tang, 1983a, [The research and application of description on sphere from ellipsoid], *Acta Geodetica et Cartographica Sinica*, 12(2): 134–151 [with English abstract].

1983b, [A direct transformation between the Gauss–Krüger projection and the Mercator projection] in Wu and Hu, 1983 [*Map Projection Papers*], 276–286.

1983c, [The inverse problem of isometric latitude] in Wu and Hu, 1983 [*Map Projec-tion Papers*], 323–328.

1983d, [On a spherical graticule for gnomonic projection] in Wu and Hu, 1983 [*Map Projection Papers*], 346–362.

1984, [The selection of spherical radius for gnomonic projection on large medium scale], *Acta Geodetica et Cartographica Sinica*, 13(2): 141–151 [with English abstract].

1985, [*Mathematical Basis for Charts*], The Navigation Guarantee Department of the Chinese Navy Headquarters.

1990, [Computation method for positioning by long-range hyperbolic navigation sys-tem], *Acta Geodetica et Cartographica Sinica*, 19(3): 179–186.

Hua Tang and Li Hongli, 1990, [The positioning calculation method for hyperbolic navigation system of two separated baselines], *Navigation of China*, 1990(2): 89–95.

Huang Guoshou, 1985 [Map projections for variable scale city maps], *Acta Geodetica et Cartographica Sinica*, 14(3): 188–195 [with English abstract].

1988, [An introduction to the space oblique Mercator projection], *Cartography* (Wuhan Technical University, Wuhan, China) (2): 12–14.

Li Changming, 1979, [On the classification of map projections], *Acta Geographica Sinica*, 34(2): 139–155. [Reprinted in Wu and Hu, 1983 [*Map Projection Papers*], 27–43].

Li Guozao, 1963, [Double azimuthal projection], *Acta Geodetica et Cartographica Sinica*, 6(4): 279–302 [with Russian abstract, reprinted (without abstract) in Wu and Hu, 1983 [*Map Projection Papers*], 58–74].

1981, [Cylindrical transverse conformal projection with two standard meridians], *Acta Geodetica et Cartographica Sinica*, 10(4): 305–312 [with English abstract, reprinted (without abstract) in Wu and Hu, 1983 [*Map Projection Papers*], 137–150. Abstract for similar paper in Budapest, International Cartographic Association, 14th World Conference, August 17–24, 1989, *Abstracts-Resumés*, 620].

1983a, [Double azimuthal projection] in Wu and Hu, 1983 [*Map Projection Papers*], 58–74.

1983b, [A modified transverse cylindrical conformal projection for 1: 1 000 000-scale topographic maps] in Wu and Hu, 1983 [*Map Projection Papers*], 151–157.

1987, [On $m = n^k$ orthogonal projections], *Acta Geodetica et Cartographica Sinica*, 16(2): 149–157 [with English abstract].

Li Guozao, Yang Qihe, and Hu Dingquan, 1993, *Ditu touying* [*Map Projections*], Chinese People's Liberation Army Press.

Li Hongli, 1989, [On the solution and inverse of rhumb line], *Navigation of China*, 1989(1): 47–52.

1993, [Double equidistant projection and its application], *Acta Geodetica et Cartographica Sinica*, 22(1): 65–73.

Li Jiaquan, 1985, [A research on transformation of finite element between conformal projections], *Acta Geodetica et Cartographica Sinica*, 14(3): 214–225 [with English abstract].

Li Yongpu, 1994, [Differential method of the transformation between plane coordinate systems], *Acta Geodetica et Cartographica Sinica*, 23(1): 37–44 [with English abstract].

Liu Fuchang, 1985, [Map projection and calculating-graphic for determining the positions and scope of beam region of communication satellite], *Acta Geodetica et Cartographica Sinica*, 14(3): 196–204 [with English abstract].

Liu Hongmo, 1985, [The numerical method of transformation of conformal projection], *Acta Geodetica et Cartographica Sinica*, 14(1): 61–72 [with English abstract, Polynomial transformations].

Liu Jiahao, 1965, [On the law of conversions of distortions in the cylindrical, azimuthal and conic projections], *Acta Geodetica et Cartographica Sinica*, 8(3): 210–217 [with Russian abstract, reprinted (without abstract) in Wu and Hu, 1983, [*Map Projection Papers*], 179–185].

1983, [On the patterns of distortion on cylindrical, azimuthal and conic projections] in Wu and Hu, 1983, [*Map Projection Papers*], 179–185.

Liu Jiahao and Li Guozao, 1963, [Pseudoazimuthal projections and their application for a whole map of China], *Acta Geodetica et Cartographica Sinica*, 6(2): 104–119 [with English abstract, reprinted (without abstract) in Wu and Hu, 1983, [*Map Projection Papers*], 200–214].

Liu Tisheng, 1957, Bei-tzin da-hsue hsue-bao (tzu-shan ke-hsue) [Construction of a cylindrical equal-area projection with two standard parallels], *Acta Scientiae Naturalis Univ. Pekinensis*, 3(2): 253–258.

Ma Junhai, 1988, [Computer program for calculation of the geographic coordinates of topo-map sheet corners and rectangular Gauss projection coordinates from sheet numbers], *Cartography* (Wuhan Technical University, Wuhan, China), (1): 21–23.

Meng Jiachun, 1959, [Study on the Projections for the Provinces' Atlas of China], *Acta Geodetica et Cartographica Sinica*, 3(4): 238–246.

    1982, [Transformation of conical projections], *Acta Geodetica et Geophysica*, (4): 189–205.

    1984, [On conical projection and its application to the general map of China], *Acta Geodetica et Geophysica*, (5): 147–160.

Shen Yongnian, 1991, [The expansion and computation formula for bivariate series of Gauss–Krüger Projection], *Acta Geodetica et Cartographica Sinica*, 20(2): 139–147 [with Russian abstract].

Shi Shusen, 1958, [Transformation of formulas of I.T. Petoval'tsev from formulas of Khristov with Gauss conformal projection with constant coefficients and their supplement (tr. from Russ. title)], *Acta Geodetica et Cartographica Sinica*, 2(2): 5–80 [with Russian abstract].

Si Xiaoyan and Hu Yuju, 1994, [Oblique camera type projection and the comparison with azimuthal projection at its centre outside of the globe], *Journal of WTUSM*, (3).

Wang Qiao and Hu Yuju, 1992, [Variable-scale map projection with rectangular frame], *Scitech. Journal of WTUSM*, (4).

    1993a, A kind of adjustable map projection with 'magnifying glass' effect, in Selected papers for English edition – *Acta Geodetica et Cartographica Sinica*, Beijing, Publishing House of Surveying and Mapping, 82–93 [English].

    1993b, [Variable-scale map projections for tourist maps and the applied software], *Cartography*, (2).

Wang Qiao *et al.*, 1996, [*Methods for Variable Scale Map Projections*], Publishing House of WTUSM.

Wu Zhongxing, 1961, [*Mathematical Cartography*], Zhengzhou Institute of Surveying and Mapping.

    1963, [The condition and tendency of development of mathematical cartography (tr. from Russian title)], *Acta Geodetica et Cartographica Sinica*, 6(1): 39–43 [with Russian abstract].

    1965, [Elementary analysis of conflicting movements in map projection. Some understanding of the work and philosophy of Chairman Mao], *Acta Geodetica et Cartographica Sinica*, 8(2): 115–124 [reprinted in Wu and Hu, 1983, [*Map Projection Papers*], 6–14].

    1979, [How to transform coordinates of points from one kind of map projection to another], *Acta Geographica Sinica*, 34(1): 55–68.

    1980, A research on the transformation of map projections in computer-aided cartography, presented at Tokyo, 10th International Cartographic Conference, unpublished manuscript.

    1989, A discussion on polyfocal projection, *Acta Geographica Sinica*, 44(1): 101–104.

    1993a, *Map Projection and Cartography Papers*, China Map Publishing House.

    1993b, [On the principle and method of exploring map projection] in Wu and Hu, 1983 [*Map Projection and Cartography Papers*], 8–19.

    1993c, [A new equivalent projection apply to eastern and western hemisphere map and continental map] in Wu, [*Map Projection and Cartography Papers*], 42–50.

Wu Zhongxing and Hu Yuju, eds, 1983a, *Ditu touying lun wenji* [*Map Projection Papers*], China Surveying and Mapping Press [27 papers by various cartographers, 12 reprinted from earlier journals].

    1983b, [Thirty years' development of map projection study in China] in [*Map Projection Papers*], 1–5.

Wu Zhongxing and Yang Qihe, 1981, [A research on the transformation of map projec-
    tions in computer-aided cartography], *Acta Geodetica et Cartographica Sinica*, 10(1):
    20–44 [with English abstract, reprinted (without abstract) in Wu and Hu, 1983,
    [*Map Projection Papers*], 287–322].

    1989, *Shuxue zhitu xue yuanli* [*Principles of Mathematical Cartography*], China Sur-
    veying and Mapping Press.

Xu Houze, 1957a, [Conversion problem from one zone to the other zone for Gauss plane
    rectangular coordination], *Geodetica Special Publication No. 5*, (pp. 29–38), Geo-
    graphic Institute, CAS.

    1957b, [On horizontal axis Mercator projection and Gauss–Krüger projection once
    again], *Geodetica Special Publication No. 5*, (pp. 39–47), Geographic Institute,
    CAS.

    1958, [General formulas for the transformation of coordinates of conformal projections
    (tr. from Russian title)], *Acta Geodetica et Cartographica Sinica*, 2(3): 186–194
    [with Russian abstract].

Yang Qihe, 1962, [Several problems on the basic mathematics for the common atlas],
    Cartography Annual Conference Papers (Vol. 2).

    1965, [The use of numerical methods for establishing optional conical projections],
    Acta Geodetica et Cartographica Sinica, 8(4): 295–318 [with Russian abstract, re-
    printed (without abstract) in Wu and Hu, 1983, [*Map Projection Papers*], 112–136].

    1979a, [A discussion of new map projections], *Bulletin of Surveying and Mapping*, (5):
    25–32.

    1979b, [*Tables of Regional Map Projections*], Zhengzhou, Zhengzhou Institute of
    Surveying and Mapping.

    1980a, [*Tables of Coordinate Transformation for Gridlines of Adjacent Zones of the
    Gauss Projection*], Peking, China Surveying and Mapping Press.

    1980b, [Research on map projection transformation], *Bulletin of Surveying and Map-
    ping*, (1): 30–33.

    1981a, [Gauss–Krüger projection and transverse Mercator projection], *Bulletin of Sur-
    veying and Mapping*, (6): 34–37.

    1981b, [Geometrical characteristics of the Gauss projection and a discussion of some
    of its problems of application], *Technical Communication of Surveying and Map-
    ping*, (2): 42–47.

    1982a, [On the numerical method for transforming the zones of Gauss projection],
    Acta Geodetica et Cartographica Sinica, 11(1): 18–31 [with English abstract].

    1982b, [A research on numerical transformation between conformal projections], *Acta
    Geodetica et Cartographica Sinica*, 11(4): 268–282 [with English abstract].

    1983a, [Parameter B_0, n_0 on the conical projection and research on its characteristics
    and application], *Bulletin of Surveying and Mapping*, (6): 41–43.

    1983b, [Research on azimuthal projections], in Wu and Hu, 1983, [*Map Projection
    Papers*], 75–95.

    1983c, [Research on the Gauss–Krüger projection family], in Wu and Hu, 1983 [*Map
    Projection Papers*], 158–172.

    1983d, [A discussion of three modified projections – modified equidistant azimuthal,
    conic and cylindrical projections], in Wu and Hu, 1983, [*Map Projection Papers*],
    186–199.

    1983e, [Research on the polyconic projection family], in Wu and Hu, 1983 [*Map
    Projection Papers*], 215–229.

1984a, [On the general formulas for the projection from ellipsoidal surface onto spherical surface and the characteristics of extremity], *Acta Geodetica et Cartographica Sinica*, 13(3): 225–236 [with English abstract].

1984b, [Research on the third kind of coordinate transformation of map projections], Zhenzhou Institute of Surveying and Mapping, *Journal*, 1(1): 109–121.

1985a, [Direct solution transformation formula between conformal conic projection and example of its application], *Bulletin of Surveying and Mapping*, (3): 42–44.

1985b, [Computation method of the variable coefficient on the inverse solution of isometric latitude], *Bulletin of Surveying and Mapping*, (5): 31–33.

1986a, [Research on the theory and application of map projection transformation], Zhengzhou Institute of Surveying and Mapping, *Journal*, 3(1): 65–73.

1986b, [The discussion and application of the constant coefficient and the variable coefficient formulae of direct and inverse solution of Gauss–Krüger projection], *Acta Geodetica et Cartographica Sinica*, 15(2): 141–154 [with English abstract].

1986c, [A study of the software for automatically setting up map mathematical foundation], First Computer-Assisted Cartography Conference, *Papers*, 167–176.

1986d, [BASIC programs for 1: 10 000, 1: 5 000, and larger-scale topographic sheet elements], *Bulletin of Surveying and Mapping*, (4): 25–28.

1987a, [*The Principles and BASIC Program for Conformal Projection Transformation*], China Surveying and Mapping Press.

1987b, [Computation method on break spacing to join sheets of adjacent zones], *Cartography*, (3): 15–16.

1987c, [The perspective azimuthal projection under variable view points], *Acta Geodetica et Cartographica Sinica*, 16(4): 306–313 [with English abstract].

1987d, The transforming and measuring of position information of map, *Proceedings*, London, International Cartographic Association, 13th World Conference.

1988, [On combinatorial azimuthal projections], *Bulletin of Surveying and Mapping*, (2): 32–34.

1989a, [Mathematical model of direct and inverse solutions to coordinate transformation for the modified polyconic projection], *Bulletin of Surveying and Mapping*, (2): 42–44.

1989b, [The mathematical method of plotting the coordinate grid of adjacent zones by using the coordinatograph and its BASIC programming], *Acta Geodetica et Cartographica Sinica*, 18(2): 108–114 [with English abstract].

1989c, A research on constant coefficient conformal projection transformation and its application, Budapest, International Cartographic Association, 14th World Conference, August 17–24, *Abstracts-Resumés*, 621.

1990a, *Ditu touying bianhuan yuanli yu fongfa* [*The Principles and Methods of Map Projection Transformation*], Chinese People's Liberation Army Press.

1990b, [A research of linear transformation on the gnomonic projection and its characteristics and application], *Acta Geodetica et Cartographica Sinica*, 19(2): 102–109 [with English abstract].

1990c, [Analytical computation method for two-point azimuthal projection and two-point equidistant projection], *Bulletin of Surveying and Mapping*, (6): 34–37.

1990d, [Polynomials for conformal projection transformation and research on its application], *Cartography*, (4): 13–16.

1991, [Mathematical model for a great- or small-circle position line and its application], *Chinese Yearbook of Cartography*, 73–76.

1992, [Affine transformation of equal-area projections and some discussion of modified equal-area projections], *Acta Geodetica et Cartographica Sinica*, 21(1): 67–77.

1993, [Analytical transformation method for digitizing data processing on a map of the world], *Chinese Yearbook of Cartography*, 35–37.

1994, [An overview of the theory and method of conformal projection]: Zhengzhou Institute of Surveying and Mapping, *Journal*, 11(2): 133–139; (3): 181–187.

1995, [Six latitudes used in surveying and mapping and their transformation equations], *Communication of Surveying and Mapping*, 18(3): 14–19.

Yang Qihe and Lu Xiao hua, 1991, The study of inverse transformation methods for topographical map data processing, '91 FIG Conference, *Papers*.

1995, The theory of position line and positioning navigation software system, '95 FIG Conference, *Papers*.

Yang Qihe and Yang Xiaomei, 1995, [Analysis of digital map processing method], *Bulletin of Surveying and Mapping*, (2): 31–34.

1997, [Three types of latitudinal function and linear interpolation algorithms used in surveying and cartography], *Acta Geodetica et Cartographica Sinica*, 26(1): 92–95.

Yang Xiaomei, Yang Qihe and Yang Yubing, 1997, [Space information positioning system and its application], *Developments in Surveying and Mapping*.

Ye Xuean, 1953, [*Map Projections*], Longmen United Publishing House.

1957, [Formulae for the coordinate transformation between zones of the Gauss–Krüger and Lambert Conformal Conic projections], *Acta Geodetica et Cartographica Sinica*, 1(2): 181–194.

1958, [Formulas for the coordinates from a zone of the Gauss projection to an adjacent zone and from the Lambert projection on a secant cone to the Gauss projection using the method of an auxiliary point on either site (continuation 2) (tr. from Russian title)], *Acta Geodetica et Cartographica Sinica*, 2(2): 121–134; (3): 35–54 [with Russian abstract].

Yu Chongwen, 1980, [*Method and Application of Mathematical Topography*], Metal Industry Publishing House.

Yuan Yinxiang, 1983, [A numerical method for the transformation of map projections by densifying with spline function], *Acta Geodetica et Cartographica Sinica*, 12(4): 300–308 [with English abstract].

Zhang Shenjia, 1988, [How to teach the chapter of 'Map projections'], *Cartography* (Wuhan Technical Univ.), (3): 40–43.

Zhang Shunqing, 1991, [A kind of sine equivalent pseudocylindrical projection family], *Cartography*, 1991(2): 50–51.

1993, [1980 national geodetic coordinate system and the transformation of topographic map mathematical foundation], *Cartography*, 1993(1): 22–26.

Zhong Yexun, 1963, [The state and capabilities of mathematical cartography], *Acta Geodetica et Cartographica Sinica*, 6(1): 39–43 [with Russian abstract].

1983a, [A polyconic projection with two spherical arcs as border meridians] in Wu and Hu, 1983, [*Map Projection Papers*], 230–254.

1983b, [On the problems of using high-order polynomials for latitude ϕ to approximate the meridians on a map projection] in Wu and Hu, 1983, [*Map Projection Papers*], 329–334.

Zhong Yexun *et al.*, 1965, [The design and analytical calculating method of polyconic projection in unequal graticules], *Acta Geodetica et Cartographica Sinica*, 8(3): 218–236 [with Russian abstract].

Zhou Chenggong, 1957, [Problems with the discovery of new map projections with the distortion ellipse on hand], *Acta Geodetica et Cartographica Sinica*, 1(1): 1–14 [with Russian abstract, reprinted (without abstract) in Wu and Hu, 1983, [*Map Projection Papers*], as [Developing new map projections with pre-defined ellipse of distortion], 15–26].

Zhu Changqing, 1997, [*Calculating Method and their Applications in Surveying and Mapping*], China Surveying and Mapping Press.

Zhu Huatong, 1986, [*Build of Geodetic Coordinate System*], China Surveying and Mapping Press.

OTHER REFERENCES

Brandenberger, G., 1985, *Koordinatentransformationen für Digitale Kartographische Daten mit Lagrange- und Spline-Interpolation* [Coordinate transformation for digital cartography with Lagrange and spline interpolation], Institute für Kartographie, ETH, Zürich.

Bugayevskiy, L.M. and Portnov, A.M., 1984, [*The Theory of Single Space Photographs*], Moscow: Nedra.

Bugayevskiy, L.M. and Snyder, J.P., 1995, *Map Projections – A Reference Manual*, Taylor & Francis, London.

Butekovich, A.V., 1964, [Study of the problem of solving geodetic computing].

Cheng, Yang, 1992, On conformal projection maintaining a desired curve on the ellipsoid without distortion, ASPRS/ACSM (American Society for Photogrammetry and Remote Sensing/American Congress on Surveying and Mapping), Summer Convention, August 3–8, Washington, D.C., Technical Papers, 3: 294–303.

   1996, The conformal space projection, *Cartography and Geographic Information Systems*, 23(1): 37–50.

Colvocoresses, A.P., 1974, Space Oblique Mercator, *Photogrammetric Engineering*, 40(8): 921–926.

Compile Group, 1979, [*Mathematics Manual*], Demo – Education Press.

Dorling, D., 1996, *Area Cartograms: Their Use and Creation*, School of Environmental Studies, University of East Anglia.

Ginzburg, G.A., 1964, *Manual of Mathematical Cartography*, Trudy TsNIIGAik 160, Moskva [in Russian].

Heristov, V.K., 1957, [Gauss–Krüger coordinate on surface of revolution ellipsoid] (Russian version).

Kadmon, N., 1978, A polyfocal projection for statistical surface, *The Cartographic Journal*, 15(1).

Lambert, J.H., 1772, [Anmerkungen und Zusätze zur Entwerfen der Land- und Himmelscharten], translated by W. Tobler as *Lamberts Notes on Maps*, Michigan Geographical Publication No. 8, Department of Geography, University of Michigan, 1972.

Snyder, J.P., 1977, A comparison of pseudocylindrical map projections, *The American Cartographer*, 4(1): 59–81.

   1978, The Space Oblique Mercator projection, *Photogrammetric Engineering and Remote Sensing*, 44(5): 585–596.

   1981, *Space Oblique Mercator Projection – Mathematical Development*, US Geological Survey Bulletin 1518.

1982, *Map Projections Used by the US Geological Survey*, US Geological Survey Bulletin 1532, 193–210, 2nd ed., 1983.

1987a, *Map Projections – A Working Manual*, US Geological Survey Professional Paper 1395, 214–219, reprinted with corrections 1989, 1994.

1987b, Magnifying-glass azimuthal map projections, *American Cartographer*, 14(1): 61–68.

1988, New equal-area map projections for non-circular regions, *American Cartographer*, 15(4): 341–355.

Tissot, N.A., 1881, *Mémoire sur la représentation des surfaces et les projections des cartes géographiques*, Gauthier-Villars, Paris.

Tobler, W.R., 1963, Geographic Area and Map Projections, *The Geographical Review*, LIII, 1: 59–78.

1965, Medieval distortions: the projections of ancient maps, *Annals, Association of American Geographers*, 56(2): 351–360.

1973, A hyperelliptical and other new pseudo cylindrical equal area map projections, *Journal of Geophysical Research*, 78(11): 1753–1759.

1976, The geometry of mental maps, in R. Golledge and G. Rushton, eds, *Essays on the Multidimensional Analysis of Perceptions and Preferences*, Columbus, Ohio State University Press, 69–81.

1977, Numerical approaches to map projections, in I. Kretschmer, ed., *Studies in Theoretical Cartography*, Vienna, Deuticke, 1977; pp. 51–64.

1986a, Pseudo-cartograms, *American Cartographer*, 13(1): 43–50.

1986b, Polycylindric map projections, *The American Cartographer*, 13(2): 117–120.

1994, Bidimensional regression, *Geographical Analysis*, 26: 186–212.

Tobler, W.R. and Chen, Z., 1986, A quadtree for global information storage, *Geographical Analysis*, 18(4): 360–371.

Urmayev, N.A., 1962, *The Fundamentals of Mathematical Cartography*, Trudy, TsNIIGAiK, (144).

ACKNOWLEDGEMENT

Our thanks to Yang Cheng (now Cheng Yang) for translating the titles in Wu and Hu, ed., 1983, *Map Projection Papers*.

Index